Statistical Theory of Heat

Wilhelm Brenig

Statistical Theory of Heat

Nonequilibrium Phenomena

With 39 Figures

Springer-Verlag Berlin Heidelberg New York
London Paris Tokyo Hong Kong

Professor Dr. Wilhelm Brenig
Physik-Department, Technische Universität München, James-Franck-Str.,
D-8046 Garching, Fed. Rep. of Germany

ISBN 3-540-51036-2 Springer-Verlag Berlin Heidelberg New York
ISBN 0-387-51036-2 Springer-Verlag New York Berlin Heidelberg

Library of Congress Cataloging-in-Publication Data.
Brenig, Wilhelm.
Statistical theory of heat : nonequilibrium phenomena / W. Brenig.
p. cm.
Sequel to: Statistische Theorie der Wärme. Bd. I. Gleichgewicht.
ISBN 0-387-51036-2 (U.S.)
1. Nonequilibrium thermodynamics. 2. Statistical mechanics.
3. Matter, Kinetic theory of. I. Brenig, Wilhelm. Statistische
Theorie der Wärme. Bd. I, Gleichgewicht. II. Title.
QC318.I7B74 1989 536'.71 – dc20 89-19698

This work is subject to copyright. All rights are reserved, whether the whole or part of the material is concerned, specifically the rights of translation, reprinting, reuse of illustrations, recitation, broadcasting, reproduction on microfilms or in other ways, and storage in data banks. Duplication of this publication or parts thereof is only permitted under the provisions of the German Copyright Law of September 9, 1965, in its version of June 24, 1985, and a copyright fee must always be paid. Violations fall under the prosecution act of the German Copyright Law.

© Springer-Verlag Berlin Heidelberg 1989
Printed in the United States of America

The use of registered names, trademarks, etc. in this publication does not imply, even in the absence of a specific statement, that such names are exempt from the relevant protective laws and regulations and therefore free for general use.

2154/3150-543210 – Printed on acid-free paper

Preface

This text on the statistical theory of nonequilibrium phenomena grew out of lecture notes for courses on advanced statistical mechanics that were held more or less regularly at the Physics Department of the Technical University in Munich. My aim in these lectures was to incorporate various developments of many-body theory made during the last 20–30 years, in particular the correlation function approach, not just as an "extra" alongside the more "classical" results; I tried to use this approach as a unifying concept for the presentation of older as well as more recent results. I think that after so many excellent review articles and advanced treatments, correlation functions and memory kernels are as much a matter of course in nonequilibrium statistical physics as partition functions are in equilibrium theory, and should be used as such in regular courses and textbooks. The relations between correlation functions and earlier vehicles for the formulation of nonequilibrium theory such as kinetic equations, master equations, Onsager's theory, etc., are discussed in detail in this volume.

Since today there is growing interest in nonlinear phenomena I have included several chapters on related problems. There is some nonlinear response theory, some results on phenomenological nonlinear equations and some microscopic applications of the nonlinear response formalism. The main focus, however, is on the linear regime.

The basic concepts of the statistical theory of equilibrium phenomena are a prerequisite for this book, which is actually a sequel to my textbook *Statistische Theorie der Wärme I. Gleichgewicht* (Springer 1983). Sometimes results from this earlier publication are quoted in the text. Since there exists no English translation, so far, almost any other textbook on equilibrium (quantum) statistical mechanics may be used instead.

Nonequilibrium statistical theory, although over a century old, is still going strong. It is a vast field of present research activity with applications not only in physical sciences but also in chemistry and biology. If this book helps to attract a number of excellent graduate students to the field it will have fulfilled its purpose.

Garching, March 1989   W. Brenig

Acknowledgements

I have profited immensely from many discussions with colleagues and friends and I should like to take this opportunity to thank Kurt Schönhammer for many readings of the manuscript from the earliest versions on and for his many stimulating and critical remarks, furthermore Lorenz Cederbaum, George Feher, Donald Fredkin, Wolfgang Götze, Mario Liu, Erwin Müller-Hartmann, Hannes Risken, Albert Schmid, Hartwig Schmidt, Franz Schwabl, Dietrich Stauffer, Harry Suhl, Dieter Vollhardt, Herbert Wagner, Franz Wegner, Peter Wölfle and Wilhelm Zwerger for stimulating discussions.

Many interesting questions and remarks, of course, also arose from discussions with graduate students in courses on statistical physics.

I am very grateful to Dr. Herbert Müller for undertaking the tedious job of transposing the original typescript of the manuscript from "ScienTEX" to "TeX" and for producing several of the figures in the text.

The cooperation with Springer-Verlag has been always friendly and efficient.

The completion of the book in its last stages was sponsored by a six months' grant from the Volkswagen Foundation.

Contents

Part I Correlation Functions and Kinetic Equations

1. **Introduction** 3
2. **General Equations of Motion of Statistical Physics** 6
   - 2.1 Quantum Statistics 6
   - 2.2 Classical Statistics 10
   - 2.3 Approximations 12
   - Problems 13
3. **Small Amplitude Perturbation Theory (Linear Response)** 14
   - Problems 20
4. **Brownian Motion (Relaxator)*** 22
   - Problems 27
5. **Brownian Motion (Oscillator)*** 29
   - Problems 32
6. **Dispersion Relations and Spectral Representations** 33
   - Problems 36
7. **Symmetry Properties of Correlation Functions** 37
   - Problems 41
8. **Detailed Balance, Fluctuations and Dissipation** 42
   - Problems 44
9. **Scattering of Particles and Light**** 45
   - Problems 50
10. **Energy Dissipation, Detailed Balance and Passivity** 52
    - 10.1 The Response of Conserved Quantities to External Forces .. 52
    - 10.2 Energy Dissipation and Passivity 53
    - Problems 55
11. **The High-Frequency Behaviour of Response Functions** 56
    - Problems 58
12. **The Low-Frequency Behaviour of Response Functions** 59
    - Problems 62
13. **Stochastic Forces, Langevin Equation** 63
    - 13.1 The Subtraction Method (Langevin) 63

	13.2 The Projection Method (Zwanzig and Mori)	66
	Problems	68
14.	**Brownian Motion: Langevin Equation***	69
	Problems	72
15.	**Nonlinear Response Theory**	73
	15.1 The General Initial Value Case	73
	15.2 Low-Frequency Perturbation Theory	75
	Problems	77
16.	**The Increase of Entropy and Irreversibility**	78
	16.1 General	78
	16.2 Linear Response	79
	16.3 Low-Frequency Response	81
	Problems	81
17.	**The Increase of Entropy: A Critical Discussion****	82
	17.1 Maxwell's Demon	82
	17.2 Gibbs' Ink Parable	83
	17.3 Zermélo's Recurrence Paradox	84
	17.4 Loschmidt's Reversibility Objection	86
	Problems	87

Part II Irreversible Thermodynamics

18.	**The Nyquist Formula**	90
	Problems	92
19.	**Thermomechanical Effects**	93
	19.1 Diffusion and the Mechanocaloric Effect ($\Delta T = 0$)	95
	19.2 Heat Diffusion and Thermomechanical Pressure Difference ($\dot{N} = 0$)	95
	Problems	98
20.	**Diffusion and Thermodiffusion**	99
	Problems	103
21.	**Thermoelectric Effects**	104
	Problems	107
22.	**Chemical Reactions**	108
	Problems	111
23.	**Typical Time Evolutions of Simple Chemical Reactions**	112
	23.1 Zero-Order Reactions	112
	23.2 First-Order Reactions	113
	23.3 Second-Order Reactions	113
	23.4 Third-Order Reactions	114
	Problems	115

24. Coupled Nonlinear Reactions 116
    24.1 Nonequilibrium Phase Transitions 116
    24.2 Kinetic Oscillations 117
    24.3 "Chaos" 120
         Problems 121
25. Chemical Fluctuations 123
         Problems 126
26. Sticking, Desorption, Condensation and Evaporation 127
         Problems 130
27. Nucleation 131
         Problems 136
28. The Oscillator with Mechanical and Thermal Attenuation* . 137
         Problems 141
29. Hydrodynamics 142
         Problems 148
30. Hydrodynamic Long-Time Tails 149
         Problems 151
31. Matter in Electromagnetic Fields 152
         Problems 157
32. Rate Equations (Master Equation, Stosszahlansatz) 158
         Problems 163
33. Kinetic Transport Equations 164
         Problems 168
34. The Dynamic Conductivity in the Relaxation Time Model . 169
    34.1 Longitudinal Excitations 169
    34.2 Transverse Excitations 170
    34.3 Discussion of $\sigma^l(k,\omega)$ and $\sigma^t(k,\omega)$ 171
    34.4 Quantum Corrections 174
         Problems 176
35. Zero Sound 177
         Problems 179
36. The Fokker–Planck Approximation 180
         Problems 183
37. Brownian Motion and Diffusion* 184
         Problems 187
38. Fokker–Planck and Langevin Equations 188
39. Transport Equations in the Hydrodynamic Regime 193
    39.1 The Hydrodynamic Approximation 193
    39.2 Diffusion of Particles and Heat 195
    39.3 The Viscosities 198
         Problems 200

40. **The Minimum Entropy Production Variational Principle** 201
    40.1 The Principle of Minimum Entropy Production 201
    40.2 The Classical Boltzmann Gas 202
    40.3 The Electron–Phonon System 206
    40.4 Fermi Liquids 210
           Problems 211

Part III Calculation of Kinetic Coefficients

41. **Approximation Methods** 214
42. **Correlation Functions for Single-Particle Problems** 217
    42.1 General 217
    42.2 Impurity Conduction (Greenwood Formula) 219
           Problems 220
43. **Perturbation Theory for Impurity Conduction** 221
           Problems 226
44. **Electron–Phonon Conduction** 227
           Problems 229
45. **Mode-Coupling Theory for Impurity Conduction** 230
    45.1 Particle Diffusion and Current Relaxation 230
    45.2 Backscattering Effects 231
    45.3 Self-Consistency Relations 234
           Problems 236
46. **Electron Localization** 237
    46.1 Breakdown of Perturbation Theory 239
    46.2 Localization and Nonergodicity 239
    46.3 Critical Behaviour and Scaling Laws 241
           Problems 242
47. **Localization and Quantum Interference*** 244
48. **Scaling Laws for Dynamic Critical Phenomena** 247
    48.1 General 247
    48.2 The Lambda Transition in Liquid Helium 250
           Problems 252
49. **Applications of Dynamic Scaling Laws** 254
    49.1 Isotropic Ferromagnets 254
    49.2 Uniaxial Antiferromagnets, Structural Phase Transitions 257
    49.3 Anisotropic Ferromagnets 258
    49.4 Liquid–Gas Transition 259
50. **Mode-Coupling Theory for Dynamic Critical Phenomena** 260
51. **Broken Symmetry and Low-Frequency Modes**** 264

52. Collision Rates ... 269
52.1 Perturbation Theory ... 269
52.2 Impurity Scattering ... 270
52.3 Chemical Reaction Rates ... 271
Problems ... 273

53. Many-Body Effects in Collision Rates ... 274
Problems ... 276

References ... 277
Subject Index ... 287

Part I
Correlation Functions and Kinetic Equations

The first part of this volume will deal with the general formalism of nonequilibrium statistical mechanics. Historically there have been different formalisms developed for different nonequilibrium phenomena. In the meantime it has turned out that the formalism of correlation, relaxation and memory functions can serve as a common basis for most phenomena. This formalism will be introduced in a reasonably general and deductive way. In order not to be too abstract and formal we consider a number of simple examples for applications of the general formalism taken from the theory of Brownian motion. Sections which are of such, more pedagogical, nature, discussing general results in terms of simple examples, are marked by a single asterisk (*). Sections with a double asterisk (**) are intended to provide additional background information and cross-relations to other fields of physics.

Formulae are numbered consecutively within each chapter. When referring back to several formulae from one section we adopt the notation (7.8,9,11), for example, instead of (7.8), (7.9) and (7.11), etc. References are provided at the end of the volume. They are referred to by numbers in square brackets in the text. Suggestions for further or complementary reading are also provided at the end.

1. Introduction

All processes in nature are irreversible.

This elementary fact is an essential ingredient of theories describing macroscopic physics (frictional motion in mechanics, viscous flow, electrical conduction, conduction of heat, diffusion, chemical reactions, etc.). Nonequilibrium "irreversible" thermodynamics provides a common framework for all such phenomena.

Reversible thermodynamics, in contrast, deals only with equilibrium states which are known to appear spontaneously in closed systems after sufficiently long times. Processes (i.e., transitions from one such state to another) can be treated only if they proceed infinitely slowly compared to the time scales for the establishment of equilibrium. The notation "thermostatics" would be appropriate for this branch of physics but is used only rarely.

Irreversible thermodynamics deals with the dynamics (or "kinetics") of macroscopic processes. Kinetic equations are at the heart of nonequilibrium theory. *Time* plays an important role in this theory in two ways. First of all, thermodynamical quantities are explicitly considered as functions of time. The basic equations contain derivatives with respect to time, and constants of dimensionality "time" (relaxation times, collision times, etc.). Secondly, the *direction* of time plays an essential role. In most other branches of physics the difference between past and future is only of minor importance. In nonequilibrium thermodynamics, however, the fundamental difference between reversible and irreversible processes, pointed out by R. Clausius and M. Planck last century, shows up explicitly in the basic equations. The term "irreversible thermodynamics" is thus very well justified.

Nonequilibrium theory is more general than reversible thermodynamics. It contains equilibrium theory as the limiting case of infinitely slow processes. The task of irreversible thermodynamics is wider and more difficult. The foundations and applications of the theory are less developed than in equilibrium thermodynamics. Many fascinating problems of irreversible processes in solids, liquids, hot plasmas and astrophysical objects, in biophysical and chemical reactions are waiting for solutions.

The *statistical* point of view, which has been so successful in equilibrium theory, plays an at least equally important role in nonequilibrium. Most statements in the introduction and second chapter of [1.1], concerning the relationship between macroscopic phenomenological and microscopic statistical concepts, remain valid for both reversible and irreversible phenomena.

Some milestones on the way to a microscopic statistical description of irreversible processes are *Boltzmann*'s "Stoßzahlansatz" in the kinetic theory of gases (1872) [1.2], the theory of Brownian motion developed by *Einstein* (1905) [1.3] and *Langevin* (1908) [1.4] and *Onsager*'s symmetry relations between coupled irreversible processes (1931) [1.5]. Einstein's relation between mobility and the diffusion constant was generalized to electric circuits within classical physics by *Schottky* (1918) [1.6], with the inclusion of quantum effects by *Nyquist* (1928) [1.7] and extended to a general relation between thermodynamical fluctuations and dissipation by *Callen* and *Welton* (1951) [1.8].

A more or less deductive derivation of the basic equation of nonequilibrium theory from quantum mechanics and statistics is in principle possible. It is, however, more involved, than, for example, the derivation of equilibrium thermodynamics, mathematically as well as conceptually. Furthermore, no general consensus appears to exist about the best formalism. Different phenomena are often treated by different formalisms, sometimes only for historical reasons. Most of the time we will use a formalism closely related to the correlation functions of linear response theory, but later we shall generalize it so that strongly nonlinear phenomena (such as chemical reactions) can be handled on a similar footing.

We illustrate our procedure by means of the block diagram of Fig. 1.1. Here we deal mainly with the upper and right-hand parts of this diagram. We start out from the basic equations of motion of quantum statistics and then introduce approximations of these equations for small amplitudes or low frequencies. This on the one hand leads us to the theory of linear response, and on the other to further basic equations of irreversible thermodynamics, such as kinetic equations. We are going to use the term "kinetic equations" rather loosely for practically all equations of thermodynamics containing time derivatives and phenomenological constants (such as kinetic coefficients). The discussion and solution of such equations forms the main content of the second part of this volume. Kinetic equations in a more restricted sense, i.e., the basic equations of so-called transport theory, are discussed in the sections at the end of this part. In the third part we then present several examples of microscopic calculations of kinetic coefficients using correlation functions.

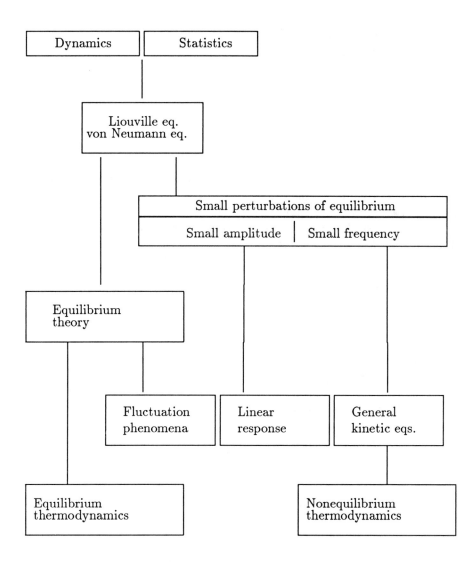

Fig. 1.1. Logical relations between subfields of statistical physics

2. General Equations of Motion of Statistical Physics

The equations of motion of statistical mechanics are first-order differential equations for the evolution of statistical ensembles in time. In quantum statistics [2.1] the *von Neumann equation* [2.2] describes the evolution of the statistical operator; in classical statistics the *Liouville equation* [2.3] describes the evolution of the statistical distribution function.

2.1 Quantum Statistics

The quantum statistical expression for the average value of an observable $A(t)$,

$$\langle A(t) \rangle = \mathrm{Tr}\{A(t)\varrho(t)\} \quad , \tag{2.1}$$

is invariant with respect to unitary transformations of the form

$$\mathrm{Tr}\{A(t)\varrho(t)\} = \mathrm{Tr}\{U^* A(t) U U^* \varrho(t) U\} \tag{2.2}$$

with $U^*U = UU^* = 1$. As the unitary operator U may depend on time the time dependence of the observable $A(t)$ and the statistical operator $\varrho(t)$ is arbitrary to some extent. Three different choices for the time dependence (the so-called "pictures" of quantum mechanics) have been particularly useful: the Schrödinger picture, the Heisenberg picture and the Dirac picture (or interaction picture).

In the *Schrödinger* picture, operators such as spatial coordinates x and momenta p as well as time-independent functions of them are constant in time:

$$\frac{dA_\mathrm{s}}{dt} = 0 \quad . \tag{2.3}$$

The subscript "s" indicates the Schrödinger picture.

Time dependences of observables in the Schrödinger picture can occur if one considers time-dependent functions. The Hamiltonian, for instance, of a system becomes time dependent if external forces $f_k^\mathrm{e}(t)$ act on the observables q_k of the system

$$H_\mathrm{s}(t) = H - \sum f_k^\mathrm{e}(t) q_k \tag{2.4}$$

even if the Hamiltonian H of the closed system and the quantities q_k are time independent.

The time dependence of state vectors $|t\rangle_s$ is governed by the Schrödinger equation

$$\frac{d|t\rangle_s}{dt} = -\frac{i}{\hbar}H_s(t)|t\rangle_s \quad . \tag{2.5}$$

If the statistical operator is written as a sum of projection operators $\sum |i,t\rangle\langle t,i|$, (2.5) leads to the *von Neumann equation*

$$\frac{d\varrho_s(t)}{dt} = -\frac{i}{\hbar}[H_s(t), \varrho_s(t)] \quad . \tag{2.6}$$

The solution of (2.5,6) can be written as a unitary transformation

$$|t\rangle_s = U(t, t_0)|t_0\rangle_s \quad , \tag{2.7}$$

and

$$\varrho_s(t) = U(t, t_0)\varrho_s(t_0)U^*(t, t_0) \quad , \tag{2.8}$$

where U obeys the differential equation

$$\frac{dU}{dt} = -\frac{i}{\hbar}H_s(t)U \tag{2.9}$$

with the initial condition $U(t_0, t_0) = 1$. The solution of this equation is simple only if H_s is time independent. Then

$$U = \exp[-i(t - t_0)H_s/\hbar] \quad . \tag{2.10}$$

The *Heisenberg* picture can be obtained from the Schrödinger picture by the transformations

$$\varrho_h = U^*(t, t_0)\varrho_s(t)U(t, t_0) = \varrho_s(t_0) \tag{2.11}$$

and

$$A_h(t) = U^*(t, t_0)A_s(t)U(t, t_0) \quad . \tag{2.12}$$

Thus the statistical operator is constant in time in the Heisenberg picture whereas physical observables change in time according to the equations of motion (indicating a time derivative by a dot)

$$\dot{A}_h(t) = \frac{i}{\hbar}[H_h(t), A_h(t)] + U^*\dot{A}_s(t)U \tag{2.13}$$

with the initial condition

$$A_h(t_0) = A_s(t_0) \quad . \tag{2.14}$$

The choice of t_0 is arbitrary. Usually one chooses t_0 as the zero of the time scale, $t_0 = 0$.

For operators that are time independent in the Schrödinger picture the second term on the r.h.s. of (2.13) vanishes, so (2.13) then simplifies to

$$\dot{A}_h = \frac{i}{\hbar}[H_h(t), A_h] \ . \qquad (2.15)$$

For the Hamilton operator, on the other hand, the first term in (2.13) vanishes, leaving

$$\dot{H}_h(t) = U^* \dot{H}_s U \ . \qquad (2.16)$$

The *Dirac picture* (or interaction picture) stands between the two other pictures. The transformation U is applied only with respect to part of the Hamiltonian H_s, the unperturbed part H_s^0. For instance, in (2.4) one may choose $H_s^0 = H$. Since H is time independent one can use (2.10), which for $t_0 = 0$ leads to a time dependence of observables in the Dirac picture of the form

$$A_d(t) = e^{iHt/\hbar} A_s(t) e^{-iHt/\hbar} \ , \qquad (2.17)$$

and, for the statistical operator,

$$\varrho_d(t) = e^{iHt/\hbar} \varrho_s(t) e^{-iHt/\hbar} \ . \qquad (2.18)$$

Taking the time derivative of these equations using (2.4) and the von Neumann equation (2.6) one obtains the differential equation

$$\dot{\varrho}_d(t) = \frac{i}{\hbar} \sum_l f_l^e(t) [q_{ld}(t), \varrho_d(t)] \ . \qquad (2.19)$$

For the integration of this equation we need initial conditions. We are going to consider two special cases.

1. The external field case: For large negative times the external forces f^e vanish and the system is in equilibrium described by a statistical operator with the temperature $T = 1/(k\beta)$ and the free energy F:

$$\varrho_d(-\infty) = \varrho_s(-\infty) = \exp[-\beta(H-F)] = \varrho \ . \qquad (2.20)$$

2. The initial value case: For $t = 0$ the system is in partial equilibrium corresponding to maximal entropy with given average values $\langle q_k(0)\rangle = Q_k(0)$. The external forces vanish for $t \geq 0$. This corresponds to a statistical operator

$$\varrho_d(t) = \exp\left\{-\beta\left[H - \sum f_l(0)q_l(0) - K(0)\right]\right\} \ , \quad t \geq 0 \ . \qquad (2.21)$$

Here the $f_l(0)$ are Lagrange parameters occurring in the determination of the statistical operator from the principle of maximum entropy under the constraint that certain averages have prescribed values $Q_k(0)$ (see e.g. [Ref. 2.4, Sect. 10]). $K(0)$ is a generalized free enthalpy. Although the initial situation described by

(2.21) is rarely established exactly, in practice it is used frequently because of its simplicity [2.5]. From a physical point of view one expects the memory of possible unphysical details of (2.21) to be lost after some (short) microscopic relaxation time.

For an ergodic system the initial value situation can be considered as a special case of the external field situation: If an external field is turned on quasi statically from a suitable initial state at $t = -\infty$ and then switched off abruptly at $t = 0$ one has exactly the initial value situation from $t = 0$ onwards.

On the other hand, the initial value case is more general than the external field case, since the Lagrange parameters $f_k(0)$ in (2.21) can be used to describe "nonmechanical", i.e. purely thermodynamical, perturbations of total equilibrium such as deviations of temperature and chemical potential from their equilibrium values. Initial conditions of the type (2.21) are therefore particularly useful if one wants to describe heat conduction and diffusion.

For the initial condition (2.20) the differential equation (2.19) can be transformed into an integral equation by integration over t:

$$\varrho_d(t) = \varrho + \frac{i}{\hbar} \sum_l \int_{-\infty}^{t} f_l^e(t')[q_l(t'), \varrho_d(t')]dt' \quad . \tag{2.22}$$

The average values of the quantities $q_k(t)$ and the deviations

$$\Delta Q_k(t) = Q_k(t) - Q_k(-\infty) = Q_k(t) - \mathrm{Tr}\{q_k \varrho\} = Q_k(t) - Q_k^0 \tag{2.23}$$

from their equilibrium values can then be expressed – making use of the cyclic invariance of the trace ($\mathrm{Tr}\{ab\} = \mathrm{Tr}\{ba\}$) – as

$$\boxed{\Delta Q_k(t) = \sum_l \int_{-\infty}^{+\infty} \chi_{kl}(t,t') f_l^e(t')dt' \quad .} \tag{2.24}$$

Here we have introduced the so-called dynamical susceptibility

$$\boxed{\chi_{kl}(t,t') = \frac{i}{\hbar} \langle [q_k(t), q_l(t')] \rangle_{t'} \Theta(t - t') \quad .} \tag{2.25}$$

The averaging is done with the statistical operator $\varrho_d(t')$.

For the initial condition (2.21) the susceptibility turns out to be the key quantity, too. We start out from the time derivative

$$\dot{Q}_k(t) = \frac{i}{\hbar} \mathrm{Tr}\{[H, q_k(t)]\varrho_d(0)\} \quad , \tag{2.26}$$

then using the cyclic invariance of the trace and the fact that $\varrho_d(0)$ commutes with $H - \sum f_l(0)q_l(0)$, see (2.21), we find

$$\dot{Q}_k(t) = - \sum_l \chi_{kl}(t,0) f_l(0) \quad , \quad t \geq 0 \quad , \tag{2.27}$$

and after integration using (2.23)

$$\Delta Q_k(t) = \Delta Q_k(0) - \int_0^t \sum_l \chi_{kl}(t',0) f_l(0) dt'; \quad t \geq 0 \quad . \tag{2.28}$$

Equations (2.24) and (2.28) may be considered as the central results of this section. They express causality in two ways. First of all the external forces or the initial deviations from equilibrium are the cause of changes of the $Q_k(t)$: If $f_k^e(t') = 0$ or $f_k(0) = 0$ then $\Delta Q_k(t) = 0$. Secondly the effect happens *after* the cause: For good reasons (see Chap. 17) we have solved an *initial* value problem – not a final value problem. The integrals on the r.h.s. of (2.22, 23, 28) involve only times *smaller* than t.

We conclude this section with a few remarks concerning the extension of the foregoing results to continuous and complex variables q_k. Continuously distributed variables can be treated in analogy to discrete ones by considering the corresponding integrals. For deviations of the energy density $h(r)$ and particle density $n(r)$ from their equilibrium values, for instance, the statistical operator (2.21) will involve integrals of the form

$$\int \beta(r)[h(r) - \mu(r)n(r)] d^3r \quad .$$

Such integrals, in principle, can be transformed into discrete Fourier sums for periodic boundary conditions in a box. It is often useful, however, to work directly with the integrals.

In Fourier transformations (and in other cases, too, for instance in the presence of magnetic fields) one has to deal with complex "forces" f_k and non-Hermitian "coordinates" q_k. The complete sums $\sum f_k q_k$, however, occurring in the Hamiltonian (2.4) or in the statistical operator (2.21) always have to be Hermitian.

2.2 Classical Statistics

The classical limit of the von Neumann equation (2.6) can be obtained using the results of e.g. [Ref. 2.4, Sect. 23]. Introducing the classical distribution function $\varrho(p,x,t)$ according to [Ref. 2.4, Eq. (23.8)], the commutator of (2.6) has to be replaced by the Poisson bracket for $\hbar \to 0$. Thus in the classical limit one finds the *Liouville equation*

$$\frac{\partial \varrho(p,x,t)}{\partial t} = -\frac{\partial(H,\varrho)}{\partial(p,x)} . \tag{2.29}$$

Here we have introduced the shorthand notation

$$(p,x) = (\ldots p_n \ldots, \ldots x_n \ldots)$$

and similarly

$$\frac{\partial(H,\varrho)}{\partial(p,x)} = \sum_n \frac{\partial H}{\partial p_n}\frac{\partial \varrho}{\partial x_n} - \frac{\partial H}{\partial x_n}\frac{\partial \varrho}{\partial p_n} . \tag{2.30}$$

Since Planck's constant has disappeared completely from (2.29) one expects that there is a derivation of the Liouville equation that involves only the laws of classical physics. Indeed, if we consider an ensemble in which I systems occupy the points (x^i, p^i) in phase space the distribution function has the form

$$\varrho(p,x,t) = \sum_i \frac{\delta(p-p^i)\delta(x-x^i)}{I} = \sum_i \varrho_i(p,x,t) . \tag{2.31}$$

This function is time dependent since the p_n^i and x_n^i depend on time according to the equations of motion

$$\dot{p}_n^i(t) = -\frac{\partial H}{\partial x_n^i} , \qquad \dot{x}_n^i(t) = \frac{\partial H}{\partial p_n^i} . \tag{2.32}$$

Differentiating (2.31) one obtains

$$\frac{\partial \varrho(p,x,t)}{\partial t} = -\sum \frac{\partial(\varrho_i \dot{p}^i)}{\partial p} + \frac{\partial(\varrho_i \dot{x}^i)}{\partial x} . \tag{2.33}$$

This equation, obviously, expresses the conservation of probability. The r.h.s. of (2.24) is just the negative divergence of probability current density in phase space. We now replace the phase space velocities $\dot{x}^i$, $\dot{p}^i$ by the r.h.s. of (2.32) and make use of the fact that the ϱ_i are delta functions in phase space. In the partial derivatives of H occurring on the r.h.s. of (2.32) the arguments (p^i, x^i) can be replaced by (p,x) after insertion into (2.24). Then this equation takes the form

$$\frac{\partial \varrho(p,x,t)}{\partial t} = \frac{\partial[\varrho(\partial H/\partial x)]}{\partial p} - \frac{\partial[\varrho(\partial H/\partial p)]}{\partial x} . \tag{2.34}$$

On differentiating the products the terms involving second derivatives of H cancel out and what is left is exactly (2.29).

The cancellation of second derivatives can be interpreted geometrically as a generalized "incompressibility" of the probability flow in phase space. Introducing the generalized velocity $v = (\dot{p}, \dot{x}) = (-\partial H/\partial x, \partial H/\partial p)$ in phase space one can write

$$\text{div } v = -\partial(\partial H/\partial x)/\partial p + \partial(\partial H/\partial p)/\partial x = 0 \quad . \tag{2.35}$$

This incompressibility of the flow may be illustrated by considering the motion of a volume element in phase space moving with the system points. The volume of this element remains constant (although its shape will change in general). See Fig. 2.1.

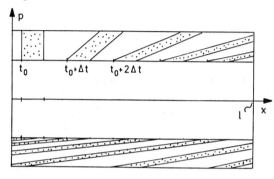

Fig. 2.1. Flow pattern of a one-dimensional ideal gas in (two-dimensional) phase space. Each point corresponds to a particle with a spatial coordinate x ($0 \leq x \leq \ell$) and a momentum p. Particles with larger p have a larger velocity $v = p/m$ and thus move faster. The area of the parallelogram containing the same particles remains constant

2.3 Approximations

The equations (2.24,28), although simple in appearance, are still far from a solution of any specific problem. The evaluation of expressions (2.25) for the susceptibility in general is, to say the least, tedious. In fact, it is fair to say that it is impossible, in practice. The equations are, however, convenient starting points for approximation schemes if the system is sufficiently close to equilibrium (or some quasi-static equilibrium). Such situations occur if the forces f^e or f are sufficiently small or sufficiently slowly varying in time. This leads to the approximation schemes of linear or slow response. A schematic view of the relations between various approximations is shown in Fig. 2.2.

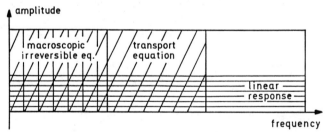

Fig. 2.2. Schematic view of the ranges of validity of approximation schemes in an amplitude-frequency diagram. Examples of macroscopic irreversible equations are hydrodynamics, heat conduction and chemical reactions. Kinetic equations: Boltzmann equation, Vlasov equation, etc. Linear response: Kubo-Greenwood-Mori formalism, etc.

PROBLEMS

2.1 Determine the interaction Hamiltonian, the external forces f_k^e and the "coordinates" q_k for
   (a) a charged particle (charge e, position x) in a uniform electric field $E(t)$,
   (b) a magnetic moment m in a uniform magnetic field $B(t)$,
   (c) electrons in an arbitrary electromagnetic field $E(r,t), B(r,t)$.

2.2 Determine the statistical operator $\varrho_d(0)$ with maximum entropy, the Lagrange parameters $f_k(0)$ and the "coordinates" q_k for
   (a) a "Brownian particle" (plant pollen in water, charged oil drop in air, electron in a semiconductor, etc.) with initial momentum $\langle p(0) \rangle = P$,
   (b) a "Brownian oscillator" with initial momentum P and initial coordinate $\langle x(0) \rangle = X$,
   (c) a gas or liquid with initial particle number density $\langle n(r,0) \rangle = n(r)$ and energy density $\langle h(r,0) \rangle, \varepsilon(r)$.

2.3 Application of the Liouville theorem (constancy of phase space volume): An ion source emits ions with energies $E_1 = 5 \pm 1\text{keV}$ from an area $F_1 = 1\text{mm}^2$ into a solid angle $\Omega_1 = 1/10\text{sterad}$. The ions are accelerated to an energy $E_2 = 10\text{MeV}$ and focused onto an area $F_2 = 1\text{cm}^2$. Calculate the final solid angle Ω_2 of the ion beam using the Liouville theorem.

3. Small Amplitude Perturbation Theory (Linear Response)

The basic equations (2.24, 28) are, of course, far from a solution of the problem of nonequilibrium statistical physics. The evaluation of the expression (2.25) for the susceptibility is difficult. In fact it is impossible in the general nonlinear case. The situation is improved a great deal for sufficiently small amplitudes of the external forces $f^e(t)$ or the Lagrange parameters $f(0)$. Then a linear approximation for the response $Q(t)$ can be adopted in which the statistical operator $\varrho_d(t)$ is replaced by the operator ϱ of total equilibrium. Furthermore, the initial deviation $Q_k(0)$ occurring in (2.28) can be expressed in terms of the $f_k(0)$ by means of the isothermal susceptibility χ^T in thermal equilibrium as

$$\Delta Q_k(0) = \sum_l \chi^T_{kl} f_l(0) \quad . \tag{3.1}$$

The notation can be made a little more compact by defining the so-called relaxation function Φ by [an i being introduced in t-space in order to avoid it in ω-space later on, for instance in (3.32)]

$$\Phi_{kl}(t) = i \left(\chi^T_{kl} - \int_0^t \chi_{kl}(t', 0) dt' \right) \Theta(t) \quad . \tag{3.2}$$

Then (2.28) can be written as

$$\Delta Q_k(t) = -i \sum_l \Phi_{kl}(t) f_l(0) \quad ; \quad t \geq 0 \quad . \tag{3.3}$$

The averaging in the dynamical suceptibility (2.25) occurring in (2.24) and (3.2) is now performed with the statistical operator ϱ of total equilibrium. The susceptibility is thus independent of the external forces f^e or Lagrange parameters f. The response ΔQ therefore is linear in the forces f^e for the external field case or in the Lagrange parameters f for the initial value case. Since $\varrho = \exp[\beta(F - H)]$ commutes with H, the dynamical susceptibility in equilibrium is invariant against translation in time, i.e.,

$$\chi_{kl}(t,t') = \chi_{kl}(t-t',0) = \chi_{kl}(t-t'). \tag{3.4}$$

Here the second equality sign serves as the definition of the function $\chi_{kl}(t)$ depending on only a single time variable. The integral on the r.h.s. of (2.24) is then a convolution of the dynamical susceptibility with the external force. The Fourier transform of (2.24) then takes the form of a product

$$\Delta Q_k(\omega) = \sum_l \chi_{kl}(\omega) f_l^e(\omega) \tag{3.5}$$

with the Fourier transform

$$\chi_{kl}(\omega) = \frac{i}{\hbar} \int_0^t \langle [q_k(t), q_l(0)] \rangle e^{i\omega t} dt \tag{3.6}$$

of the dynamical susceptibility.

An alternative expression for the relaxation function can be derived by a direct application of thermodynamic perturbation theory, i.e., a power expansion of

$$\varrho_d = \text{const} \times \exp\left\{-\beta\left[H - \sum_l f_l(0) q_l(0)\right]\right\} \tag{3.7}$$

in powers of $f(0)$. In quantum mechanics one has to bear in mind that the $q(0)$ do not in general commute with H. The "quantum Taylor series" up to the first order takes the form (see e.g. [Ref. 3.1, Sect. 26])

$$\varrho_d = \left(1 + \sum_l \int_0^\beta \Delta q_l(\alpha)\, d\alpha\, f_l(0)\right) \varrho + O(f^2) \ . \tag{3.8}$$

Here we have introduced the notation

$$a(\alpha) = \exp(-\alpha H) a(0) \exp(\alpha H) \tag{3.9}$$

for arbitrary operators a. [For the evolution of the same operator in time we keep – though not quite consistent with (3.9) – the notation $a(t)$.] In (3.7) the deviations $\Delta q(\alpha)$ from the equilibrium values Q^0 occur instead of the $q(\alpha)$ themselves: this takes care of the normalisation condition $\text{Tr}\{\varrho_d\} = 1$ and determines the "const" in (3.7) up to first order in $f(0)$.

For the evolution of averages such as $\Delta Q_k(t) = \text{Tr}\{\Delta q_k(t) \varrho_d\}$ in time one obtains, using (3.8),

$$\Delta Q_k(t) = \sum_l \beta \langle \Delta q_k(t); \Delta q_l(0) \rangle f_l(0) \quad . \tag{3.10}$$

Here we have introduced a symbol which will occur frequently from now on:

$$\boxed{\beta \langle a; b \rangle = \int_0^\beta \langle ab(\alpha) \rangle d\alpha = \int_0^\beta \mathrm{Tr}\{ab(\alpha)\varrho\} d\alpha} \quad . \tag{3.11}$$

This symbol is reminiscent of the Dirac expression for a scalar product in Hilbert space. We shall, indeed, see later on that the bracket symbol introduced in (3.10) has the properties of a generalised scalar product (in what is called a Liouville space). If we identify a relaxation function Φ now as

$$\boxed{\Phi_{kl}(t) = \mathrm{i}\beta \langle \Delta q_k(t); \Delta q_l(0) \rangle \Theta(t)} \tag{3.12}$$

we can write (3.10) again in the form (3.3). If one compares (3.2) and (3.3,10) in the limit of $t \to 0$ one finds

$$\lim_{t \to 0} \Phi_{kl}(t) = \mathrm{i}\beta \langle \Delta q_k; \Delta q_l \rangle = \mathrm{i}\chi_{kl}^T \quad . \tag{3.13}$$

The relaxation function can therefore be considered as a generalisation of the isothermal susceptibility of equilibrium thermodynamics to nonequilibrium processes. Another generalisation is, of course, the dynamical susceptibility $\chi(t)$ or its Fourier transform (3.6). In the static limit we define

$$\chi_{kl}(\omega = 0) = \int_0^\infty \chi_{kl}(t, 0) dt = \chi_{kl}^0 \quad . \tag{3.14}$$

If this result is combined with (3.2) one finds

$$\lim_{t \to 0} \Phi_{kl}(t) = \mathrm{i}(\chi_{kl}^T - \chi_{kl}^0) \quad . \tag{3.15}$$

Since in the external field case (2.20) the system is thermally isolated during the action of the external forces it cannot come to an equilibrium with the heat bath. The quasi-static isolated susceptibility χ^0 therefore will be different from the isothermal susceptibility χ^T. If the system comes to equilibrium internally (i.e., if it is "ergodic") one expects χ^0 to be equal to the adiabatic thermodynamic susceptibility. This can easily be checked. On the one hand, if the limit of $\Phi(t)$ for t going to infinity exists, it can be obtained from

$$\lim_{t \to \infty} \Phi(t) = \lim_{\eta \to 0} \eta \int_0^\infty e^{-\eta t} \Phi(t)\, dt \quad , \quad \eta \geq 0 \quad . \tag{3.16}$$

Now since

$$\lim_{\eta \to 0} \eta \int_0^\infty e^{-\eta t} e^{-i\omega t} dt = \begin{cases} 1 & \text{if } \omega = 0 \\ 0 & \text{if } \omega \neq 0 \end{cases} \quad (3.17)$$

only the zero-frequency Fourier components of $\Delta q_k(t)$ remain in (3.12). Denoting them by Δq_k^0 we can write

$$\chi_{kl}^0 = \chi_{kl}^T - \beta \langle \Delta q_k^0 \Delta q_l^0 \rangle \quad . \quad (3.18)$$

Here we have replaced the symmetric correlation (3.10) by the ordinary product correlation since the zero-frequency components commute with the Hamiltonian. If the system is ergodic the eigenvalues of H are nondegenerate and the zero-frequency part of q_k can be written as

$$q_k^0 = \sum \langle \nu | q_k | \mu \rangle | \mu \rangle \langle \mu | \quad , \quad (3.19)$$

where $|\mu\rangle$ are the eigenstates of H with the eigenvalues E_μ. Equation (3.19) says that q_k^0 is a function of H. Now for a macroscopic system the fluctuations of H are small compared to the average energy. We therefore expect a linear relation between Δq_k^0 and ΔH of the form

$$\Delta q_k^0 = \frac{\langle \Delta q_k^0 \Delta H \rangle}{\langle \Delta H^2 \rangle} \Delta H \quad . \quad (3.20)$$

Inserting this into (3.18) we find (note that $\langle \Delta q_k^0 \Delta H \rangle = \langle \Delta q_k \Delta H \rangle$)

$$\chi_{kl}^0 = \chi_{kl}^T - \frac{\beta \langle \Delta q_k \Delta H \rangle \langle \Delta H \Delta q_l \rangle}{\langle \Delta H \Delta H \rangle} \quad . \quad (3.21)$$

Now from thermodynamics one derives (see Problem 3.3)

$$\chi_{kl}^S = \chi_{kl}^T - \frac{T}{C_f} \left(\frac{\partial Q_k}{\partial T}\right)_f \left(\frac{\partial Q_l}{\partial T}\right)_f \quad . \quad (3.22)$$

Where C_f is the specific heat for constant forces $f_k = 0$. Because of the general relations between second derivatives of the generalized free enthalpy K and thermodynamical fluctuations

$$C_f = \beta \langle \Delta H^2 \rangle \quad , \quad (\partial Q_k / \partial T)_f = \beta \langle \Delta q_k \Delta H \rangle \quad , \quad (3.23)$$

one has $\chi_k^0 = \chi_k^S$ as expected above.

If the system is internally nonergodic one has further constants C of motion besides H. Then (3.20) will have further additive terms on the r.h.s. proportional to ΔC. Consequently (3.21) will have further subtraction terms on the r.h.s. and χ_{kl}^0 will in general be smaller than χ_{kl}^S.

Nonergodicity in general will lead to a nonzero limit of $\Phi(t)$ for t going to infinity: the system does not relax back to its isothermal equilibrium value $\Phi = 0$.

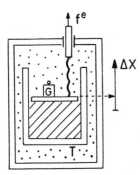

Fig. 3.1. External force acting on a piston compressing a gas in a heat bath at temperature T. The time dependence of $f^e(t)$ and piston displacement $\Delta X(t)$ corresponding to the "initial value case" (2.21) are indicated in Fig. 3.2

A particularly simple case of nonergodicity occurs if the isothermal and adiabatic susceptibilities are different. In this case equilibrium between the system and its heat bath is not established.

As an example we consider a gas in a heat bath kept at a pressure P_0 by a weight G. The external force f^e is exerted by a spring, see Fig. 3.1.

Initially the piston will be lifted up by the external force isothermally and quasi statically to a position $\Delta X(0) = \chi^T f^e(0)$ (Fig. 3.2). Then the force is switched off and the system relaxes back adiabatically to the final position $\Delta X(\infty) = (\chi^T - \chi^S) f(0)$.

A similar example is the relaxation of a ferromagnet after switching off a magnetic field, or a ferroelectric sample in an electric field, etc. Further examples of nonergodic behaviour occur for conserved quantities. We will see in Sect. 10.1 that the linear response of conserved quantitites to external forces vanishes. Thus for such quantities $\chi^0 = 0$ and the relaxation function approaches $i\chi^T$ for large times. In fact it is equal to it for all times.

In all these examples the nonergodic behaviour occurs only for a single variable, or at most very few. Examples with infinitely many nonergodic variables are ideal gases or harmonic solids. In these cases all energies of gas particles or normal modes are conserved. In principle every system has an infinite number of conserved quantities: the projection operators projecting onto the eigenstates of H. All these conserved observables, however, are not accessible individually

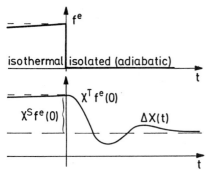

Fig. 3.2. External force and shift of the position of the piston in Fig. 3.1 from its equilibrium value as a function of time. The external force is increased infinitely slowly to its value at $t = 0$ and then switched off abruptly. The piston follows this force isothermally up to its value at $t = 0$ and then relaxes adiabatically back to a new equilibrium position at large t with a temperature somewhat larger than T

in macroscopic experiments (except perhaps vibrational normal modes in solids). Physically accessible variables are usually linear combinations of the conserved variables, which relax to zero because of the dispersion of energies in the original conserved quantities. For instance, even in harmonic solids, variables pertaining to a single point (or small localized regions) in coordinate space behave ergodically and relax to their equilibrium values. A nontrivial example where local excitations behave nonergodically occurs for electrons in strong random potentials in the context of Anderson localisation.

The two expressions (3.2,12) can be compared more directly if one writes (3.12) as

$$\Phi_{kl}(t) = i\beta \left[\langle \Delta q_k(0); \Delta q_l(0) \rangle + \int_0^t \langle \dot{q}_k(t'); q_l(0) \rangle dt' \right] \Theta(t) \quad . \tag{3.24}$$

In the second term on the r.h.s. of this equation we have omitted the Δ's, since in the time derivative the constant term $Q_k(-\infty)Q_l(-\infty)$ drops out. Looking at (3.2) one expects the relation

$$\chi_{kl}(t,0) = -\beta \langle \dot{q}_k(t); q_l(0) \rangle \Theta(t) = \beta \langle q_k(t), \dot{q}_l(0) \rangle \Theta(t) \quad , \tag{3.25}$$

where the second equation follows from the first one because of the invariance of equilibrium expectation values under time translations. A direct proof of (3.25) is possible using an important symmetry relation of equilibrium expectation values of products of operators [3.2, 3] following from the cyclic invariance of the trace:

$$\langle a(t)b(0) \rangle = \langle b(\beta)a(t) \rangle = \langle b(0)a(t - i\hbar\beta) \rangle \quad . \tag{3.26}$$

The second equation again is a consequence of time translation invariance. Similarly one has

$$\langle b(0)a(t) \rangle = \langle a(t)b(\beta) \rangle = \langle a(t - i\hbar\beta)b(0) \rangle \quad . \tag{3.27}$$

Now using (3.9) and (2.17) one can write

$$b(\beta) = b(0) - \int_0^\beta [H, b(\alpha)] d\alpha = b(0) - \frac{\hbar}{i} \int_0^\beta \dot{b}(\alpha) d\alpha \quad , \tag{3.28}$$

the dot indicating a time derivative. If one inserts (3.28) into (3.27) and uses (3.11), one does indeed arrive at (3.25). Needless to say, using (3.24) and (3.2) as derived from (2.28) one can obtain the expression (3.12) for the relaxation function without the use of thermodynamic perturbation theory, e.g. (3.8). The Fourier transform of (3.26, 27), as we shall see later, is the famous detailed balance relation (Callen and Welton (1951)). [3.4].

Equation (3.25) yields general expressions for dynamical susceptibilities in classical statistics without commutators or Poisson brackets. A glance at the expression for the bracket symbol (3.11) shows that it approaches the expectation value of the product in the classical limit:

$$\langle a; b \rangle \to \langle ab \rangle \quad \text{(classical limit)} \quad . \tag{3.29}$$

Thus the dynamical suceptibilities in the classical limit can be written as

$$\chi_{kl}(t) \to -\beta \langle \dot{q}_k(t) q_l(0) \rangle \Theta(t) = \beta \langle q_k(t) \dot{q}_l(0) \rangle \quad \text{(classical limit)} \quad . \tag{3.30}$$

If one differentiates the relaxation function with respect to t one finds, using (3.2),

$$\chi_{kl}(t) = \chi_{kl}^{\mathrm{T}} \delta(t) + \mathrm{i} d\Phi_{kl}(t)/dt \tag{3.31}$$

or after Fourier transformation

$$\boxed{\chi_{kl}(\omega) = \chi_{kl}^{\mathrm{T}} + \omega \Phi_{kl}(\omega) \quad .} \tag{3.32}$$

The central result contained in (3.3, 5, 6, 12) is Onsager's famous "regression theorem": The regression of (small) *spontaneous fluctuations* in equilibrium follows the laws of the relaxation of forced *average deviations* from their equilibrium values.

Summing up foregoing and anticipating later results one can say that (a) the relaxation of weakly coupled nonequilibrium averages, (b) the response to (classical) external forces, (c) the correlation of equilibrium fluctuations and (d) the inelastic scattering of quantum objects such as light quanta, neutrons and electrons can all be described in terms of the same type of correlation functions. It is thus not surprising that a vast number of experimental results and theoretical investigations are expressed and formulated in terms of these correlation functions; for example, lineshape studies of electronic, infrared and Raman spectra; inelastic scattering results of light, neutrons and electrons; lineshapes in magnetic and dielectric experiments; acoustic and transport data; and many others.

PROBLEMS

3.1 Electrons of (gas) atoms are in equilibrium at a temperature T. They are then exposed to a (weak) uniform electric field $E(t)$ in the x direction. Use linear response theory to calculate the response of the x coordinate of the electrons. Evaluate the expressions for the dynamical susceptibility in terms of the eigenstates of the unperturbed electrons and compare the results for the polarisibility $\alpha = e^2 \chi$ in the limit $T \to 0$ with the results of quantum mechanical dispersion theory.

3.2 A harmonic crystal with the Hamiltonian $H = \sum \omega_k b_k^* b_k$ is exposed to an external perturbation $-f(t)q$ where $q = \sum q_k(b_k + b_k^*)$; b_k (b_k^*) are the usual annihilation (creation) operators for bosons with commutation rules $[b_k, b_q] = [b_k^*, b_q^*] = 0$ and $[b_k, b_q^*] = \delta_{kq}$. Calculate the dynamical susceptibility, the relaxation function $\Phi(t)$ and their Fourier transforms. In particular, for $q_k = \bar{q}/\sqrt{\omega_k v}$ (v the volume of the periodicity box), corresponding to a perturbation strongly localized in space, evaluate $\Phi(t)$ explicitly for a Debye spectrum of the harmonic frequencies ω_k.

3.3 Using thermodynamic relations, express the quantity $(\chi^T - \chi^S)$ occurring in the large time asymptotic value (3.22) of the relaxation function for the example of Fig. 3.1 in terms of the thermal expansion coefficient α.

4. Brownian Motion (Relaxator) *

In the next two chapters we are going to apply the general formalism to two very simple examples. The aim is twofold: first of all, pedagogical. We want to demonstrate how some well-known results manifest themselves in the framework of the formalism of the foregoing sections. Secondly, we want to use the equations of Brownian motion as a heuristic starting point for more general kinetic equations.

As a first example we consider the relaxation of a single variable $Q(t)$ to its equilibrium value, which we define as zero. At time $t = 0$ we assume a (small) deviation $Q(0) \neq 0$ from this equilibrium value. We describe the relaxation of this deviation back to equilibrium under three simplifying conditions: (i) linear response, (ii) classical approximation for the relaxation function, (iii) existence of the first-order phenomenological differential equation in time. This leads us to the equation of the so-called relaxator

$$\dot{Q}(t) = -\gamma Q(t) \quad . \tag{4.1}$$

In general the phenomenological constant γ could be a function of $Q(t)$. We assume that γ for $Q \rightarrow 0$ approaches a certain nonvanishing limit which then because of condition (i) occurs in (4.1). The solution of (4.1) with the given initial value $Q(0)$,

$$Q(t) = Q(0)e^{-\gamma t} = Q(0)e^{-t/\tau} \quad , \tag{4.2}$$

then describes an exponential relaxation towards equilibrium with a relaxation time $\tau = 1/\gamma$.

Let us consider a few examples of equations of the type (4.1).

a) Particles moving in a medium under the influence of *friction*, for instance electrons in a conducting medium, or "Brownian particles" (plant pollen in water, oil drops or dust particles in air), obey (4.1) if $Q(t)$ is identified with the average (drift) momentum $P(t)$ of the particles.

b) The overdamped oscillator. In this case the equation of motion for the position $X(t)$ of the oscillator takes the form

$$m(\ddot{X} + \varrho\dot{X}) + f(X) = 0 \quad . \tag{4.3}$$

This equation is of second order in t, but for sufficiently strong damping the acceleration term $m\ddot{X}$ can be neglected. In the linear regime one then arrives at

$$\dot{X} = -[f'(0)/(m\varrho)]X \quad . \tag{4.4}$$

This is an equation of type (4.1) if Q is identified with X and

$$\gamma = f'(0)/(m\varrho) = \omega_0^2/\varrho \quad ; \tag{4.5}$$

ω_0 is the vibrational frequency of the undamped oscillator.

c) *An electrical* example of an overdamped oscillator can be found in the *RC element*. It consists of a capacitor with capacity C whose plates are connected via an Ohmic resistor (resistance R), neglecting all inductances, see Fig. 4.1.

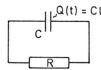

Fig. 4.1. Schematic for an RC circuit

In the absence of external voltages one has the equation

$$R\dot{Q} = -Q/C \quad . \tag{4.6}$$

This is again an equation of type (4.1) with a relaxation time

$$\tau = 1/\gamma = RC \quad . \tag{4.7}$$

d) *A thermodynamic* example is the *heat transfer* between two systems of different temperature. In equilibrium two systems in thermal contact have the same temperature T. Suppose this equilibrium is disturbed and a deviation ΔT from the equilibrium temperature exists at $t = 0$. This deviation will also relax to zero according to an equation of type (4.1) if ΔT is identified with Q. In contrast to the examples a, b, c, then Q is a purely thermodynamical quantity. With the aid of the specific heat C, however, it can be transformed into the energy difference $e = C\Delta T$, i.e. a mechanical quantity.

We now turn to (3.10) in order to determine the relaxation of the correlations $\langle q(t); q(0)\rangle$. According to the simplifying condition (ii), it is sufficient to consider

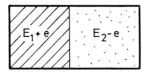

Fig. 4.2. Two systems with energies $E_1 + e$ and $E_2 - e$ in thermal contact; $E_{1,2}$ the equilibrium values of the energies. For sufficiently small thermal conductivity of the contact the two systems have homogeneous deviations $\Delta T_{1,2} = e/C_{1,2}$ of their temperatures from the equilibrium value

the classical limit (3.21), i.e. $\langle q(t)q(0)\rangle$. To fix our attention we consider the example a), i.e. a Brownian particle, and identify q with the momentum p of the particle.

We do not determine the Lagrange parameter $f(0)$ explicitly (see Problem 2.2a) but only use the fact that the time dependence of the correlation $\langle p(t)p(0)\rangle$ is proportional to a macroscopic deviation $P(t)$ from equilibrium, the proportionality constant being determined directly from the initial condition:

$$\langle p(t)p(0)\rangle = \langle p(0)^2\rangle e^{-\gamma t} \quad , \quad t \geq 0 \quad . \tag{4.8}$$

The initial value then follows directly from the equipartition theorem $\langle p(0)^2\rangle = mkT$. Now equilibrium expectation values are invariant against translations in time

$$\langle p(t)p(0)\rangle = \langle p(t+\tau)p(\tau)\rangle = \langle p(0)p(-t)\rangle \quad . \tag{4.9}$$

(The second equation is a special case of the first one for $\tau = -t$.) Thus one can generalize (4.9) to

$$\langle p(t')p(t'')\rangle = mkT\exp(-\gamma|t'-t''|) \quad ; \quad t', t'' \text{ arbitrary} \quad . \tag{4.10}$$

The remarkable fact expressed by this equation is that the microscopic thermal fluctuations of the momentum $p(t)$ of a Brownian particle have a "memory" over times of the order of the macroscopic relaxation time $\tau = 1/\gamma$, although (4.1) at first sight does not contain any memory effects: $\dot{Q}(t)$ is directly determined by $Q(t)$ and not the earlier history of $Q(t-\tau)$.

In order to determine the correlation function (4.10) from a given $p(t)$ one has to replace the ensemble average by an average over time

$$\langle p(t)p(0)\rangle = \int_0^{t_0} p(t+t')p(t')dt'/t_0 \tag{4.11}$$

where t_0 has to be sufficiently large compared to τ. For practical purposes a discretised version of (4.11) is preferable

$$\langle p(t)p(0)\rangle = \sum_{i=1}^{I} p(t+t_i)p(t_i)/I \quad . \tag{4.12}$$

Here the t_i are smoothly (for instance equidistantly, $t_i = it_0/I$) distributed "observation times".

One should bear in mind that the correlation function (4.10) determines $p(t)$ only "on the average". So curves $p(t)$ which look rather different in detail can have the same correlation (4.10). Figure 4.3 exhibits two limiting cases corresponding to either a "light" Brownian particle suffering a few strong collisions after times of the order of $\Delta\tau \approx \tau$, or a "heavy" particle suffering many weak collisions with $\Delta\tau \ll \tau$. The so-called collision time $\Delta\tau$ (the average *distance*

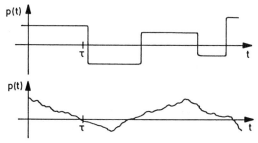

Fig. 4.3. Typical curves $p(t)$ for the thermal fluctuations of a Brownian particle. Upper curve: "light" particle, strong momentum changes after a collision time $\Delta\tau \approx \tau$. Lower curve: "heavy" particle, many small momentum changes after collision times, few large changes after times of the order of τ

between successive statistically independent collisions) – in contrast to the *duration* $\delta\tau$ of a single collision, as we shall see later on – does not show up in general in the correlation function (4.10).

Experimentally it is easier to determine the position than the momentum of a Brownian particle. An important result concerning the statistical evolution of position in time can be derived by integrating (4.10) over t' and t'' from 0 to t using $p = m\dot{x}$, i.e.

$$x(t) - x(0) = \int_0^t p(t')dt'/m \quad . \tag{4.13}$$

After some intermediate steps one arrives at

$$\langle [x(t) - x(0)]^2 \rangle = \frac{2kT}{m\gamma}[t - (1 - e^{-\gamma t})/\gamma] \quad ; \quad t \geq 0 \quad . \tag{4.14}$$

This tells us how the average mean square displacement of the position from its initial value $x(0)$ evolves in time. See Fig. 4.4.

Two limiting cases of this result are easy to interpret: the cases where the time t is either large or small compared to the relaxation time τ. Then one obtains by expansion of (4.14)

$$\langle [x(t) - x(0)]^2 \rangle = (kT/m) \begin{cases} T^2 & 0 \leq t \ll \tau, \\ 2t/\gamma & t \gg \tau. \end{cases} \tag{4.15}$$

For the interpretation of these limiting results let us consider the upper part of Fig. 4.3. For times t small compared to the collision time $\Delta\tau \approx \tau$ the

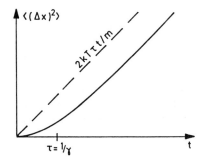

Fig. 4.4. Dependence of mean square position displacement of a Brownian particle on time

motion of the particle can be considered as approximately free. Thus one has $x(t) - x(0) = \Delta x(t) = vt$, v being the thermal velocity. The equipartition law then leads to

$$\langle (\Delta x)^2 \rangle = \langle v^2 \rangle t^2 = kTt^2/m \tag{4.16}$$

in agreement with (4.15).

For times t large compared to τ, on the other hand, one can subdivide Δx into approximately $N = t/\tau$ statistically independent parts $\Delta_n x$, each of which describes an approximately free motion:

$$\Delta x = \sum_{n=1}^{N} \Delta_n x \quad . \tag{4.17}$$

Because of the statistical independence of the terms in this sum the mean squared average will be

$$\langle (\Delta x)^2 \rangle = \sum \langle \Delta_n x \Delta_m x \rangle = \sum \langle (\Delta_n x)^2 \rangle = N \langle (\Delta_n x)^2 \rangle \quad . \tag{4.18}$$

Using $\langle (\Delta_n x)^2 \rangle = \langle v^2 \rangle \tau^2$ and $N = t/\tau$ one finds

$$\langle (\Delta x)^2 \rangle = N \langle v^2 \rangle \tau^2 = kT\tau t/m \tag{4.19}$$

again in qualitative agreement with the large t limit of (4.15).

After these more pedagogical considerations let us turn to the second aim of this (and the next) chapter: the development of a "canonical" form of kinetic equations. Let us consider the relaxation function (3.12) in particular. In our special case

$$\Phi(t) = i\beta \langle p(t)p(0) \rangle \Theta(t) \quad . \tag{4.20}$$

Differentiating (4.8) one sees that the relaxation function obeys the differential equation

$$(-id/dt - i\gamma)\Phi(t) = \chi^T \delta(t) \tag{4.21}$$

with the isothermal susceptibility

$$\chi^T = \beta \langle p^2 \rangle = m \quad . \tag{4.22}$$

In Fourier space (4.21) may be written as

$$\Phi(\omega) = -(i\gamma + \omega)^{-1}\chi^T \quad , \tag{4.23}$$

so the relaxation function $\Phi(\omega)$ has a simple pole in ω-space at $\omega = -i\gamma$ and behaves asymptotically for large ω as $-\chi^T/\omega$. Such a simple structure, of course, cannot be correct in general. It is useful, however, to write the relaxation function in general as

$$\Phi(\omega) = [N(\omega) - \omega]^{-1} \qquad (4.24)$$

in terms of a memory function $N(\omega)$. Introducing its Fourier transform $N(t)$ one can write (4.24) in t-space as

$$-id\Phi(t)/dt - \int_0^t N(t-t')\Phi(t')dt' = \chi^T \delta(t) \quad . \qquad (4.25)$$

The rate of change of the relaxation function at time t is not determined by $\Phi(t)$ alone. It remembers the earlier history of $\Phi(t)$. Only if the memory of $N(t)$ is short, more precisely if $N(t)$ has a decay length $\delta\tau \ll \tau$, does (4.25) reduce approximately to (4.21) for $t \gg \delta\tau$. The relaxation constant γ then is given in terms of the memory kernel $N(t)$ as

$$\int_0^t N(t)dt = N(\omega = 0) = -i\gamma \quad . \qquad (4.26)$$

Physically, in our special case of Brownian motion the memory decay length corresponds to the duration $\delta\tau$ of collisions of the Brownian particle with the particles of the heat bath. In the upper curve of Fig. 4.3 $\delta\tau$ would correspond to the width of the transitions between the various steps of $p(t)$. Whenever this width is small compared to the average length τ of the steps one can expect the approximate validity of (4.21) for large times, or (4.23) for low frequencies.

The existence of two (or more) well-separated time scales is typical for many situations of irreversible thermodynamics. Mathematically it usually allows the derivation of kinetic equations which are "local in time" (or so-called Markovian equations) such as (4.1) or (4.21).

PROBLEMS

4.1 Solve (4.1) with an inhomogeneity $i(t)$, i.e. $\dot{Q}(t)+Q(t) = i(t)$, either directly in t-space (for instance by "variation of constants" or in ω-space. Compare the results. Consider, in particular the cases $i(t) = i_0\delta(t)$ and $i(t) = i_0\cos\omega t$ or $i(t) = i_0\exp(-i\omega t)$.

4.2 The inhomogeneity of the foregoing problem is proportional to the external "force" $f^e(t)$: $i(t) = af^e(t)$. Determine the constant a by considering the static case $dQ/dt = 0$ for which $Q = \chi^T f^e$. With the corresponding value of a determine the dynamical susceptibility $\chi(\omega) = f^e(\omega)/Q(\omega)$ by Fourier transformation of $\dot{Q} + \gamma Q = af^e$.

4.3 Consider the case of Brownian motion in a medium with the isothermal susceptibility (4.22). What is the dimensionality of the external "force" (compare Problem 2.2a)? Determine the dynamical susceptibility from the relaxation function (4.23) using (3.32), i.e. $\chi(\omega) = \chi^T + \omega\Phi(\omega)$, and compare the result with Problem 4.2.

4.4 A Brownian particle (radius $r = 10^{-6}$cm, molecular weight 10^4) moves in a medium with viscocity η (for instance water with $\eta = 0.018\,\mathrm{g\,cm^{-1}s^{-1}}$). How far has it moved away from its initial position on the average at room temperature in a day? For the calculation of the relaxation time τ use Stokes' formula $dP/dt = -6\pi \eta r P/m$. What is the value of τ?

5. Brownian Motion (Oscillator) *

A second simple and important example is the Brownian motion of a harmonic oscillator [5.1–3]. Calling the displacement of the oscillator from its equilibrium position ($X_0 = 0$) $X(t)$, one obtains the equation of motion in the presence of an external force f^e

$$m(\ddot{X} + \omega_0^2 X + \gamma \dot{X}) = f^e(t) \quad . \tag{5.1}$$

The solution of the homogeneous equation [$f^e = 0$, with given initial values $X(0)$ and $\dot{X}(0)$] reads

$$X(t) = e^{-\gamma t/2}\{X(0)\cos(\omega_r t) + [\dot{X}(0) + \gamma X(0)/2]\sin(\omega_r t)/\omega_r\} \tag{5.2}$$

with the resonance frequency

$$\omega_r = \sqrt{\omega_0^2 - \gamma^2/4} \quad . \tag{5.3}$$

As in the previous chapter, we want to determine the correlation function $\langle x(t)x(0)\rangle$. Since position and velocity at equal times (for instance at $t = 0$) are uncorrelated one has the initial condition

$$\langle \dot{x}(0)x(0)\rangle = 0 \quad . \tag{5.4}$$

Using the regression theorem together with (5.2, 4) one finds

$$\langle x(t)x(0)\rangle = \langle x(0)^2\rangle e^{-\gamma t/2}[\cos(\omega_r t) + \gamma \sin(\omega_r t)/(2\omega_r)] \quad . \tag{5.5}$$

The value at $t = 0$ follows from the second initial condition, the equipartition law, which now reads

$$\langle x(0)^2\rangle = kT/(m\omega_0^2) \quad . \tag{5.6}$$

It is again instructive to derive the same results using the dynamical susceptibility $\chi(\omega) = X(\omega)/f^e(\omega)$. Fourier transformation of (5.1) yields

$$\chi(\omega) = 1/[m(\omega_0^2 - \omega^2 - i\omega\gamma)] \quad . \tag{5.7}$$

From this the relaxation function can again be determined using (3.32), i.e. $\chi = \chi^T + \omega\Phi$ with $\chi^T = 1/m\omega_0^2$. One finds

$$\Phi(\omega) = \frac{\omega + i\gamma}{m\omega_0^2(\omega_0^2 - \omega^2 - i\gamma\omega)} \quad . \tag{5.8}$$

From this one can calculate $\Phi(t)$ by Fourier back transformation. One obtains (most simply by application of Cauchy's theorem in the complex ω-plane), see Problem 5.1, after some intermediate steps

$$\Phi(t) = i\langle x(t)x(0)\rangle\Theta(t)/(kT) \quad , \tag{5.9}$$

where $\langle x(t)x(0)\rangle$ is given by (5.5).

If the numerator and denominator of (5.8) are divided by $\omega+i\gamma$, the relaxation function can be cast into the general form (4.24) with

$$N(\omega) = \frac{\omega_0^2}{\omega + i\gamma} \quad . \tag{5.10}$$

The main difference compared to the relaxator of the previous chapter is that, now, the memory function has a nontrivial frequency dependence, except in the limit of the overdamped oscillator [see (4.5) – note the difference in notation between γ and ρ – and Problems 5.2, 5.3].

The relaxation function of the oscillator, in contrast to the relaxator, has *two* poles in the complex ω-plane. This is, of course, a direct consequence of the occurrence of *second*-order time derivatives in the equation of motion (5.1).

Now, in analogy to Hamilton's equations in mechanics, the second-order differential equation (5.1) can be transformed into a system of first-order equations by introducing the two variables $Q_1(t) = X(t)$ and $Q_2 = P(t) = mdX(t)/dt$. The single second-order equation (5.1) is then equivalent to the two first-order equations $dX/dt = p/m$ and $dP/dt = -m\omega_0^2 X - \gamma P$, or, in matrix notation,

$$\frac{dQ_k}{dt} = -\sum \gamma_{kl} Q_l \tag{5.11}$$

with

$$\{\gamma\} = \begin{pmatrix} 0 & -1/m \\ m\omega_0^2 & \gamma \end{pmatrix} \quad . \tag{5.12}$$

The analogue of (4.21) now also becomes a matrix equation

$$\sum_l (-i\delta_{kl} d/dt - i\gamma_{kl})\Phi_{lm}(t) = \chi_{km}^T \delta(t) \tag{5.13}$$

with the isothermal susceptibility matrix

$$\{\chi_{km}^T\} = \beta\{\langle q_k q_m\rangle\} = \begin{pmatrix} 1/m\omega_0^2 & 0 \\ 0 & m \end{pmatrix} \quad . \tag{5.14}$$

Equations (4.23–25) also become matrix equations. For instance, (4.23) takes the form

$$\Phi_{km}(\omega) = -\sum [(\mathrm{i}\gamma+\omega)^{-1}]_{kl}\chi_{lm}^T \quad . \tag{5.15}$$

The essential point now is that (4.26) corresponds to a frequency-independent matrix $N_{kl}(\omega) = -\mathrm{i}\gamma_{kl}$. Thus by enlarging the number of variables we have eliminated the memory effects contained in (5.10). This situation is typical for irreversible thermodynamics. In order to have Markovian kinetic equations such as (5.13) with constant "kinetic coefficients" γ_{kl} one has to chose a sufficiently large set of independent variables Q_k. More precisely, one has to include in this set all independent "slow" variables (in contrast to the fast variables of the rest of the system, the "heat bath"). In our special case of the oscillator there are two independent slow variables.

Finally we mention another version of (5.11) that has certain advantages. One expresses the Q_k on the r.h.s. of (5.11) in terms of the Lagrange parameters f_k (physically speaking the thermodynamically conjugate "forces") in a linear approximation

$$Q_k = \sum_l \chi_{kl}^T f_l \quad . \tag{5.16}$$

Then, introducing the "mobility" matrix

$$\{\mu_{km}\} = \left\{\sum_l \gamma_{kl}\chi_{lm}^T\right\} = \begin{pmatrix} 0 & -1 \\ 1 & m\gamma \end{pmatrix} \quad , \tag{5.17}$$

one can write (5.11) as

$$\boxed{dQ_k/dt = -\sum_l \mu_{kl}f_l \quad .} \tag{5.18}$$

The advantage of this equation over (5.11) is twofold:

(i) The matrix μ has simpler symmetry properties than γ. This is not an accident of our special model but has its general root in time reversal invariance. It is a special case of Onsager's famous symmetry relations of kinetic coefficients, which take a simple form only if the kinetic equations are written as a linear relation between "velocities" dQ_k/dt and "forces" f_k.

(ii) Equation (5.18) remains valid even beyond the linear regime; in our case for the anharmonic oscillator. Then the second equation for instance reads $dP/dt = -f(X) - \gamma P$. Generally speaking kinetic equations of the form (5.18) allow arbitrary (nonlinear) thermodynamic relations between forces and coordinates. In fact we shall see that practically all equations of irreversible thermodynamics (nonlinear hydrodynamics, Boltzmann's equations, diffusion, chemical rate equations, etc.) are of the form (5.18).

From the $\Phi_{kl}(\omega)$ one can determine the correlation functions $\langle q_k(t); q_l(0)\rangle$ by Fourier back transformation. For the oscillator one can, because of $q_1(t) =$

$m dq_2(t)/dt$, calculate all other correlation functions from (5.9) by differentiation with respect to t. The matrix formulation (5.15) thus contains no new information beyond (5.8). Nevertheless it is instructive to derive (5.8) once more from (5.15) (Problem 5.5).

PROBLEMS

5.1 Calculate the Fourier transform of (5.8) and verify (5.5,6,9).

5.2 Discuss the position of the poles $\omega = \pm\omega_r + i\gamma/2$ of (5.7,8) as a function of γ in the complex ω-plane. Consider in particular the case of strong damping $\gamma \gg 2\omega_0$. In this case one has $\omega = i\gamma/2 - \omega_0^2/\gamma + \cdots$; insert this expansion into (5.5) and expand the r.h.s. up to terms of first order in ω_0^2/γ.

5.3 The results of the foregoing problem contain the limiting cases of both an overdamped oscillator and the free particle ($\omega_0 = 0$). Which limit leads to the results of the overdamped oscillator? Which of the two poles discussed in the foregoing problem gives no contribution to the correlation function any more and what is the contribution of the other pole to (5.5)? Compare the result for $\langle x(t)x(0)\rangle$ with the considerations of this chapter, in particular (5.5).

5.4 In the limit $\omega_0 \to 0$ the correlation function $\langle x(t)x(0)\rangle$ diverges because of (5.6). The mean square displacement $\langle [x(t) - x(0)]^2 \rangle$, however, remains finite. Use the expansion obtained in Problem 5.2 and discuss the limit $\omega_0 \to 0$. Compare the result with (4.14).

5.5 Calculate $\Phi_{kl}(\omega)$ from (5.15) and compare it with (5.8).

6. Dispersion Relations and Spectral Representations

The two response functions $\chi_{kl}(t)$ and $\Phi_{kl}(t)$ by definition [(2.25) and (3.12)] vanish for $t < 0$. Physically speaking this expresses "causality": the "effect" $Q(t)$ is influenced only by "causes" $f^e(t')$ and $f(0)$, at times t' and 0, respectively, before t. Mathematically one can use this fact to continue the Fourier transform of the correlation functions [e.g. (3.6)] into the (upper) complex ω half plane without endangering the convergence of the Fourier integrals. The Fourier integrals on the real ω-axis then can be obtained as limits of the integrals in the complex plane from above. If one introduces

$$\chi_{kl}(z) = \frac{i}{\hbar} \int_0^\infty \langle [q_k(t), q_l(0)] \rangle e^{izt} \, dt \quad ; \quad \text{Im}\{z\} \geq 0 \tag{6.1}$$

and

$$\Phi_{kl}(z) = i\beta \int_0^\infty \langle \Delta q_k(t); \Delta q_l(0) \rangle e^{izt} \, dt \quad ; \quad \text{Im}\{z\} \geq 0 \tag{6.2}$$

then these two integrals are analytic functions in z for $\text{Im}\{z\} > 0$ and continuous on the real axis $z = \omega$ from above. This fact can be expressed more explicitly by writing the integrands as a Fourier integral for all t. Introducing

$$\frac{1}{\hbar} \langle [q_k(t), q_l(0)] \rangle = \frac{1}{\pi} \int_{-\infty}^{+\infty} \chi_{kl}''(\omega) e^{-i\omega t} \, d\omega \tag{6.3}$$

and

$$\beta \langle \Delta q_k(t); \Delta q_l(0) \rangle = \frac{1}{\pi} \int_{-\infty}^{+\infty} \Phi_{kl}''(\omega) e^{-i\omega t} \, d\omega \tag{6.4}$$

(with ω being real), one can carry out the t integrals in (6.1,2). One finds

$$\chi_{kl}(z) = \frac{1}{\pi} \int_{-\infty}^{+\infty} \frac{\chi_{kl}''(\omega)}{\omega - z} \, d\omega \tag{6.5}$$

and a completely equivalent equation for $\Phi_{kl}(z)$. Such representations of functions in the complex plane by Cauchy integrals along the real axis are called spectral representations, and $\chi''(\omega)$ and $\Phi''(\omega)$ are known as the spectral functions of $\chi(z)$ and $\Phi(z)$.

The spectral respresentation (6.5) can now also be considered in the lower z half plane. It defines an analytic function in this half plane, too. The limit of this function on the real axis from below, however, does not agree with the limit from above. The discontinuity of the spectral representation (6.5) in crossing the real axis can be read off from (6.5) using

$$1/(\omega - x \mp i0) = P/(\omega - x) \mp i\pi\delta(\omega - x) \quad , \tag{6.6}$$

leading to

$$\chi(\omega \pm i0) = \chi'(\omega) \pm i\chi''(\omega) \tag{6.7}$$

and a completely equivalent equation for $\Phi(\omega \pm i0)$.

The continuous part $\chi'(\omega)$ is given by the principal value integral

$$\chi'(\omega) = P\frac{1}{\pi} \int_{-\infty}^{+\infty} \frac{\chi''(\omega')}{\omega' - \omega} d\omega' \quad . \tag{6.8}$$

This is one half of the famous "dispersion relations" first derived by *Kramers* and *Kronig* [6.1]. The other half can be obtained by applying Cauchy's theorem to the function $\chi(z)$ (Problem 6.1.).

The functions $\chi(z)$ and $\Phi(z)$ in the lower half plane can be written as analytical continuations of the Fourier transforms of the so-called "advanced" correlation functions in time. Together with the "retarded" functions (6.1) one can write

$$\chi_{kl}(z) = \pm\frac{i}{\hbar} \int_{-\infty}^{+\infty} \langle[q_k(t), q_l(0)]\rangle \Theta(t) e^{izt} dt \quad ; \quad \text{Im}\{z\} \gtrless 0 \tag{6.9}$$

and a completely equivalent equation for $\Phi_{kl}(z)$.

For further considerations it is useful to express the spectral functions $\chi''(\omega)$ and $\Phi''(\omega)$ in terms of the Fourier transform of the simple product correlation $s_{kl}(t) = \langle \Delta q_k(t) \Delta q_l(0) \rangle$, namely

$$\langle \Delta q_k(t) \Delta q_l(0) \rangle = \frac{1}{\pi} \int_{-\infty}^{+\infty} s_{kl}(\omega) e^{-i\omega t} d\omega \quad . \tag{6.10}$$

Taking into account that, because of time translation invariance

$$\langle [q_k(t), q_l(0)] \rangle = s_{kl}(t) - s_{lk}(-t) \tag{6.11}$$

and (Problem 6.2.)

$$\beta \langle \Delta q_k(t); \Delta q_l(0) \rangle = \int_0^\beta s_{kl}(t + \hbar\alpha/i) d\alpha \quad , \tag{6.12}$$

one finds immediately

$$\chi''_{kl}(\omega) = [s_{kl}(\omega) - s_{lk}(-\omega)]/\hbar \quad , \tag{6.13}$$

and after some intermediate steps

$$\Phi''_{kl}(\omega) = s_{kl}(\omega)(1 - e^{-\beta\hbar\omega})/\hbar\omega \quad . \tag{6.14}$$

One can now generalize the considerations of Chap. 3, in particular (3.31,32), from retarded to advanced correlation functions and continue (3.32) into the whole complex plane as

$$\chi_{kl}(z) = \chi^T_{kl} + z\Phi_{kl}(z) \quad . \tag{6.15}$$

Using (6.7) one then finds for the spectral functions

$$\chi''_{kl}(\omega) = \omega \Phi''_{kl}(\omega) \quad . \tag{6.16}$$

If this relation is combined with (6.13, 14) one obtains

$$\boxed{s_{kl}(-\omega) = s_{lk}(\omega) e^{-\beta\hbar\omega} \quad .} \tag{6.17}$$

This symmetry property of the product correlations is one of the many versions of detailed balance. We are going to consider it in more detail in the next two chapters.

It is instructive to evaluate the r.h.s. of (6.10) in the Heisenberg representation. Introducing the eigenstates $|\mu\rangle$ of H with eigenvalues E_μ and using

$$\langle\mu|q_k(t)|\nu\rangle = \langle\mu|q_k|\nu\rangle \exp[i(E_\mu - E_\nu)t/\hbar] \quad , \tag{6.18}$$

one finds (F, the free energy)

$$s_{kl}(\omega) = \pi \sum \langle\mu|\Delta q_k|\nu\rangle\langle\nu|\Delta q_l|\mu\rangle \\ \times \exp[-\beta(E_\mu - F)]\delta((E_\nu - E_\mu)/\hbar - \omega) \quad . \tag{6.19}$$

Let us use this equation to derive once more the relation (3.18) for the static susceptibilities. For this purpose we decompose (6.19) into its zero-frequency part and the rest. The zero-frequency part is obtained from (6.19) by keeping only the terms for which $E_\mu = E_\nu$ (for ergodic systems this means $\mu = \nu$). This leads to [notation as in (3.18)]

$$s_{kl}(\omega) = \pi\langle\Delta q^0_k \Delta q^0_l\rangle\delta(\omega) + s^r_{kl}(\omega) = \pi s^0_{kl}\delta(\omega) + s^r_{kl}(\omega) \quad . \tag{6.20}$$

When this expression is inserted into (6.14) one finds, after expanding the exponential function for small ω – apart from a factor β – the same zero frequency part for $\Phi''_{kl}(\omega)$ as in (6.20). Inserted into the spectral representation for Φ, [compare (6.5)], this leads to a pole term in $\Phi(z)$ at $z = 0$ and a corresponding decomposition of (6.15)

$$\chi_{kl}(z) = \chi^T_{kl} - \beta s^0_{kl} + z\Phi^r_{kl}(z) \quad . \tag{6.21}$$

In the limit $z \to 0$, together with (6.20), this yields exactly (3.18). Note also that, according to (6.13), the zero-frequency parts of s have a tendency to cancel in χ; in particular the diagonal term χ_{kk} has no zero-frequency part.

PROBLEMS

6.1 *Kramers-Kronig relations.* Apply Cauchy's integral theorem

$$\int_{C_\pm} \frac{\chi(\omega)}{\omega - z} = \pm 2\pi i \chi(z)$$

and use for the two curves $C_+(C_-)$ the two straight lines infinitesimally below (above) the real axis plus the infinite upper (lower) semicircles. Decompose the frequency denominators according to (6.6) to derive

$$P \int \frac{\chi(\omega' \pm i0)}{\omega' - \omega} d\omega' = \pm i\pi \chi(\omega \pm i0) \quad .$$

Convince yourself that the difference of these two equations is (6.8). What is the sum?

6.2 Verify (6.12) using the definition (3.11) of the bracket symbol and Heisenberg's form of the time dependence of operators (2.10,12 or 17) together with the cyclic invariance of the trace.

6.3 Calculate $\chi(z)$ from (6.5) for the spectral functions

$$\chi''(x) = \begin{cases} 1/(1+x^2) \\ \pi\Theta(1-x)\Theta(1+x)/2 \\ 2\sqrt{1-x^2}\,\Theta(1-x)\Theta(1+x) \end{cases}$$

where $x = \omega/\omega_0$ is a dimensionless frequency. Plot the real and imaginary parts of χ.

7. Symmetry Properties of Correlation Functions

In this chapter we are going to derive three symmetry relations of the correlation functions that follow either directly from their definition or from the invariance of nature under translations or reversal of time. There are further symmetry properties that follow from the invariance of nature under translations in space, Galilean invariance and possible symmetry properties of the equilibrium state (homogeneity and isotropy of liquids and gases or lattice symmetry of solids), which we will consider only in examples later on.

Complex conjugation. The behaviour of correlation functions under complex conjugation follows directly from their definition. We start out from the simple product correlation $s_{kl}(t)$ introduced in connection with (6.10). Making use of $(\mathrm{Tr}\{A\})^* = \mathrm{Tr}\{A^*\}$ and the cyclic invariance of the trace one finds

$$\langle q_k(t) q_l(0) \rangle^* = \langle q_l^*(0) q_k^*(t) \rangle \quad . \tag{7.1}$$

If this relation is combined with the definition (2.25) of susceptibilities one obtains the symmetry property

$$[\chi_{kl}(t)]^* = \chi_{\bar{k}\bar{l}}(t) \quad , \tag{7.2}$$

where we have introduced the shorthand notation $q_k^* = q_{\bar{k}}$ for the labels of Hermitian conjugate operators. The corresponding symmetry property for the Kubo relaxation function can then be read off most easily from (3.2), namely

$$[\Phi_{kl}(t)]^* = -\Phi_{\bar{k}\bar{l}}(t) \quad . \tag{7.3}$$

These results may now be used together with the definitions of Fourier transforms (6.3,4,5 and 9) to derive the corresponding relations in Fourier space. One obtains

$$[\chi''_{kl}(\omega)]^* = -\chi''_{\bar{k}\bar{l}}(-\omega) \quad , \tag{7.4}$$

$$[\Phi''_{kl}(\omega)]^* = \Phi''_{\bar{k}\bar{l}}(-\omega) \tag{7.5}$$

and

$$[\chi_{kl}(z)]^* = \chi_{\bar{k}\bar{l}}(-z^*) \quad , \tag{7.6}$$

$$[\Phi_{kl}(z)]^* = -\Phi_{\bar{k}\bar{l}}(-z^*) \quad . \tag{7.7}$$

Time translation invariance. The correlation functions of linear response theory are equilibrium averages, taken with a time-independent statistical operator that commutes with the Hamiltonian H. As a consequence of this and the cyclic invariance of the trace the product correlations obey

$$\langle q_k(t)q_l(0)\rangle = \langle q_k(t+t')q_l(t')\rangle = \langle q_k(0)q_l(-t)\rangle \ . \tag{7.8}$$

The second equation is a special case of the first for $t' = -t$.

For the commutator and bracket symbol one obtains

$$\langle [q_k(t), q_l(0)]\rangle = -\langle [q_l(-t), q_k(0)]\rangle \tag{7.9}$$

and

$$\langle q_k(t); q_l(0)\rangle = \langle q_l(-t); q_k(0)\rangle \ . \tag{7.10}$$

The Fourier transformation of these relations according to (6.3,4) leads to

$$\chi''_{kl}(\omega) = -\chi''_{lk} \quad , \quad \Phi''_{kl}(\omega) = \Phi''_{lk}(-\omega) \tag{7.11}$$

and, using the spectral representation (6.5) (including the one for Φ),

$$\chi_{kl}(z) = \chi_{lk}(-z) \quad \text{and} \quad \Phi_{kl}(z) = -\Phi_{lk}(-z) \ . \tag{7.12}$$

The last two equations obviously connect points in the upper frequency half plane with those in the lower half plane. After Fourier back transformation according to (6.9) this leads to relations between the retarded and advanced correlation functions in t-space of the form

$$[\chi_{kl}(t)]_{\text{ret}} = [\chi_{lk}(-t)]_{\text{adv}} \tag{7.13}$$

and

$$[\Phi_{kl}(t)]_{\text{ret}} = -[\Phi_{lk}(-t)]_{\text{adv}} \ . \tag{7.14}$$

These relations can, of course, also be derived from (7.9,10) after multiplication with $\Theta(t)$.

Time reversal invariance. The operation τ of time reversal is defined in physical terms as the change of the sign of time t accompanied by the reversal of the time evolution of all processes. Mathematically one writes

$$\tau q_k(t)\tau^{-1} = q_{\underset{k}{-}}(-t) \ . \tag{7.15}$$

This transformation is particularly simple for operators which have a definite parity under time reversal. There are "coordinate-like" operators with parity +1, such as the spatial coordinates of particles, the energy, energy density, particle number, particle number density, etc., and "momentum-like" operators with parity -1, such as the momentum, spin, current density, energy current density, etc.

For operators with definite parity one has

$$q_{\underset{k}{\tau}} = \sigma_k q_k = \begin{Bmatrix} +1 \\ -1 \end{Bmatrix} \cdot q_k \quad \text{for} \quad \begin{cases} \text{coordinate-like operators}, \\ \text{momentum-like operators}. \end{cases} \quad (7.16)$$

An example of an operator without definite parity is the number density in momentum space $n(p)$ or the phase space density $n(p,r)$ with, for instance, $\tau n(p)\tau^{-1} = n(-p)$.

In order to keep the commutation relations between coordinates and momentum invariant under time reversal one has to chose τ as an antilinear operator, obeying [7.1, 2]

$$\tau a = a^* \tau \quad , \quad (7.17)$$

where a is an arbitrary complex number.

Instead of Hermitian conjugation one can then define an "anti-conjugation" by (von Neumann's notation for vectors and scalar products in this case is somewhat more convenient than Dirac's)

$$(\varphi, \tau\psi) = (\hat{\tau}\varphi, \psi)^* \quad . \quad (7.18)$$

In order to keep the norm of vectors in Hilbert space invariant under the operation τ one has to chose τ as an "antiunitary" operator with

$$\tau\hat{\tau} = \hat{\tau}\tau = 1 \quad . \quad (7.19)$$

Scalar products then go over into their complex conjugate under the transformation τ

$$(\tau\varphi, \tau\psi) = (\varphi, \psi)^* \quad . \quad (7.20)$$

Let us now consider the time reversal operation on a trace

$$\text{Tr}\{\tau A \tau^{-1}\} = \sum(\varphi_n, \tau A \tau^{-1} \varphi_n) = \sum(\hat{\tau}\varphi_n, A\hat{\tau}\varphi_n)^* = \text{Tr}\{A^*\} \quad . \quad (7.21)$$

Here the φ_n and consequently $\tau\varphi_n$ and $\hat{\tau}\varphi_n$ are assumed to form a complete orthonormal system in Hilbert space. Note that (7.21) implies that the *cyclic invariance of the trace does not hold for antiunitary operators* τ.

Now we assume that the Hamiltonian H and the statistical operator ϱ are invariant under time reversal:

$$\tau H \tau - 1 = H, \quad \text{or} \quad [\tau, H] = 0 \quad \text{and similarly} \quad [\tau, \varrho] = 0 \quad . \quad (7.22)$$

Then one obtains, using (7.21),

$$\text{Tr}\{\tau q_l(0) q_k(t) \tau^{-1} \varrho\} = \text{Tr}\{q_k^*(t) q_l^*(0) \varrho\} \quad . \quad (7.23)$$

Rather than using this relation directly, we combine it with time translation invariance (7.8) to obtain

$$\langle q_{\bar{k}}(t)q_{\bar{l}}(0)\rangle = \langle q_{\underset{l}{-}}(t)q_{\underset{k}{-}}(0)\rangle \quad . \tag{7.24}$$

Since these symmetry properties involve only the indices, they can be carried over straightforwardly to all other correlation functions. We collect the results together with the earlier ones of this chapter in Table 7.1.

Table 7.1. Symmetry relations for correlation functions

Complex conjugation	
$[\chi_{kl}(t)]^* = \chi_{\bar{k}\bar{l}}(t)$	$[\Phi_{kl}(t)]^* = -\Phi_{\bar{k}\bar{l}}(t)$
$[\chi''_{kl}(\omega)]^* = -\chi''_{\bar{k}\bar{l}}(-\omega)$	$[\Phi''_{kl}(\omega)]^* = \Phi''_{\bar{k}\bar{l}}(-\omega)$
$[\chi_{kl}(z)]^* = \chi_{\bar{k}\bar{l}}(-z^*)$	$[\Phi_{kl}(z)]^* = -\Phi_{\bar{k}\bar{l}}(-z^*)$
Time translation	
$[\chi_{kl}(-t)]_{\text{ret}} = [\chi_{lk}(t)]_{\text{adv}}$	$[\Phi_{kl}(-t)]_{\text{ret}} = [\Phi_{lk}(t)]_{\text{adv}}$
$\chi''_{kl}(\omega) = -\chi''_{lk}(-\omega)$	$\Phi''_{kl}(\omega) = \Phi''_{lk}(-\omega)$
$\chi_{kl}(z) = \chi_{lk}(-z)$	$\Phi_{kl}(z) = -\Phi_{lk}(-z)$
Time reversal + translation	
$\chi_{\bar{k}\bar{l}}(t) = \chi_{\underset{l\ k}{-\ -}}(t)$	$\Phi_{\bar{k}\bar{l}}(t) = \Phi_{\underset{l\ k}{-\ -}}(t)$
$\chi''_{\bar{k}\bar{l}}(\omega) = \chi''_{\underset{l\ k}{-\ -}}(\omega)$	$\Phi''_{\bar{k}\bar{l}}(\omega) = \Phi''_{\underset{l\ k}{-\ -}}(\omega)$
$\chi_{\bar{k}\bar{l}}(z) = \chi_{\underset{l\ k}{-\ -}}(z)$	$\Phi_{\bar{k}\bar{l}}(z) = \Phi_{\underset{l\ k}{-\ -}}(z)$

The symmetry relations following from time reversal invariance form the basis of Onsager's symmetry relations between kinetic coefficients. We will consider them in more detail later on. They become particularly simple for Hermitian operators with definite time reversal parity. In this case one has, for instance,

$$\chi_{kl} = \sigma_k \sigma_l \chi_{kl} \quad \text{valid for all variables } t, z \text{ and } \omega \quad . \tag{7.25}$$

The same symmetry relation holds for the relaxation function Φ_{kl}, and, because of (3.32), also for the static isothermal susceptibility:

$$\chi_{kl}^T = \sigma_k \sigma_l \chi_{kl}^T \quad . \tag{7.26}$$

On the other hand all static thermodynamic susceptibilities have to be symmetric, since they are second derivatives of the free energy:

$$\chi_{kl}^T = \chi_{kl}^T \quad . \tag{7.27}$$

The obvious consequence of (7.26, 27) is

$$\chi_{kl}^T = 0 \quad \text{if } q_k \text{ and } q_l \text{ have opposite time reversal parity} \quad . \tag{7.28}$$

The symmetry relations can be generalized to the case when time reversal symmetry is broken by a magnetic field B. Then one has instead of (7.22)

$$\tau H(B) \tau^{-1} = H(-B) \quad . \tag{7.29}$$

In the time reversed correlation functions similarly the sign of the magnetic field has to be reversed.

PROBLEMS

7.1 Check the symmetry relations for the relaxation functions of the oscillator, in particular time reversal invariance, using (5.12, 14, 15, 17).

7.2 Discuss the symmetry properties for correlation functions of densities [such as particle number density $n(r,t)$, energy density $\varepsilon(r,t)$, etc.] following from spatial symmetries, in particular homogeneity or lattice translation symmetry.

8. Detailed Balance, Fluctuations and Dissipation

Relations between spontaneous and forced deviations from equilibrium occur over and over again in statistical mechanics. It starts already in equilibrium situations with the well-known relations between fluctuations and static susceptibilities [Ref. 5.1, Sects. 14,49,50]. In qualitative terms: the larger the response to external perturbations, the larger the spontaneous fluctuations.

Another example from equilibrium theory is Planck's formula

$$m\omega_0^2 \langle x^2 \rangle = \hbar\omega_0 [1/2 + 1/(e^{\beta\hbar\omega_0} - 1)] \tag{8.1}$$

for the amplitude fluctuations of a harmonic oscillator. Keeping in mind the relation $m\omega_0^2 = f$ between the force constant f and eigenfrequency ω_0 of the oscillator one has again: small force constant – large susceptibility – large amplitude fluctuations. Note in addition that this holds not only for the (classical) thermal fluctuations at large temperatures $kT \gg \hbar\omega$, but also for the (quantum mechanical) zero-point fluctuations at $T = 0$.

Einstein's relation $D = BkT$ between the diffusion constant D and mobility B is the first one connecting a fluctuation quantity D with a dissipative one B. We have discussed the connections between diffusion and the position fluctuations of Brownian particles in Chap. 4, in particular (4.14). Einstein's relation was extended to a relation between noise and dissipation in electric circuits by Schottky and Nyquist. Callen and Welton have put it into its general form, which we are now going to derive.

The starting point of this celebrated fluctuation-dissipation theorem, as it is now called, is an innocent looking symmetry relation between product correlations that we have come across several times already [compare (3.26,27) and (6.17), which is essentially the Fourier transform of (3.26)]. We write down (3.26) for the product correlation $s_{kl}(t)$, making use of time translation invariance, together with its Fourier transform (6.17):

$$\boxed{\begin{array}{c} s_{kl}(-t) = s_{lk}(t - i\hbar\beta) \quad , \\[6pt] \hline \\[-6pt] s_{kl}(-\omega) = s_{lk}(\omega)e^{-\beta\hbar\omega} \quad . \end{array}} \tag{8.2}$$

Because of its importance it may be of interest to derive the second half of (8.2) once again directly from the representation (6.19) of $s_{kl}(\omega)$. For convenience we write out (6.19) again:

$$s_{kl}(\omega) = \pi \sum \langle \mu | \Delta q_k | \nu \rangle \langle \nu | \Delta q_l | \mu \rangle \times \exp[-\beta(E_\mu - F)] \delta((E_\nu - E_\mu)/\hbar - \omega) \quad . \tag{6.19}$$

If in this expression ω is replaced by $-\omega$ the frequency delta function remains unchanged if the indices μ and ν are interchanged. In the double sum this amounts to interchanging k and l and to multiplying the exponential by $\exp[\beta(E_\mu - E_\nu)]$. Because of the delta function in (6.19) the energy difference in this exponential can be replaced by $-\hbar\omega$. This leads directly to the second half of (8.2).

Equation (8.2) is said to express *detailed balance* in inelastic processes with an energy gain of the thermodynamic system ($\hbar\omega$ positive) and energy loss of the system ($\hbar\omega$ negative). The loss processes are always suppressed by the Boltzmann factor $\exp(-\beta\hbar\omega)$, measuring the probability of having thermal excitations around which can be annihilated. In particular, at zero temperature there can be only gain processes, because the system is in its ground state. We shall discuss detailed balance in scattering processes in the next chapter. The equations (6.13,14) can now be used to express the fluctuation quantities s_{kl} in terms of the dissipative quantities χ''_{kl} and Φ''_{kl}:

$$s_{kl}(\omega) = \hbar\omega[1 + n(\omega)]\Phi''_{kl}(\omega) \quad , \tag{8.3}$$

similarly

$$s_{lk}(-\omega) = \hbar\omega n(\omega)\Phi''_{kl}(\omega) \tag{8.4}$$

and a symmetric version

$$[s_{kl}(\omega) + s_{lk}(-\omega)]/2 = \hbar\omega[1/2 + n(\omega)]\Phi''_{kl}(\omega) \quad . \tag{8.5}$$

Equations (8.3–5) are different versions of the fluctuation–dissipation theorem. Since χ''_{kl} according to (6.9) involves the commutator of q_k and q_l and the l.h.s. of (8.5) the anticommutator, (8.5) is often written explicitly as a relation between the commutator and anticommutator of the two fluctuating variables.

It is instructive to consider the two limiting cases of very low and very high frequencies. They can be obtained from an expansion of the average energy

$$\hbar\omega[1/2 + n(\omega)] = \epsilon(\omega, T) = \begin{cases} \hbar|\omega|/2 + e^{-\beta\hbar|\omega|} & \hbar|\omega| \gg kT \\ kT & \hbar|\omega| \ll kT \end{cases} \tag{8.6}$$

of an oscillator with eigenfrequency ω in contact with a heat bath at temperature T; note that $\epsilon(\omega, T)$ is even in ω. In general one has only the inequality

$$\hbar\omega[1/2 + n(\omega)] \geq kT \quad . \tag{8.7}$$

If the spectral function Φ'' is large only in the classical regime $\hbar\omega \ll kT$ one can write approximately

$$s_{kl}(\omega) \approx s_{lk}(-\omega) \approx kT\Phi''_{kl} \quad . \tag{8.8}$$

On the other hand, in the quantum limit one has

$$s_{kl}(\omega) = \hbar|\omega|[\Theta(\omega) + \exp(-\beta\hbar|\omega|)]\Phi''_{kl}(\omega) \quad . \tag{8.9}$$

The product correlation then has practically only spectral contributions in the gain regime of positive frequencies.

For illustration let us consider the example of the amplitude fluctuations of a damped oscillator with a spectral function $\Phi''(\omega)$, compare (5.8). If $x(t)$ is the coordinate of the oscillator, then, in particular

$$\langle x(t)x(0) + x(0)x(t)\rangle \to 2\langle x^2\rangle \quad \text{for } t \to 0 \quad . \tag{8.10}$$

Combining this with (6.10) and (8.5) one finds

$$\langle x^2\rangle = \frac{1}{\pi}\int_{-\infty}^{+\infty} \Phi''(\omega)\hbar\omega[1/2 + n(\omega)]d\omega \quad . \tag{8.11}$$

This is a generalization of (8.1) that allows one to treat fluctuation phenomena for which quantum effects as well as damping phenomena play a role.

PROBLEMS

8.1 Starting from (8.11) calculate the first nonvanishing quantum mechanical correction to $\langle x^2\rangle$ from the classical value $\langle x^2\rangle_{\text{cl}}$ by expanding $\epsilon(\omega, T)$ in (8.6) in powers of $\beta\hbar\omega$. Express the result in terms of $\langle p^2\rangle_{\text{cl}}$.

8.2 *Wiener-Kinchin Theorem.* This theorem relates the mean square fluctuation of the Fourier transform of a statistical variable to the Fourier transform of the correlation function. Let $q(t)$ be a statistically fluctuating variable and

$$q(\omega_n) = \int_{-t_0/2}^{+t_0/2} q(t)e^{i\omega_n t}dt/(2\pi t_0)^{1/2}$$

($\omega_n = 2\pi n/t_0$, n integer). Demonstrate that, as long as t_0 is large compared to the decay time τ of the correlation function, one has

$$\langle q(\omega_m)q(-\omega_n)\rangle = \int_{-\infty}^{+\infty} \langle q(t)q(0)\rangle e^{i\omega_n t}dt\delta_{m,n}/(2\pi) \quad .$$

9. Scattering of Particles and Light **

The elastic and inelastic scattering of particles (such as electrons, neutrons and light atoms) and light quanta by many-particle systems is an important technique for determining the structure and dynamics of solids and liquids. We do not want to go into all the details of such scattering experiments but restrict ourselves to those aspects which are interesting in the context of nonequilibrium statistical mechanics.

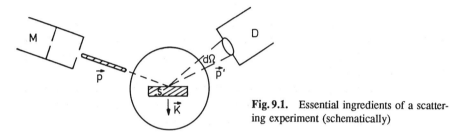

Fig. 9.1. Essential ingredients of a scattering experiment (schematically)

Figure 9.1. illustrates the general setup of a scattering experiment. Particles or quanta of a definite momentum p are produced by a machine M and then scattered by the system S. They are counted by a detector D and their energy ε' and angle Ω (and spin, if necessary) after the scattering event are measured. The momentum $\hbar k$ and energy $\hbar \omega$ transferred to the system are then determined by the conservation laws of momentum and energy:

$$p = p' + \hbar k \quad , \tag{9.1}$$

$$\varepsilon(p) = \varepsilon(p') + \hbar \omega \quad , \tag{9.2}$$

$$\varepsilon(p) = \begin{cases} p^2/2m & \text{particles} \\ |p|c & \text{light} \end{cases} \quad . \tag{9.3}$$

The scattering intensity is usually measured by a scattering cross section, which can be calculated as the modulus squared of a scattering amplitude. Suppose the scattering amplitude for the scattering by the nth particle of the system is $t_n(p, p')$. If one neglects multiple scattering events (an approximation which is closely related to linear response theory) the scattering amplitude for the whole systems is obtained by summing up the amplitudes of all particles at the positions

r_n with the appropriate phase factors $\exp(i\mathbf{k} \cdot \mathbf{r}_n)$:

$$T_{p,\mu,p',\mu'} = \langle \mu | \sum_n t_n(p,p') \exp(i\mathbf{k} \cdot \mathbf{r}_n) | \mu' \rangle \quad ; \quad (9.4)$$

$\mu(\mu')$ are the initial (final) quantum states of the system in the scattering process. Normally the system is not in a definite quantum state before the scattering, but (at best) in a thermodynamic equilibirum state. Furthermore, the final state of the system usually remains unknown. Thus one measures only an average scattering cross section, which is averaged over the initial states with statistical weights $\varrho_\mu = \exp[\beta(F - E_\mu)]$ and summed over the final states μ'. The final result for the so-called double differential cross-section can then be written in the form

$$\boxed{\frac{d^2\sigma}{d\Omega\, d\hbar\omega} = \frac{d\sigma}{d\Omega} \frac{s(k,\omega)}{\pi}} \quad . \quad (9.5)$$

Here $(d\sigma/d\Omega)$ is the differential cross section for the scattering by a single particle and $s(k,\omega)$ the scattering function. The scattering function depends only on the internal properties of the system and is closely related to the spectral functions introduced in Chap. 6.

The scattering of electrons is particularly simple. They interact with the *charges* of the system. In this case $d\sigma/d\Omega$ is the Rutherford cross section and the scattering function (denoted by a label ϱ) is given by

$$s_\varrho(k,\omega) = \pi \sum |\langle \mu | q_n \exp(i\mathbf{k} \cdot \mathbf{r}_n) | \mu' \rangle|^2 \delta((E_\mu - E_{\mu'})/\hbar + \omega) \varrho_\mu / N \quad . \quad (9.6)$$

We have divided this quantity by the number N of atoms of the system since we define (9.5) as the cross section per atom. The q_n are the "charge numbers" of the scattering centres of the system: $q = 1$ for electrons, $q = -Z$ for the nuclei. If one introduces the Fourier transform

$$\varrho_k = \sum_n q_n \exp(i\mathbf{k} \cdot \mathbf{r}_n)/\sqrt{N} \quad (9.7)$$

of the charge density and uses the completeness of the eigenstates μ, one can carry out the double sum in (9.6). The result can be written as

$$s_\varrho(k,\omega) = \int \langle \varrho_k(t) \varrho_k^*(0) \rangle e^{i\omega t} dt / 2 \quad . \quad (9.8)$$

Neutrons are scattered on the one hand by nuclei, on the other by the magnetic moments of the atomic electrons. The nuclear scattering amplitude normally contains a part depending on the nuclear spin:

$$t_n = a_c + a_i (\mathbf{s} \cdot \mathbf{I}_n) \quad (9.9)$$

where s (I) is the spin of the neutron (nucleus) and a_c and a_i are the coherent and incoherent scattering lengths, respectively. The scattering cross section contains, after averaging, two contributions: one with the spectral function

$$s_n(k,\omega) = \int \langle n_k(t) n_k^*(0)\rangle e^{i\omega t} dt/2 \tag{9.10}$$

of the density fluctuations, with the Fourier transform

$$n_k = \sum \exp(i\mathbf{k}\cdot \mathbf{r}_n)/\sqrt{N} \tag{9.11}$$

of the number density of nuclei.

The second contribution contains, in principle, the spectral function of the nuclear spin density fluctuations. Since, however, the spins of different nuclei are uncorrelated at those temperatures which are normally reached, the off-diagonal terms in the double sum of the spectral function average out to zero. This destroys the spatial coherence of contributions from different nuclei and leaves only the incoherent diagonal terms of the double sum, the so-called incoherent spectral function

$$s_i(k,\omega) = \int \sum_n \langle \exp[i\mathbf{k}\cdot \mathbf{r}_n(t)]\exp[-i\mathbf{k}\cdot \mathbf{r}_n(0)]\rangle \exp(i\omega t) dt/2N \quad. \tag{9.12}$$

If the system contains two or more isotopes with different scattering lengths there are further incoherent terms, which, however, we are not going to consider in this short overview.

We have mentioned already that nucleons are also magnetically scattered. In this case the spectral function of electronic spin fluctuations can be measured, as might be expected:

$$s_\sigma(k,\omega) = (\hat{k}_i \hat{k}_j - \delta_{ij}) \int \langle s_{ik}(t) s_{jk}^*(0) \rangle e^{i\omega t} dt/2 \quad. \tag{9.13}$$

Here s_{ik} is the Fourier transform of the spin density

$$s_{ik} = \sum_n \sigma_{in} \exp(i\mathbf{k}\cdot \mathbf{r}_n)/\sqrt{N} \tag{9.14}$$

and σ_{in} the ith component of the Pauli spin matrix of the nth electron. The prefactor in (9.13) contains the components $\hat{k}_i = k_i/|k|$ of the unit vectors in the direction of the k vector. It projects out the transverse part of the spin fluctuations and follows from the directional dependence of the magnetic dipole-dipole interaction in Fourier space (Problem 9.3.).

Light quanta are mainly scattered by the electrons. Depending on the wavelength of the light one distinguishes X-ray scattering, Raman and Brillouin scattering.

For X-rays the scattering properties of an atom are described in terms of a form factor $f(k)$. As long as the electrons follow the nuclear motion rigidly,

X-ray scattering measures again the density fluctuations of the nuclei, (9.10,11).

In Raman and Brillouin scattering the wavelength of the light is large compared to the atomic dimensions. The dependence of scattering strength on the momentum transfer is therefore irrelevant. On the other hand, one has usually strong dependences of the atomic scattering cross sections on the frequency of the incident light, in particular in the vicinity of resonance frequencies of the atoms (or molecules). In the simplest way this can be taken into account in a semi-phenomenological theory, describing the propagation of the light wave by a wave equation

$$\left(\Delta + \frac{\omega_0^2}{c^2}\varepsilon_0\right) E_i = -\sum_j \frac{\omega_0^2}{c^2} \delta\varepsilon_{ij} E_j \quad . \tag{9.15}$$

Here ω_0 is the frequency of the incident light and $\delta\varepsilon_{ij}$ the deviation of the dielectric tensor from its equilibrium value $\varepsilon_0\delta_{ij}$. Particularly simple is the situation for isotropic liquids. In this case one has to consider only the density dependence of the dielectric function. For small deviations one can write

$$\delta\varepsilon_{ij} = (\partial\varepsilon/\partial n)\delta n\, \delta_{ij} \quad . \tag{9.16}$$

The fluctuations of the dielectric function thus can be traced back to the density fluctuations again. The cross section is again proportional to the spectral function s_n (9.10) of the density fluctuations. In general, however, one measures by light scattering directly the spectral function of the fluctuations of the dielectric function

$$s_\varepsilon(k,\omega) = \int \langle \varepsilon_k(t)\varepsilon_k^*(0)\rangle e^{i\omega t} dt/2 \quad . \tag{9.17}$$

Here ε_k is the Fourier component of the fluctuation

$$\varepsilon_k = \int \sum e_i \delta\varepsilon_{ij}(r) e'_j \exp(i\mathbf{k}\cdot\mathbf{r}) d^3r/\sqrt{V} \tag{9.18}$$

of the dielectric function, e_i (e'_j) are the components of the polarisation vectors of the incident (scattered) light beam, and V the volume of the system. The results are collected in Table 9.1. The scattering amplitudes are in most cases of the order of nuclear radii (a_c, a_i) or of the classical electron radius, $e^2/(m_e c^2)$. The corresponding scattering cross sections thus are of the order of 10^{-24}cm^2. The neglect of multiple scattering thus is well justified in most cases. An exception is the scattering of electrons with the much larger Rutherford scattering cross section. In particular for low energy electron scattering, multiple scattering effects play an important role.

Information which is relevant in the context of thermodynamics is contained in the spectral functions $s(k,\omega)$. As can be seen from their definitions (9.6,etc.) they are of exactly the same kind as occurs in the linear response theory for

Table 9.1. Scattering cross sections and spectral functions for the scattering of particles and light

Process	$d\sigma/d\Omega$	Spectral function
Electron scattering	$\left(\frac{2m_e e^2}{\hbar^2 k^2}\right)^2$	$s_\varrho(k,\omega)$
Coherent neutron scattering	$\frac{|p'|}{|p|} a_c^2$	$s_n(k,\omega)$
Incoherent neutron scattering	$\frac{|p'|}{|p|} a_i^2 I(I+1)$	$s_i(k,\omega)$
Magnetic neutron scattering	$\frac{|p'|}{|p|} \left(\frac{1.91 e^2}{m_e c^2}\right)^2$	$s_\sigma(k,\omega)$
X-ray scattering	$\frac{8\pi}{3}\left(\frac{Zf(k)e^2}{m_e c^2}\right)^2$	$s_n(k,\omega)$
Brillouin, Raman scattering	$\left(\frac{pp'}{2\hbar^2}\right)^2 \sqrt{\varepsilon_0}$	$s_\varepsilon(k,\omega)$

(classical) external fields, compare (6.19). They therefore obey the same symmetry relations (Chaps. 7,8), in particular the detailed balance condition (8.2). This relation acquires a very direct physical meaning in the context of scattering: processes with positive ω (known as Stokes processes) correspond to an energy gain of the system, those with negative ω (anti-Stokes processes) to an energy loss of the system. Detailed balance thus says that in equilibrium loss processes are always suppressed compared to gain processes by the Boltzmann factor $\exp(-\beta\hbar\omega)$. The requirement of initial equilibrium of the system is, indeed, essential for the validity of detailed balance. A well-known and important counter example is the *laser*. Laser action requires a so-called "inversion" of the excitation probabilities, in some way just the opposite of equilibrium. Another counter example would be a mixture of ortho and para hydrogen, which can easily be taken away from equilibrium for a long time just by heating. Although "upward" processes (with positive ω) are more probable than "downward" processes this does not necessarily mean that an incoming particle in a single collision can only lose energy on the average. One has to bear in mind that in a scattering process ω cannot take arbitrary values. It is kinematically restricted by the conservation laws (9.1–3). Particularly for particles with very low incident energy, upward processes with large ω are forbidden, because they would require negative final energies $\varepsilon(p')$ (Problem 9.1.). So, on the average one expects a cooling of hot and a heating of cold incident beams in qualitative accord with the second law of thermodynamics.

The kinematics can be taken into account more directly by introducing in $s(k,\omega)$ for k and ω the quantities $p - p'$ and $\varepsilon(p) - \varepsilon(p')$ from the conservation laws. The spectral functions then become functions of p and p' and the detailed balance relation (8.2) can be written as

$$S(p',p)\exp[-\beta\varepsilon(p)] = S(p,p')\exp[-\beta\varepsilon(p')] \tag{9.19}$$

with

$$S(p',p) = s(p - p', \varepsilon(p) - \varepsilon(p')) \quad . \tag{9.20}$$

Finally, let us mention a generalization of detailed balance (9.19) to cases where the simple single scattering approximation (9.4) for the transition matrix does not hold any more. Such cases occur, as mentioned above, for low energy electron scattering or, also, for the scattering of atoms or molecules by solids or liquids. The generalization can be obtained by a combined application of time reversal invariance and the equilibrium properties of the initial state.

Time reversal invariance for the T-matrix implies

$$T_{p\mu,p'\mu'} = \left(T_{\bar{p}'\bar{\mu}',\bar{p}\bar{\mu}}\right)^* \quad . \tag{9.21}$$

If we consider, for instance, average transition rates r_{pq}, proportional to the modulus squared of the T-matrix (9.21) averaged over the initial states μ of the system with the weights ϱ_μ and summed over all final states μ', a line of argument completely equivalent to the one leading to detailed balance [9.1] now leads to

$$r_{p'p}\exp[-\varepsilon(p)] = r_{\bar{p}\bar{p}'}\exp[-\varepsilon(p')] \quad . \tag{9.22}$$

This equation is, in many cases, just as good as (9.19). It is often also just called "detailed balance".

PROBLEMS

9.1 Neutrons with an incident energy $\varepsilon = p^2/2m$ are scattered through an angle Θ. Determine the momentum transfer $\hbar k$ as a function of scattering angle Θ and energy transfer $\hbar\omega$. Plot the possible energy and momentum transfers in the (ω, k)-plane with the incident moentum p as the parameter.

9.2 Determine the wavelength $\lambda = 2\pi/k$ corresponding to a momentum tranfer $\hbar k$ for X-rays of an energy of 10keV for a scattering angle $\Theta = 60°$ and an energy transfer $\hbar\omega \ll 10\text{keV}$. How large would the energy have to be for neutrons to transfer the same momentum in an elastic scattering event with the same scattering angle?

9.3 Calculate the differential scattering cross section for a neutron by an electron in the first-order Born approximation. Use an interaction Hamiltonian $H_i = -\boldsymbol{\mu}_n \cdot \boldsymbol{B}_e$ with $\boldsymbol{B}_e = \nabla \times \nabla \times (\boldsymbol{\mu}_e/r)$. Compare the result with (9.13).

10. Energy Dissipation, Detailed Balance and Passivity

Another consequence of the detailed balance relation for the dynamics of many-particle systems is an inequality for the energy dissipation of external forces. Before we derive this inequality we have to look a little more closely at the response of conserved quantities (the energy being one such quantity).

10.1 The Response of Conserved Quantities to External Forces

According to (2.22) and a slight generalization of (2.24,25), the response of an arbitrary observable $A(t)$ to an external force $f^e(t)$ is given quite generally by

$$\langle A(t)\rangle = A_0 + \frac{i}{\hbar}\sum_l \int_{-\infty}^{t} \langle [A(t), q_l(t')]\rangle f_l^e(t')dt' \quad . \tag{10.1}$$

The average in this equation is taken with the statistical operator $\varrho_d(t)$, $A_0 = \langle A(-\infty)\rangle$ is the expectation value of A before switching on the external field and

$$A(t) = \exp(iHt/\hbar)A(0)\exp(-iHt/\hbar) \tag{10.2}$$

is the time dependence of the operator $A(t)$ (H being the Hamiltonian of the closed system).

If A is a conserved quantity it commutes with H. Obviously then it is constant in time in the closed system. As a consequence of (10.1) it turns out, in addition, that there is even *no linear response* of a *conserved quantity* to external forces in an open system: in linear response theory the average on the r.h.s. of (10.1) is done with the statistical operator $\varrho = \varrho(H)$ of the equilibrium state before switching on the external forces. The observable A then also commutes with ϱ and, taking into account the cyclic invariance of the trace, one finds that the susceptibility occurring in (10.1) vanishes in zeroth order in the external forces:

$$\chi_{Ak}(t) = \frac{i}{\hbar}\langle [A(t), q_k(0)]\rangle \Theta(t) = 0 \quad . \tag{10.3}$$

The corresponding relaxation function Φ_{Ak} then can be easily expressed in terms of the static isothermal susceptibility (Problem 10.2.).

The quadratic response for constants A will, in general, be different from zero. To evaluate it, one has to calculate the expectation value on the r.h.s. of (10.1) to first order in the external forces f^e.

10.2 Energy Dissipation and Passivity

This procedure now becomes particularly simple if A is equal to the Hamiltonian H. In this case the commutator occurring in the susceptibility can be evaluated quite generally using the equations of motion

$$\dot{q}_l(t) = \frac{i}{\hbar}[H, q_l(t)] \ . \tag{10.4}$$

Inserting this into (10.1) one finds for $A = H$

$$\langle H(t) \rangle = H_0 + \sum \int_{-\infty}^{t} \dot{Q}_k(t) f_k^e(t) dt \ . \tag{10.5}$$

This is obviously nothing but the time integral of the differential energy conservation law

$$\frac{d\langle H(t) \rangle}{dt} = \sum \dot{Q}_k(t) f_k^e(t) \tag{10.6}$$

(Problem 10.1.), saying that the rate of change of the energy of the system is equal to the work done by the external forces per unit time.

The total energy of the external forces dissipated by the system is obtained by taking the integral in (10.5) for $t \to \infty$. After introducing the Fourier transform of the r.h.s. of (10.5) one finds

$$\langle H(\infty) - H(-\infty) \rangle = \sum \int_{-\infty}^{+\infty} \frac{d\omega}{2\pi}(-i\omega) Q_k(\omega) f_k^e(-\omega) \ . \tag{10.7}$$

This result is correct to all orders in the external forces. To second order in these forces it is sufficient to use for $Q_k(\omega)$ the first-order (linear response) result (3.5). Inserting this into (10.7) one obtains

$$\langle H(\infty) - H(-\infty) \rangle = \sum \int_{-\infty}^{+\infty} \frac{d\omega}{2\pi}(-i\omega) f_k^e(-\omega) \chi_{kl}(\omega) f_l^e(\omega) \ . \tag{10.8}$$

For the susceptibility we use the spectral representation (6.5) with z infinitesimally above the real ω-axis: $z = \omega + i0$ [for the integration variable in (6.5) one had then better introduce a new notation, say ω']. The spectral function then can be expressed in terms of the product correlation function by means of (6.13). For the sake of compactness, we introduce the notation

$$s(\omega) = \sum f_k^e(-\omega) s_{kl}(\omega) f_l^e(\omega) \ . \tag{10.9}$$

For the discussion of the symmetry and reality of this quantity it is useful to express it in terms of

$$W(\omega) = \sum q_l f_l^e(\omega) \tag{10.10}$$

and $W(-\omega)$ respectively. Since the Hamiltonian is Hermitian one has

$$\sum q_k f_k^e(t) = \sum q_k^* f_k^e(t)^* \quad . \tag{10.11}$$

After Fourier transformation and use of (10.10) this leads to

$$W(\omega)^* = W(-\omega) \quad . \tag{10.12}$$

We insert this into (10.9) combined with (6.19) and obtain

$$s(\omega) = \pi \sum |\langle \mu|W(\omega)^*|\nu\rangle|^2 \delta(\omega - \omega_{\mu\nu}) \varrho_\mu \geq 0 \quad ,$$
$$\omega_{\mu\nu} = (E_\nu - E_\mu)/\hbar \quad . \tag{10.13}$$

The second term on the r.h.s. of (6.13) together with (10.9) and (10.13) after interchanging the summation indices k and l finally leads to a term

$$\chi''(\omega) = \frac{s(\omega) - s(-\omega)}{\hbar} \quad . \tag{10.14}$$

This function is real and odd in ω. Its contribution to the energy dissipation is obtained after multiplying it by ω and integrating over all ω.

Besides the δ-function part of the spectral representation there is a principal value part, compare (6.7,8). This can be obtained from (10.13), replacing the δ-function by $P/(\omega_{\mu\nu} - \omega)$ and *adding* a corresponding term with ω replaced by $-\omega$. In contrast to (10.14) this yields an *even* function in ω. Multiplication by ω and integration over ω yields a net contribution to the energy dissipation equal to zero. This is fortunate, as otherwise (because of the factor $-i$ in the integrand) the net contribution would be imaginary. So, finally one obtains for the energy dissipation

$$\langle H(\infty) - H(-\infty)\rangle = \int \omega \chi''(\omega) d\omega / 2\pi \quad . \tag{10.15}$$

The r.h.s. of this equation is, first of all, real, as it should be. Moreover, one can show that it is non-negative. The proof uses again the detailed balance relation (8.2). If one inserts this into (10.9) and interchanges the summation indices k and l on one side of the resulting equation one finds

$$s(-\omega) = s(\omega) e^{-\beta\hbar\omega} \quad . \tag{10.16}$$

If one inserts this into (19.14) and keeps (10.13) in mind, one obtains the inequality

$$\boxed{\omega \chi''(\omega) \geq 0} \quad . \tag{10.17}$$

Together with (10.15) this means that the energy of an equilibrium state can only be *increased* by weak external forces. This property of equilibrium states is called *passivity*. Linear response theory deals with *linear passive systems*. In this sense an equilibrium state is a kind of generalization of a quantum mechanical ground state. (However, remember in this context the discussion at the end of the previous chapter, where it was pointed out that an equilibrium state at nonzero temperature may lose energy by inelastic scattering).

The inequality (10.17) can also be expressed more directly, without introducing $W(\omega)$ and $\chi''(\omega)$. It is customary to express the k sum in (10.8), for example, in terms of the Hermitian conjugate quantities, using (10.11). If we use the notation $q_k^* = q_{\bar{k}}$ as introduced in Chap. 7, we find

$$\boxed{\begin{aligned}&\langle H(-\infty) - H(-\infty)\rangle \\ &= \sum \int_{-\infty}^{+\infty} f_k^e(\omega)^* \omega \chi''_{kl}(\omega) f_l^e(\omega) d\omega/(2\pi) \geq 0\end{aligned}} \qquad (10.18)$$

and, since this is valid for arbitrary $f_k^e(\omega)$, the "inequality"

$$\boxed{\text{``} \omega \chi''_{kl}(\omega) = \Phi''_{kl}(\omega) \geq 0 \text{''}} \qquad (10.19)$$

The meaning of this "inequality" is nothing but (10.17,18), namely: if one multiplies (10.19) by $a_k^* a_l$ and sums over k and l the result (for arbitrary complex numbers a_k) will be a non-negative real number.

We shall see later on that passivity (10.18,19) is intimately connected with irreversibility, i.e. the increase of entropy in closed systems. In fact we shall see that the increase of entropy can be derived from the inequality (10.19).

PROBLEMS

10.1 Use the von Neumann equation (2.6) or (2.19) to derive the differential law of energy conservation $d\langle H(t)\rangle = \sum d\langle q_k(t)\rangle/dt\, f_k^e(t)$.

10.2 Prove that the relaxation function for a conserved quantity A is given by $\Phi_{Ak}(z) = -\chi_{Ak}^T/z$.

10.3 If you did not know that a friction constant γ, as occurring in (4.23) or (5.8), has to be positive, how could you derive it from passivity (10.19)?

11. The High-Frequency Behaviour of Response Functions

The behaviour of response functions $\chi_{kl}(z)$ and $\Phi_{kl}(z)$ at high frequencies z can be most easily obtained from the spectral representation (6.5), i.e.

$$\Phi_{kl}(z) = \frac{1}{\pi} \int_{-\infty}^{+\infty} \frac{\Phi''_{kl}(\omega)}{\omega - z} d\omega \quad , \tag{11.1}$$

by a power expansion in descending powers of z. This leads to a series

$$\Phi_{kl}(z) = -\frac{1}{z} \sum_{n=0}^{\infty} \frac{\Phi_{kl}^{(n)}}{z^n} \quad . \tag{11.2}$$

The expansion coefficients $\Phi_{kl}^{(n)}$ can be expressed in terms of the nth moments of the spectral function. Using (11.1) one finds

$$\Phi_{kl}^{(n)} = \frac{1}{\pi} \int_{-\infty}^{+\infty} \omega^n \Phi''_{kl}(\omega) d\omega \quad . \tag{11.3}$$

With the aid of (6.16) the r.h.s. of this equation for $n \geq 1$ can also be expressed in terms of the spectral function of the dynamical susceptibility

$$\Phi_{kl}^{(n)} = \frac{1}{\pi} \int \omega^{(n-1)} \chi''_{kl}(\omega) d\omega \quad ; \quad n \geq 1 \quad . \tag{11.4}$$

For $n = 0$ one has to keep in mind possible singularities of Φ at zero frequency (compare the discussion in Chap. 3). Taking into account (3.14) and (6.5) one finds for the isolated static susceptibility (equal to the adiabatic one, for ergodic systems)

$$\frac{1}{\pi} \int \frac{\chi''_{kl}(\omega)}{\omega} d\omega = \chi_{kl}^S \quad . \tag{11.5}$$

The r.h.s. of (11.3), on the other hand, for $n = 0$ yields the isothermal susceptibility. Using (3.13) and (6.4) one finds

$$\Phi_{kl}^{(0)} = \frac{1}{\pi} \int \Phi''_{kl}(\omega) d\omega = \chi_{kl}^T \quad . \tag{11.6}$$

The higher moments of the spectral function can also be obtained from (6.4) by differentiating the l.h.s. of this equation with respect to t and putting $t = 0$:

$$\beta i^n \langle q_k^{(n)}; q_l \rangle = \Phi_{kl}^{(n)} \quad . \tag{11.7}$$

Here $q_k^{(n)}$ is the nth time derivative of $q_k(t)$ for $t = 0$. Using (6.3) and (11.4) the l.h.s. of this equation can also be expressed in terms of commutators. We write down the resulting equations for $n = 1, 2, 3$ explicitly:

$$\Phi_{kl}^{(1)} = i\beta \langle \dot{q}_k; q_l \rangle = \langle [q_k, q_l] \rangle / \hbar \quad , \tag{11.8}$$

$$\Phi_{kl}^{(2)} = \beta \langle \ddot{q}_k; q_l \rangle = i \langle [\dot{q}_k, q_l] \rangle / \hbar \quad , \tag{11.9}$$

$$\Phi_{kl}^{(3)} = i\beta \langle \dddot{q}_k; \dot{q}_l \rangle = \langle [\dot{q}_k, \dot{q}_l] \rangle / \hbar \quad . \tag{11.10}$$

In the last two equations we have expressed time derivatives of the q_k in terms of those of the q_l using time translation invariance.

Relations of the form (11.5, 6, 8–10) for the moments of spectral functions are called *sum rules*. In addition to those derived so far one often considers directly the moments of the spectral functions $s_{kl}(\omega)$ of the product correlation functions (6.10).

The sum rules become particularly simple for a single variable, say, q. In this case the indices k, l can be omitted. The first five sum rules then read

$$\boxed{\frac{2}{\pi\hbar} \int \frac{P}{\omega} s(\omega) d\omega = \frac{1}{\pi} \int \frac{\chi''(\omega)}{\omega} d\omega = \chi^0 \\ = \chi^S, \text{ ergodic systems}, } \tag{11.11}$$

$$\boxed{\frac{2}{\pi\hbar} \int \frac{1}{\omega} s(\omega) d\omega = \frac{1}{\pi} \int \Phi''(\omega) d\omega = \chi^T \quad , } \tag{11.12}$$

$$\frac{1}{\pi} \int s(\omega) d\omega = \frac{1}{\pi} \int \varepsilon(\omega, T) \Phi''(\omega) d\omega = \langle q^2 \rangle \quad , \tag{11.13}$$

$$\frac{2}{\pi\hbar} \int \omega s(\omega) d\omega = \frac{1}{\pi} \int \omega^2 \Phi''(\omega) d\omega = \frac{i}{\hbar} \langle [\dot{q}, q] \rangle \quad , \tag{11.14}$$

$$\frac{1}{\pi} \int \omega^2 s(\omega) d\omega = \frac{1}{\pi} \int \omega 2\varepsilon(\omega, T) \Phi''(\omega) d\omega = \langle \dot{q}^2 \rangle \quad . \tag{11.15}$$

If q is the spatial coordinate of a particle with mass m, the r.h.s. of (11.14) can be evaluated exactly using the commutation relations

$$\frac{i}{\hbar}[\dot{q}, q] = \frac{i}{\hbar m}[p, q] = \frac{1}{m} \quad . \tag{11.16}$$

In fact (11.14) is then nothing but the generalization of the well-known *f-sum rule* for dipole oscillator strengths to nonzero temperatures.

The sum rules listed so far are exact. In the classical limit ($\langle a;b\rangle \approx \langle ab\rangle \approx \langle ba\rangle$) further approximately valid relations can be derived.

As preparation for the next chapter we discuss now a reorganisation of the expansion (11.2) into a continued fraction (all quantities on the r.h.s. of the next equation are now matrices)

$$\Phi_{kl}(z) = \left(\frac{1}{N(z)-z}\chi^T\right)_{kl} \qquad (11.17)$$

Here we have already taken into account (11.6). Because of this the high-frequency behaviour of $N(z)$ can be written as

$$N(z) = N^{(0)} + N^{(1)}(z) \quad, \qquad (11.18)$$

where $N^{(0)}$ is a constant and $N^{(1)}(z)$ has an expansion for large z with descending powers of z, starting with a term proportional to $1/z$. The continued fraction expansion corresponds to an ansatz

$$\begin{aligned}N^{(1)}(z) &= \frac{1}{N^{(2)}-z-\cdots}N^{(1)} \\ &= -\frac{N^{(1)}}{z} - \frac{N^{(2)}N^{(1)}}{z^2} - \cdots \quad.\end{aligned} \qquad (11.19)$$

The expansion coefficients $N^{(n)}$ can be expressed in terms of the coefficients of the original expansion (11.2) by comparing both expansions. This yields

$$N^{(0)}\chi^T = \Phi^{(1)} \quad, \qquad (11.20)$$

$$N^{(2)}\chi^T = -N^{(0)}\Phi^{(1)} + \Phi^{(2)} \quad, \qquad (11.21)$$

$$N^{(2)}N^{(1)}\chi^T = -N^{(1)}\Phi^{(1)} + N^{(0)}\Phi^{(2)} - \Phi^{(3)} \quad. \qquad (11.22)$$

The continued fraction expansion usually converges much better than the original expansion (11.2). For the same number of moments $\Phi^{(n)}$ of the spectral function, (11.17,19) is a better approximation for the response function than the original expansion (11.2) terminated at maximal n.

PROBLEMS

11.1 Discuss the content of the five sum rules (11.11–15) for the dipole susceptibility, where q is the spatial coordinate of a particle. Give an estimate for χ^0, $\langle q^2\rangle$ and $\langle \dot{q}^2\rangle$ in terms of ω_0 and m if the spectral function is approximately proportional to a sum $\delta(\omega-\omega_0)+\delta(\omega+\omega_0)$.

12. The Low-Frequency Behaviour of Response Functions

The matrix $N(z)$ introduced in the last section is also well suited for the discussion of the low-frequency behavior of the response functions. Of course, $N(z)$ is nothing but the analytic continuation of the memory function $N(\omega)$ introduced in (4.24) or, in matrix form, in the context of (5.15) in the upper frequency half plane. It may be considered as a frequency-dependent generalization of kinetic coefficients. The traditional kinetic coefficients are given by the $z \to 0$ limit of $N(z)$. Nore precisely, since $N(z)$, like the response functions, may be discontinuous across the real axis, one has to consider $N(+i0)$.

In the context of (5.17,18) we had introduced another set of kinetic coefficients, called μ_{kl}. They have simpler symmetry relations and can be taken over into the nonlinear regime. It is useful to introduce in analogy to $N(z)$ the analytic continuation $M(z)$ of μ. Then (11.17) takes the two forms

$$\Phi(z) = \chi^T \frac{1}{M(z) - \chi^T z} \chi^T = \frac{1}{N(z) - z} \chi^T \quad . \tag{12.1}$$

Here again all quantities are matrices and

$$M(z) = N(z)\chi^T \quad . \tag{12.2}$$

If one considers the relaxation function $\Phi(z)$ in analogy to the Green's function of many-body theory (or of field theory), $M(z)$ corresponds to the self-energy (or the so-called "mass operator"), the isothermal susceptibility χ^T usually being equal to 1 in those theories.

Combining the equation $\chi(z) = \chi^T + z\Phi(z)$ with (12.1) one can also express $\chi(z)$ in terms of N or M:

$$\chi(z) = N(z)\Phi(z) = \chi^T + \chi^T \frac{z}{M(z) - \chi^T z} \chi^T \quad . \tag{12.3}$$

The z in the numerator on the r.h.s. of this equation may now be combined with either the left or right χ^T. If then a term $M(z)$ is added to and subtracted from the respective $\chi^T z$ one obtains two equivalent forms for $\chi(z)$:

$$\chi(z) = M(z)\frac{1}{M(z) - \chi^T z}\chi^T = \chi^T\frac{1}{M(z) - \chi^T z}M(z) \quad . \tag{12.4}$$

According to (3.5) the dynamical susceptibility $\chi(z)$ determines the linear response of the deviation $\Delta Q(z)$ of the coordinate $Q(z)$ from its equilibrium value Q_0 to an external force $f^e(z)$. A particularly simple form of kinetic equations are obtained if one introduces an "internal" force $f(z)$ (in the linear response regime) as

$$\Delta Q(z) = \chi(z)f^e(z) = \chi^T f(z) \quad , \tag{12.5}$$

where $f(z)$ is just the Lagrange parameter of a quasi-static equilibrium state with $\langle q \rangle = Q(z)$. If $\chi(z)$ in (12.5) is replaced by the second of the two expressions in (12.4) one finds

$$f(z) = \frac{1}{M(z) - \chi^T z}M(z)f^e(z) \quad . \tag{12.6}$$

Multiplying this equation by the denominator of the r.h.s. finally leads to

$$\chi^T z f(z) = M(z)[f(z) - f^e(z)] \quad . \tag{12.7}$$

We now transform this equation back into t-space and spell out the matrix indices explicitly:

$$\dot Q_k(t) = -i\sum_l \int_{-\infty}^{+\infty} M_{kl}(t - t')[f_l(t') - f_l^e(t')]dt' \quad . \tag{12.8}$$

Here the t' integration, for the time being, is carried out generally to infinity. In fact, as we shall see, one can replace the upper limit by t. This is allowed, since $M(t)$, like the response functions $\chi(t)$ and $\Phi(t)$, is a retarded function which vanishes for $t \geq 0$. Physically speaking this is again a consequence of causality and will be discussed in more detail in the next chapter.

If one considers external forces that are switched on adiabatically and then switched off abruptly at $t = 0$, the internal and external force in the integral of (12.8) cancel for $t' \leq 0$, leading to

$$\dot Q_k(t) = -i\sum_l \int_0^{t+0} M_{kl}(t - t')f_l(t')dt' \quad . \tag{12.9}$$

Here we have added a positive infinitesimal $+0$ in the upper limit of the integral in order not to lose any singular terms $\propto \delta(t - t')$ in $M(t)$.

Keeping in mind (12.2,5) one can also write

$$\dot Q_k(t) = -i\sum_l \int_0^{t+0} N_{kl}(t - t')\Delta Q_l(t')dt' \quad . \tag{12.10}$$

This equation is valid only in linear response theory, whereas, as we shall see, (12.9) can essentially be taken over to the nonlinear regime.

The equations (12.8,9) are integrodifferential equations: The "velocities" $dQ(t)/dt$ do not respond instantaneously to the forces $f(t)$ but with a memory. The decay time of this memory is identical to the decay time of the functions $N(t)$ or $M(t)$. Hence the term "memory kernel" or "memory function", which we introduced in Chap. 4.

The integrodifferential equations simplify considerably – they become differential equations – for processes with a decay time $\delta\tau$ of the memory functions that is small compared to the relaxation time τ of the response functions. As already indicated in Chap. 4, one introduces so-called *kinetic coefficients*

$$\nu_{kl} = i \int_0^\infty N_{kl}(t)dt = iN_{kl}(\omega = +i0) \qquad (12.11)$$

and/or

$$\mu_{kl} = i \int_0^\infty M_{kl}(t)dt = iM_{kl}(\omega = +i0) \quad . \qquad (12.12)$$

Then, for times t sufficiently large compared to $\delta\tau$, the r.h.s. of (12.9,10) can be integrated out to

$$\dot{Q}_k(t) = -\sum \mu_{kl} f_l(t) = -\sum \nu_{kl} \Delta Q_l(t) \quad . \qquad (12.13)$$

This approximation corresponds to a behaviour

$$\Phi(\omega) = \frac{1}{N(+i0) - \omega} \chi^T \quad ; \quad \omega\delta\tau \ll 1 \qquad (12.14)$$

of the relaxation function for low frequencies.

The validity of the approximations (12.13,14) obviously requires the existence of two well-separated time scales: one, say τ, being given directly by the *magnitude* ν_{kl} of the memory functions $N(+i0)$, and another one, say $\delta\tau \ll \tau$, being given by the *decay time* of the memory functions $N(t)$. One then can distinguish so-called *slow variables* [namely the q_k occurring explicitly in the kinetic equations (12.13)] and all other variables, the *fast variables*. The fast variables are often called bath variables, since they can be considered as the constituents of the heat bath for the slow variables.

An important first step in setting up kinetic equations therefore is always to pick a suitable set of slow variables. Very useful guidelines for picking such variables are conservation laws. We shall discuss this in the second part of this volume in the context of hydrodynamics and transport theory. Length scales in collision processes often can be converted into corresponding time scales. In Brownian motion, for instance, see Chap. 4, one has a long time scale τ corresponding to the relaxation time of the Brownian particle (related to the mean free path). This time is large compared to the *duration* of a collision, if the mean free path is large compared to the range of the interaction potential between the Brownian particle and the particles of the bath.

A more mathematically oriented procedure for picking a complete set of slow variables consists in starting with a plausible set of q_k's. One then calculates the corresponding memory functions with the procedure described in the next chapter. If these functions still contain slowly varying terms one can often remove them by including the time derivatives of the slow variables in a new enlarged set. This procedure can easily be studied for the example of a Brownian oscillator (Problem 12.1).

For completeness we mention that equations of the form (12.13), where memory effects play no role, are said to obey *Markovian* behaviour or to be of *Markov* type.

PROBLEMS

12.1 Calculate the memory functions $N(\omega)$ for the Brownian oscillator (Chap. 5) and discuss their characteristic frequency scales for
   i) the single variable $X(t)$,
   ii) after including the time derivative $dX(t)/dt$ as a second slow variable.

13. Stochastic Forces, Langevin Equation

The kinetic coefficients introduced in the previous chapter can be expressed in terms of correlations in time of the time derivatives $\dot{q}_k$ as discovered by *Langevin* [13.1] in the context of Brownian motion. This and the following two chapters will be devoted to what is sometimes called the second fluctuation–dissipation theorem. It is a generalization of Langevin's method leading to equations which ultimately can be used for microscopic calculations of kinetic coefficients. We are going to provide examples for such calculations in the third part of this volume. For the time being we just derive the corresponding formalism.

It will often be useful to work in time space, rather than frequency space. In order to have a compact notation for the integrodifferential equations occurring in this context we introduce a shorthand matrix notation in (k,t)-space. We consider, for instance, relaxation functions

$$\Phi_{kl}(t,t') = i\beta \langle q_k(t); q_l(t')\rangle \Theta(t-t') \tag{13.1}$$

and self-energies $M_{kl}(t,t')$ etc. as generalized matrices Φ, M, etc. in this space with a multiplication rule

$$(AB)_{kl}(t,t') = \sum_m \int_{-\infty}^{+\infty} A_{km}(t,t'') B_{ml}(t'',t') dt'' \quad . \tag{13.2}$$

In order to have a notation as close as possible to Fourier space with complex frequencies z we introduce the differential operator z with the matrix elements

$$z_{kl}(t,t') = i\delta_{kl} d\delta(t-t')/dt \tag{13.3}$$

and the isothermal susceptibility matrix (not to be confused with the dynamical susceptibility χ)

$$\chi^T_{kl}(t,t') = \chi^T_{kl} \delta(t-t') \quad . \tag{13.4}$$

13.1 The Subtraction Method (Langevin)

There are essentially two methods to derive equations for the self-energies: the so-called subtraction and projection methods. The subtraction method is a fairly direct consequence of the definition (12.1,3) of M. It does not introduce any new concepts, and leads only to implicit equations for M. The projection method is more fancy. It introduces new concepts like the Liouville super operator and a scalar product in operator space. It eventually leads to explicit equations for M.

A convenient starting point for the subtraction formalism is (12.1), which, after multiplication with the denominator $(M - \chi^T z)(\chi^T)^{-1}$ takes the two equivalent forms

$$\left(M \frac{1}{\chi^T} - z\right) \Phi = \Phi \left(\frac{1}{\chi^T} M - z\right) = \chi^T \quad . \tag{13.5}$$

Because of the notation (13.1,2,3) introduced above, these two equations look the same in z- and t-space. If one uses t-space, bearing in mind (3.12), the differentiation of z according to (13.3) yields two terms. One from the derivative of the Θ function in (3.12) and one from the derivative of $q_k(t)$. The remainder can be written as

$$\left(M \frac{1}{\chi^T} \Phi\right)_{kl}(t, t') = -\beta \langle \dot{q}_k(t); q_l(t') \rangle \Theta(t - t') = \chi_{kl}(t, t') \tag{13.6}$$

and, equivalently,

$$\left(\Phi \frac{1}{\chi^T} M\right)_{kl}(t, t') = \beta \langle q_k(t); \dot{q}_l(t') \rangle \Theta(t - t') = \chi_{kl}(t, t') \quad . \tag{13.7}$$

As already mentioned in the previous chapter in the context of (12.8), the self-energy $M(t, t')$ is a retarded function (i.e. nonzero only for $t \geq t'$). We shall present a general proof for this in Chap. 19 in connection with a spectral representation of $M(z)$. For the time being we content ourselves with a consistency argument. We introduce the ansatz of retarded functions M into (13.6,7) and derive an equation for M that is consistent with this ansatz.

If M is retarded the integration over the intermediate times on the l.h.s. of (13.6,7) runs only from t' to t. The resulting equations look simpler and can be interpreted in simple physical terms if one introduces (in generalization of Langevin's ideas) so-called "*fluctuating*"or "*stochastic*" *forces* $F_k(t)$ by means of the equations

$$F_k(t) = \dot{q}_k(t) + i \sum \int_{t'}^{t} \left(M \frac{1}{\chi^T}\right)_{km}(t, t'') q_m(t'') dt'' \quad , \quad t' < t \tag{13.8}$$

and

$$F_l(t') = \dot{q}_l(t') - i \sum \int_{t'}^{t} q_m(t'') \left(\frac{1}{\chi^T} M\right)_{ml}(t'', t') dt'' \quad , \quad t' < t \quad . \tag{13.9}$$

Then, after introducing nonequilibrium averages by $Q_k(t) = \overline{q_k(t)}$ and $dQ_k(t)/dt = \overline{dq_k(t)/dt}$, one sees that one can write (12.10) in the simple form

$$\overline{F_k(t)} = 0 \quad . \tag{13.10}$$

Furthermore the two equations (13.6,7) can be combined to

$$\langle F_k(t); q_l(t')\rangle \Theta(t-t') = 0 \qquad (13.11)$$

and

$$\langle q_k(t); F_l(t')\rangle \Theta(t-t') = 0 \quad . \qquad (13.12)$$

Differentiating (13.11) and (13.12) with respect to t' and t, respectively, one obtains the desired expressions for M. The simplest way is to insert (12.1) for Φ in the l.h.s. of (13.6,7) and then multiply again by the numerator of (12.1) to obtain

$$M = \chi[(1/\chi^T)M - z] = [M(1/\chi^T) - z]\chi \quad . \qquad (13.13)$$

If here for χ the expressions from the middle of (13.6,7) are used and the differentiation contained in z is carried out one arrives at

$$\boxed{M_{kl}(t,t') = i\beta\langle \dot{q}_k; q_l\rangle \delta(t-t') - i\beta\langle F_k(t); \dot{q}_l(t')\rangle \Theta(t-t')} \qquad (13.14)$$

and

$$\boxed{M_{kl}(t,t') = i\beta\langle \dot{q}_k; q_l\rangle \delta(t-t') - i\beta\langle \dot{q}_k; F_l(t')\rangle \Theta(t-t')} \quad . \qquad (13.15)$$

There is also a more symmetric version

$$\boxed{M_{kl}(t,t') = i\beta\langle \dot{q}_k; q_l\rangle \delta(t-t') - i\beta\langle F_k(t); F_l(t')\rangle \Theta(t-t')} \quad . \qquad (13.16)$$

This version agrees with (13.14,15) since the contribution added in the second term on the r.h.s. of this equation vanishes because of (13.11,12). Finally one notices, that the r.h.s. of (13.14,15,16) is a retarded function. This is consistent with our original ansatz for M.

The comparison of the r.h.s. of the above expressions for M with (11.8,20) shows that the first term on the r.h.s. is just the limit of the self-energy at infinite frequency. This is also a direct consequence of the occurrence of the δ-function in the first term of the r.h.s. of (13.14,15,16).

The foregoing results of this section may be summarized by saying that if one introduces stochastic forces $F(t)$ by (13.8,9), they have the following properties:
a) For small deviations from equilibrium their average vanishes (13.10).
b) Their equilibrium correlations with the coordinates vanish (13.11,12).

c) The equilibrium correlations of the stochastic forces with themselves yield the memory functions according to (13.16) and thus the kinetic coefficients according to (12.12).

The equations of this section are generally valid, and, within the framework of linear response theory, exact. Their derivation, however, is formal and not very transparent from the physical point of view. In the next section, therefore we consider as a simple case again the example of Brownian motion, for which a physical interpretation of the formalism is relatively easy.

13.2 The Projection Method (Zwanzig and Mori)

Since the stochastic forces according to (13.8,9) contain terms with the self-energy in them, the r.h.s. of (13.14–16) contain such terms as well. These equations, therefore, do not determine the the quantities M explicitly. We shall see that this is no handicap for a microscopic determination of M by means of perturbation theory or by iteration. There is, however, a formalism leading to explicit expressions for the memory functions M. This is the projection formalism developed by *Zwanzig* [13.2] and *Mori* [13.3]. We present a brief introduction to this formalism.

The formalism is based on the Liouville operator L. This operator acts in the space of the dynamical variables of the system and thus is sometimes called a super operator. In classical physics this is the space of functions of canonical coordinates and momenta, in quantum mechanics the space of *operators* (whereas the operators act in the Hilbert space of state vectors). For details consider Problem 13.1.

The equations of motion in terms of L take the form

$$\dot{q}_k(t) = iLq_k(t) \tag{13.17}$$

with the formal solution

$$q_k(t) = e^{iL(t-t')}q_k(t') \quad . \tag{13.18}$$

A further important concept in the projection formalism is the scalar product $\langle a; b^* \rangle$ between *operators* a and b (Problem 13.2) and the projection operator P onto the initial values $q_k(t')$, defined by

$$Pa = \sum \Delta q_k(t')(\chi^T)^{-1}_{kl}\langle \Delta q_l(t'); a\rangle \beta \tag{13.19}$$

for arbitrary operators a. One can easily verify the following two relations:

$$P\Delta q_k(t') = \Delta q_k(t') \tag{13.20}$$

and

$$P^2 = P \quad . \tag{13.21}$$

With the aid of the complement $Q = 1 - P$ every operator a can be decomposed into its "matrix elements" in the subspace of initial slow variables $q_k(t')$ and its complement

$$a = (P + Q)a(P + Q) = a_{PP} + a_{PQ} + a_{QP} + a_{QQ} \quad . \tag{13.22}$$

With the above definitions the relaxation function can now be written as

$$\Phi_{kl}(t,t') = \beta \langle \Delta q_k(t'); R_{PP} \Delta q_l(r') \rangle \tag{13.23}$$

with the so-called resolvent operator

$$R(t-t') = \mathrm{i} e^{-\mathrm{i} L(t-t')} \Theta(t-t') \quad . \tag{13.24}$$

If this equation is differentiated with respect to t the result, using again the differential operator z, takes the form

$$(L - z)R = 1 \quad . \tag{13.25}$$

The decomposition of this equation according to (13.22) then reads

$$\begin{aligned}(L-z)_{PP} R_{PP} + L_{PQ} R_{QP} &= 1 \ , \\ L_{QP} R_{PP} + (L-z)_{QQ} R_{QP} &= 0 \ .\end{aligned} \tag{13.26}$$

After eliminating R_{QP} one finally obtains

$$\left[(L-z)_{PP} - L_{PQ} \frac{1}{(L_{QQ}-z)} L_{QP} \right] R_{PP} = 1 \quad . \tag{13.27}$$

We now multiply this equation by $\langle \Delta q_k(t');$ from the left and by $\Delta q_m(t') \rangle$ from the right. Keeping in mind (13.23) we then find an explicit expression for the self-energy M by comparing (13.27) with (13.5). The self-energy is essentially given by the matrix elements of the term in the square bracket on the l.h.s. of (13.27). More precisely, if one takes into account (13.17,20) (and time translation invariance) and uses, in analogy to (13.24),

$$(L_{QQ} - z)^{-1}(t,t') = \mathrm{e}^{-\mathrm{i}QLQ(t-t')} \Theta(t-t') \ , \tag{13.28}$$

one finds

$$\boxed{\begin{aligned} M_{kl}(t,t') =& \mathrm{i}\beta \langle \dot{q}_k(t); \Delta q_l(t) \rangle \delta(t-t') \\ &- \mathrm{i}\beta \langle \mathrm{e}^{\mathrm{i}QLQ(t-t')} Q\dot{q}_k(t); \dot{q}_l(t') \rangle \Theta(t-t') \quad . \end{aligned}} \tag{13.29}$$

So finally the equation for M takes exactly the form (13.14), where, however, the stochastic force is now given by

$$\boxed{F_k(t) = e^{iQLQ(t-t')} Q\dot{q}_k(t')} \qquad (13.30)$$

With this definition of the stochastic force, (13.11) is valid, too. This is a direct consequence of $Q\Delta q_k(t') = 0$, which is equivalent to (13.20). Similarly, (13.10) is valid if the nonequilibrium average of F is explicitly spelt out in linear response theory (Problem 13.3).

At first sight one might think that an expression as weird and formal as (13.30) cannot be of any practical use. But, in fact, we shall demonstrate in the third part of this volume that it is, indeed, a possible starting point for microscopic calculations of kinetic coefficients. In particular, it is a convenient expression for the introduction of the so-called mode–mode coupling approximation in such calculations.

Subtraction and projection both have the same effect: In physical terms they remove from the time derivatives $dq_k(t)/dt$ the slowly varying part, leaving only the high-frequency part (of the complementary space Q, physically speaking, of the "heat bath"). We shall discuss this point in more detail in the next chapter.

PROBLEMS

13.1 Determine the matrix elements $L_{\mu\nu,\kappa\lambda}$ of the Liouville operator L such that $d\langle\mu|a|\nu\rangle/dt = i\langle\mu|[H,a]|\nu\rangle/\hbar = i\sum L_{\mu\nu,\kappa\lambda}\langle\kappa|a|\lambda\rangle$.

13.2 Prove the reality and positivity properties of the "scalar product of operators" $\langle a;b\rangle^* = \langle b^*;a^*\rangle$ and $\langle a;a^*\rangle \geq 0$.

13.3 Using thermodynamic perturbation theory (3.8) prove the expansion $\bar{F} = \langle F\rangle + \beta\sum\langle F;\Delta q_k\rangle f_k + \cdots$ and hence (13.10).

14. Brownian Motion: Langevin Equation*

As an illustration of the rather abstract results of the previous chapter we discuss a theory of Brownian motion going back to *Langevin* [14.1]. We consider a Brownian particle that has a momentum $p(0)$ at time $t = 0$. The collisions of this particle with other particles of the heat bath will lead to random fluctuations of its momentum that cannot be predicted in detail. On the average, however, it must relax to its equilibrium value $\langle p \rangle = 0$.

As in Chap. 4 we describe this relaxation in terms of a differential equation $d\overline{p(t)}/dt + \gamma \overline{p(t)} = 0$ with a friction coefficient γ. Now, Langevin describes this situation as follows. He introduces a so-called stochastic force $F(t)$ by means of the equation

$$F(t) = \dot{p}(t) + \gamma p(t) \quad ; \quad t > 0 \quad . \tag{14.1}$$

The idea behind this definition is that, whenever the Brownian particle has a momentum different from zero the force $\dot{p}$ cannot fluctuate randomly around zero: it has to have a systematic part which leads to its relaxation to zero. For this systematic part Langevin uses the ansatz $-\gamma p(t)$. Then according to (14.1) the average of $F(t)$ has to be zero

$$\overline{F(t)} = 0 \quad ; \quad t > 0 \quad . \tag{14.2}$$

Equations (14.1,2) have to be considered in analogy to (13.8,10). How about the other two equations (13.11,14)? In our special case (13.11) obviously reads

$$\langle F(t)p(0) \rangle \Theta(t) = 0 \quad . \tag{14.3}$$

If this equation is assumed to hold also for $t = 0$, one has to supplement (14.1) by

$$F(0) = \dot{p}(0) \quad . \tag{14.4}$$

Then $\langle F(0)p(0) \rangle = \langle \dot{p}(0)p(0) \rangle = 0$.

Equation (14.3) says that the stochastic forces are uncorrelated with the momenta. We shall demonstrate that this is equivalent to the regression ansatz

$$\langle p(t)p(0) \rangle = \langle p(0)^2 \rangle e^{-\gamma t} \quad , \tag{14.5}$$

see (4.10), for the momentum correlation.

On the one hand, differentiation of (14.5) and multiplication by $\Theta(t)$ paying attention to (14.1,4) yields directly (14.3). On the other hand, one can multiply (14.1) by $\exp(\gamma t)$ and integrate from zero to t to obtain

$$p(t) = e^{-\gamma t}\left(p(0) + \int_0^t e^{\gamma t'} F(t')dt'\right) \quad ; \quad t \geq 0 \quad . \tag{14.6}$$

If this equation is multiplied by $p(0)$ and the equilibrium average is taken, one finds (14.5) (for $t \geq 0$) precisely if (14.3) is taken into account.

Next we turn to (13.14). It can, in principle be obtained by differentiation of (14.3). Let us, instead, start directly from (14.5). Differentiation with respect to t leads to

$$\langle p(t)\dot{p}(0)\rangle = -\langle \dot{p}(t)p(0)\rangle = \text{sign}\{t\}\langle p(t)p(0)\rangle \quad . \tag{14.7}$$

Differentiating once more and taking into account

$$d\,\text{sign}\{t\}/dt = 2\delta(t) \tag{14.8}$$

one finds

$$\langle \dot{p}(t)\dot{p}(0)\rangle = -\gamma\,\text{sign}\{t\}\langle \dot{p}(t)p(0)\rangle + 2\gamma\langle p(0)^2\rangle\delta(t) \quad . \tag{14.9}$$

If we now extend the definition (14.1) of $F(t)$ to all t by

$$F(t) = \dot{p}(t) + \gamma\,\text{sign}\{t\}\,p(t) \tag{14.10}$$

and keep in mind (14.7) we obtain

$$\langle F(t)F(0)\rangle = 2\gamma\langle p(0)^2\rangle\delta(t) \quad . \tag{14.11}$$

On the other hand one can again start from (14.6). To simplify the procedure somewhat, we consider the limit of large t. More precisely we generalize the lower limit of the integration in (14.6) to, say, t_0, and consider the limit of t_0 going to minus infinity. Then (14.6) simplifies to

$$p(t) = e^{-\gamma t}\int_{-\infty}^t e^{\gamma t'} F(t')dt' \quad . \tag{14.12}$$

If this expression is multiplied by a similar one for $p(t')$ and the equilibrium average is taken, one recovers (14.5) precisely if (14.11) is valid.

Now let us compare (14.11) with (13.14). First of all one should not be misled – by the occurrence of the δ-function in (14.11) – to compare the r.h.s. of (14.11) with the first term on the r.h.s. of (13.14). In fact this term always vanishes if there is only a single slow variable, since $\langle \dot{q}; q\rangle = 0$. However, the second term on the r.h.s. of (13.14) in general will not be proportional to a δ-function. On the other hand, the r.h.s. of (14.11) is only an approximation for a function with a nonzero width, which is small compared to $1/\gamma$. As already discussed in Chap. 4, (14.5) will have to be modified for short times of the order of the duration $\delta\tau$ of the collisions of the Brownian particle. Thus the r.h.s. of (14.11) will roughly have a width $\delta\tau$. One then has to be particularly careful if (14.11) is multiplied by the discontinuous functions $\Theta(\pm t)$ in order to obtain the two parts of (13.14,15) for the memory function. In the limit of vanishing width $\delta\tau$ then each of these parts contains only half of the δ-function on the r.h.s. of (14.11) according to the rule $2\Theta(t)\delta(t) = \delta(t)$.

Our considerations clearly demonstrate the effect of adding the term γp to the time derivative of p in the definition of the fluctuating force F: it subtracts out from dp/dt all low-frequency components, not only in the average (14.2) but also in the correlations with p (14.7) and with dp/dt (14.9).

The derivation of (14.7,9) is straightforward if (14.5) is known. When *Langevin* wrote his pioneering paper on Brownian motion in 1908 this was, of course, not the case. He, instead, used (14.3,11) as plausible assumptions and derived (14.5) from them.

For further insight into the physical meaning of (14.2,3,11) we discuss a simple model. We consider the interaction of the Brownian particle with the heat bath as a succession of infinitely short statistically independent collisions. If Δp_i is the momentum change in such a collision, the equation of motion is $dp(t)/dt = \sum \Delta p_i \delta(t - t_i)$. Now, following Langevin we introduce momentum changes f_i with vanishing average by means of the equation

$$\dot{p}(t) + \text{sign}\{t\}\gamma p(t) = F(t) = \sum_i f_i \delta(t - t_i) \quad . \tag{14.13}$$

Since the average of the l.h.s. of this equation vanishes, the average of the f_i vanishes as well. Furthermore, again following Langevin, we assume that the f_i are statistically independent of the momenta and of each other:

$$\langle f_i p(0) \rangle = 0 \quad ; \quad \langle f_i f_k \rangle = \langle f_i^2 \rangle \delta_{ik} \quad . \tag{14.14}$$

Then, obviously, one has

$$\langle F(t)p(0) \rangle = \sum_i \langle f_i p(0) \rangle \langle \delta(t - t_i) \rangle = 0 \quad . \tag{14.15}$$

Furthermore, one immediate consequence of (14.13,14) is $\langle F(t)F(0) \rangle = C\delta(t)$. In order to determine the constant C we consider

$$\left\langle \left[\int_0^t F(t')dt' \right]^2 \right\rangle = t \int \langle F(t')F(0) \rangle dt' = Ct \quad . \tag{14.16}$$

The l.h.s. of this equation is now obtained by direct integration as

$$\left\langle \left(\sum_{0 \le t_i \le t} f_i \right)^2 \right\rangle = \sum_{0 \le t_i \le t} \langle f_i^2 \rangle \quad . \tag{14.17}$$

Thus one has

$$C = \sum_{0 \le t_i \le t} \langle f_i^2 \rangle / t \quad . \tag{14.18}$$

If now the integration interval t is taken to be small compared to $1/\gamma$, one obtains by an approximate integration of the l.h.s. of (14.13)

$$p(t) = (1 - \gamma t)p(0) + \sum_{t_i \le t} f_i \quad . \tag{14.19}$$

Taking the square of both sides of this equation and averaging then yields (as $\gamma t \ll 1$)

$$\langle p^2 \rangle 2\gamma t = \sum_{t_i \le t} \langle f_i^2 \rangle \quad . \tag{14.20}$$

If this is finally combined with (14.18) one finds $C = 2\gamma \langle p^2 \rangle$ in complete agreement with (14.11). This completes our statistical interpretation of Langevin's equation.

It is also obvious that any nonzero duration of the collisions will broaden the δ-function in (14.13) and consequently in (14.11) as well. As mentioned already in Chap. 4 the duration $\delta \tau$ should not be confused with the mean collision time $\Delta \tau$ (the average distance between two successive collisions) and, of course, still less with the relaxation time τ itself.

For the diffusion of electrons in solids one has for instance $\delta \tau \ll \Delta \tau \approx \tau$. For self-diffusion in liquids one usually has $\delta \tau \approx \Delta \tau \ll \tau$ ($\delta \tau$ in this case is of the order of the vibrational frequency ω_0 of the particles in the attractive well of the interaction potential). On the other hand, for the diffusion of large Brownian particles in a liquid one expects $\Delta \tau \ll \delta \tau \ll \tau$ since the Brownian particle will be hit by many different particles of the heat bath at approximately the same time. Essential for the validity of (14.11) in all these cases is, however, only the condition $\delta \tau \ll \tau$.

PROBLEMS

14.1 From (4.23) and (14.12) in Fourier space $[p(\omega) = F(\omega)/(\gamma - i\omega)]$ derive (14.11) in Fourier space.

15. Nonlinear Response Theory

In this chapter we intend to abandon the restrictions of linear response theory. First of all let us remember that the initial equations (2.24) and (2.27,28) are generally valid. It is interesting, that, starting from these equations one can derive a formalism which is surprisingly close to the formalism of linear response theory. In particular, the kinetic equations in terms of memory kernels (12.9) and the equations (13.14,15,16) for these memory kernel can be taken over to the general case without major changes.

15.1 The General Initial Value Case

We intend to restrict ourselves to closed systems, and thus consider again the "initial value case" of Chap. 2, in particular (2.21): a number of variables $q_k(0)$ have average values $Q_k(0)$. The statistical operator with maximum entropy fulfilling these conditions is given by

$$\rho_d(0) = \exp\left\{-\beta\left[H - \sum_l f_l(0) q_l(0) - K(0)\right]\right\} \qquad (15.1)$$

with Lagrange parameters $f_l(0)$ which have to be chosen so as to fulfill the conditions $\langle q_k(0)\rangle = Q_k(0)$. Since there are supposed to be no external forces this statistical operator remains the same in the interaction picture for all times $t \geq 0$.

In Chap. 2 an exact consequence of (15.1) was derived for the time derivative of $Q_k(t)$, see (2.27):

$$\dot{Q}_k(t) = -\sum_l \chi_{kl}(t,0) f_l(0) \quad , \quad t \geq 0 \quad . \qquad (15.2)$$

This equation can now be used to define a memory kernel in complete analogy to (13.5) by

$$\chi_{kl}(t,t') = [M(M - \chi^T z)^{-1} \chi^T]_{kl}(t,t') \quad . \qquad (15.3)$$

From this definition one can proceed in almost complete analogy to the linear response case to derive generalized kinetic equations as well as equations for the memory kernels. To start with the kinetic equations we introduce as in

Chap. 13 the "internal forces" $f_l(t)$ of a quasi-static equilibrium corresponding to a statistical operator

$$\rho_{d0}(t) = \exp\left\{-\beta\left[H - \sum f_l(t)q_l(t) - K(t)\right]\right\} \quad . \tag{15.4}$$

Then after writing $dQ_k(t)/dt = \sum \chi_{kl}^T(t)df_l(t)/dt$ in the l.h.s. of (15.1) and inserting (15.3) for the r.h.s. we obtain with the same algebra as in Chap. 13 the kinetic equation

$$\dot{Q}_k(t) = -i\sum_l \int_0^{t+0} M_{kl}(t,t')f_l(t')dt' \quad , \quad t \geq 0 \quad . \tag{15.5}$$

To derive equations for M we proceed again in analogy to Chap. 13. We multiply (15.3) by $[(\chi^T)^{-1}M - z]^{-1}$ to obtain

$$M_{kl}(t,t') = \frac{1}{\hbar}\langle[q_k(t),q_l(t)]\rangle\delta(t-t')$$
$$- \frac{1}{\hbar}\langle[q_k(t),F_l(t')]\rangle\Theta(t-t') \tag{15.6}$$

with the stochastic force

$$F_l(t') = \dot{q}_l(t') - i\sum_m \int_{t'}^t q_m(t'')\left(\frac{1}{\chi^T}M\right)_{kl}(t'',t')dt'' \quad , \quad t' < t \quad . \tag{15.7}$$

This expression is, in fact, identical to (13.9). Equation (15.6) may be considered as the general form of (13.15) valid in the nonlinear regime. It becomes identical to (13.15) in the linear regime if Kubo's relation (3.25) between commutators and bracket symbols is used. If one wants to bring (15.6) even closer to (13.15) one has to generalize the bracket symbol to the quasi-static equilibrium case (15.1). We do this by generalizing (3.9) to

$$a(\alpha) = \exp[-\alpha I(0)]a(0)\exp[\alpha I(0)] \tag{15.8}$$

with the "enthalpy operator"

$$I(t) = H - \sum f_l(t)q_l(t) \tag{15.9}$$

taken at time $t = 0$. The bracket symbol then is defined as in (3.11).

Relations such as (3.25) now do not hold any longer since (3.28) does not hold any more: the propagation in "imaginary time" α proceeds no longer with the Hamiltonian but with the enthalpy (15.9). However, a generalization of (3.25) can easily be derived, namely

$$i\langle [a,b]\rangle/\hbar = -\beta\langle \delta\dot{a}; b\rangle = \beta\langle a; \delta\dot{b}\rangle \quad , \tag{15.10}$$

where $\delta\dot{a}$ is an abbreviation for

$$\delta\dot{a} = i[I(0),a]/\hbar = \dot{a} - \underline{\dot{a}} \tag{15.11}$$

and $\underline{\dot{a}}$ stands for

$$\underline{\dot{a}} = -i\sum_l [a, q_l(0)] f_l(0)/\hbar \quad . \tag{15.12}$$

Keeping in mind (2.27) one has

$$\langle \dot{a}\rangle = \langle \underline{\dot{a}}\rangle = d\langle a\rangle/dt = dA(t)/dt \quad . \tag{15.13}$$

[Remember: here as well as everywhere else above, the expectation values are taken with the statistical operator (15.1) of quasi-static equilibrium.] Equation (15.13) gives a justification for denoting the l.h.s. of (15.12) in analogy to a time derivative.

If we now combine (15.10) with (15.6) we finally obtain an equation for the memory kernels

$$M_{kl}(t,t') = i\beta\langle \delta\dot{q}_k; q_l\rangle\delta(t-t') - i\beta\langle \delta\dot{q}_k(t); F_l(t')\rangle\Theta(t-t') \quad . \tag{15.14}$$

This is as close as one can get to (13.15) of linear response theory. Now, can all this accumulation of implicit, nonlinear, nonlocal definitions lead to any useful results, or is it just some swindle which tries to make complicated things look "simple" (i.e. similar to the linear theory)?

Well, one has to admit that so far nobody has been able to make any use of the completely general set of equations above. But one can indeed use them as a convenient starting point for nonlinear kinetic equations which are "local in time" (better: instantaneous or Markovian) as we shall see in the next section.

15.2 Low-Frequency Perturbation Theory

The approximations leading to Markovian kinetic equations have been discussed several times before. Consider, for instance, the context of (4.25). In our context it means that the memory kernel $M_{kl}(t,t')$ is large only in a narrow interval of order $\delta\tau$ near $t=t'$. Then again, as in (4.26), we can take $f(t)$ outside the t' integral, leading to

$$\dot{Q}_k(t) = -\sum_l \mu_{kl} f_l(t) \quad . \tag{15.15}$$

The main difference from the linear case is the absence of translational invariance and the fact that the averaging is done with the statistical operator (15.1), involving the $f_l(0)$. Thus (15.15) is not a Markovian equation yet. It would be if the averaging were done with the statistical operator (15.4). This, on the other hand, is in accord with the approximation (15.15): one expects the memory of the initial conditions to be lost after a time of the order of $\delta\tau$ so that it does not matter if one uses (15.1) or (15.4). The final result for μ_{kl} can then be written as

$$\mu_{kl}(t, f_m(t)) = i \int_{-\infty}^{t+0} M_{kl}(t, t')dt' \qquad (15.16)$$

with the additional prescription that in (15.14) one should replace $I(0)$ everywhere by $I(t)$.

Figure 15.1 displays once again the order of magnitude of the relevant amplitudes and time scales.

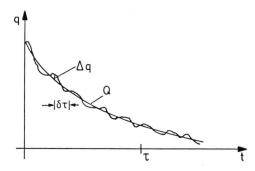

Fig. 15.1. The amplitudes of mean deviations Q_k from complete equilibrium $Q_k = 0$ and of the thermal fluctuations Δq_k around these mean deviations. $\delta\tau$ is the time after which the short time memory is lost, and τ is the relaxation time towards complete equilibrium

In irreversible thermodynamics a notation has become popular that can be traced back to Onsager. In this notation the forces f are expressed in terms of entropy derivatives. For quasi-static equilibrium the first and second laws of themodynamics yield

$$dE = TdS + \sum f_k dQ_k \quad . \qquad (15.17)$$

For a *closed* system with $dE = 0$ this yields

$$f_k = -T\left(\frac{\partial S}{\partial Q_k}\right) \quad . \qquad (15.18)$$

Onsager's notation is obtained by introducing the kinetic coefficients $L_{kl} = T\mu_{kl}$. Then (15.15) takes the form

$$\boxed{\frac{dQ_k}{dt} = \sum L_{kl} \frac{\partial s}{\partial Q_l} \quad .} \qquad (15.19)$$

This is Onsager's famous relation between "fluxes" dQ/dt and "forces" $\partial S/\partial Q$.

It may be considered as the thermodynamic generalization of Hamilton's equations of mechanics.

The kinetic coefficients L_{kl} in general are functions of the Q_k. With respect to these dependencies one may distinguish three different categories.

i) The L_{kl} are independent of the Q_k. Even in this case (15.19) is nonlinear in general since the forces in general are nonlinear functions of the coordinates.

A simple example, the anharmonic oscillator, has already been discussed in the context of (5.18). Comparing these nonlinearities with the linear theory one can emphasize two things. First of all, the nonlinearities concern only the *reversible* part of the forces, and, secondly, the connection between the velocities dQ/dt and *forces* is linear.

A further prominent example of this category is hydrodynamics (with constant viscosities and heat conductivities). The nonlinearities come from the (reversible) acceleration term in Euler's equations, and (in particular for gases) from the nonlinear dependence of pressure on density.

ii) In this category we may consider cases where the kinetic coefficients have some, but not too large dependences on the forces. For instance, friction coefficients usually have some dependence on the velocities, viscocities depend on pressure, temperature, etc., diffusion constants may depend on density and so on.

iii) The third category will then contain examples with strong nonlinearities in the irreversible parts. Typical examples are chemical reactions. The "coordinates" in this case are the concentrations c_k of the reaction partners, and the forces are the corresponding chemical potentials μ_k. If one writes the kinetic equations for the chemical reaction in the form (15.19) the Onsager coefficients are not even approximately constant in the region of interest. Nevertheless the "apparently linear" form of (15.19) is useful in certain contexts. In particular, it comes out of the microscopic theory along the lines of the first section of this chapter, as we shall demonstrate in the third part of this volume.

PROBLEMS

15.1 Prove: $I(t) - I(0) = -\int F(t')f(t')dt'$ (i.e. the work done by fluctuating forces).

16. The Increase of Entropy and Irreversibility

The increase of entropy in time is a very basic property of nonequilibrium processes. We discuss it on three different levels of generality.

16.1 General

The basic extremum property of the information entropy [Ref. 16.1, Sect. 10] leads to a corresponding extremum property of the time dependence of the thermodynamic entropy. Consider a system that at time $t = 0$ is in a quasi-static equilibrium state (2.21). Then, if $\varrho(t)$ is the corresponding quasi-static equilibrium operator (15.4) at time t, one has the general inequality

$$\text{Tr}\{\varrho(0)[\ln \varrho(t) - \ln \varrho(0)]\} \leq 0 \quad . \tag{16.1}$$

Taking from equilibrium statistics the relation $K = E - TS - \sum f_k Q_k$ and putting $\Delta H = H - E$ one can write (16.1) as

$$-k \ln \varrho(t) = S(t) + \left[\Delta H - \sum f_k(t) \Delta q_k(t)\right]/T \quad . \tag{16.2}$$

Inserting this into (16.1) one obtains the inequality

$$S(t) \geq S(0) \quad . \tag{16.3}$$

If one is interested in more details than in this rough criterion one can use the general kinetic equation (15.5) to obtain an equation for the change of entropy in time. One uses the fact that $S(t)$ depends on t only via the $f_k(t)$ or the $Q_k(t)$. Choosing the Q_k as the independent variables one has for closed systems with $dE/dt = 0$

$$\frac{dS}{dt} = \sum \frac{\partial S}{\partial Q_k} \frac{dQ_k}{dt} = -\beta \sum f_k \dot{Q}_k \quad . \tag{16.4}$$

Now if dQ_k/dt in this equation is expressed in terms of the kinetic equation (15.5) one finds

$$S(t) - S(0) = \frac{i}{T} \sum \int_0^{t+0} f_k(t') M_{kl}(t', t'') f_l(t'') dt' dt'' \geq 0 \quad . \tag{16.5}$$

If one considers on the r.h.s. of this equation the singular contribution $M^{(0)}$ of the memory function, see (15.6), i.e.

$$M_{kl}^{(0)}(t, t') = \frac{1}{\hbar} \langle [q_k(t), q_l(t)] \rangle \delta(t - t') \quad , \tag{16.6}$$

one sees that, because of its antisymmetry in k and l, it does not contribute to the entropy change. It may therefore be called the reversible part of the kinetic coefficients. Let us now consider the specializations and simplifications which occur for either small-amplitude or low-frequency processes.

16.2 Linear Response

For small-amplitude (linear-response) processes, (16.5) can be simplified taking the translational invariance of the memory function into account. Introducing the Fourier transforms

$$f_k(\omega) = \int_0^\infty f_k(t)e^{i\omega t}dt \quad , \tag{16.7}$$

one finds for the total entropy change

$$S(\infty) - S(0) = \frac{1}{\pi}\sum \int_{-\infty}^{+\infty} f_k(-\omega)M_{kl}(\omega + i0)f_l(\omega)d\omega \geq 0 \quad . \tag{16.8}$$

Interestingly enough this inequality can be derived independently from (16.3) by linking it with the inequality (10.17,18,19) expressing the "passivity" of open systems. To do this, one first of all notices that because of (12.1) the symmetries of M are essentially the same as those of Φ. In particular, time translation invariance (Table 7.1) implies $M_{kl}(z) = -M_{lk}(-z)$. If this is used in (16.8), the integration variable ω is changed into $-\omega$ and the two summation indices k and l are interchanged, one sees that $M_{kl}(\omega + i0)$ in (16.8) can be replaced by $-M_{kl}(\omega - i0)$, or, more symmetrically,

$$S(\infty) - S(0) = \frac{1}{\pi}\sum \int_{-\infty}^{+\infty} f_k(-\omega)M''_{kl}(\omega + i0)f_l(\omega)d\omega \geq 0 \quad , \tag{16.9}$$

where

$$\begin{aligned} 2M''_{kl}(\omega) &= M_{kl}(\omega + i0) - M_{kl}(\omega - i0) \\ &= \left\{\chi^T\left[\Phi^{-1}(\omega + i0) - \Phi^{-1}(\omega - i0)\right]\chi^T\right\}_{kl} \\ &= \left[\chi^T\Phi^{-1}(\omega + i0)\Phi''(\omega)\Phi^{-1}(\omega - i0)\chi^T\right]_{kl} \quad . \end{aligned} \tag{16.10}$$

Now, in order to make contact with the inequality (10.19), we define the complex numbers a_k introduced in the context of that inequality by

$$a_k = \sum_{l,m}\Phi^{-1}_{kl}(\omega - i0)\chi^T_{lm}f_m(\omega) \quad . \tag{16.11}$$

Then, choosing from the symmetries of Φ (Table 7.1) a combination of complex conjugation and time translation invariance, one has first of all

$$\Phi_{kl}(\omega + i0) = \Phi_{\bar{l}\bar{k}}(\omega - i0)^* \qquad (16.12)$$

and secondly the integrand of (16.9) can be written as

$$\sum fk(-\omega)M''_{kl}(\omega)f_l(\omega) = \sum a_k^* \Phi''_{\bar{k}l}(\omega)a_l \geq 0 \ . \qquad (16.13)$$

So, in this sense the detailed balance relation (10.16) can be considered as the common root of both passivity and irreversibility in linear response. Note, however, that irreversibility is more general than detailed balance.

Let us now dwell a little more on the analytical properties of the memory function $M_{kl}(z)$. We take as a starting point that the function

$$\Phi(z) = \sum a_k^* \Phi_{\bar{k}l}(z)a_l \qquad (16.14)$$

(the a_k now being arbitrary complex numbers) has the spectral representation

$$\Phi(z) = \frac{1}{\pi} \int \frac{\Phi''(\omega)[(\omega - x) - iy]}{(\omega - x)^2 + y^2} d\omega \qquad (16.15)$$

with a spectral function $\Phi''(\omega)$ which, because of (10.19), can never become negative. Equation (16.15) now says that Im $\{\Phi(z)\} \neq 0$ for $y \neq 0$ and thus also $\Phi(z) \neq 0$ for all z off the real axis. As a consequence, the inverse matrix of $\Phi_{kl}(z)$ has no singularities off the real axis. Thus finally

$$M_{kl}(z) = z\chi_{kl}^T + [\chi^T \Phi^{-1}(z)\chi^T]_{kl} \qquad (16.16)$$

is an analytic function in the complex z plane except at the real axis. The discontinuity of $M(z)$ across the real axis has already been determined in (16.10). The asymptotic behaviour of M for large z follows from (11.18,19), namely

$$M_{kl}(z) = M_{kl}^{(0)} + M_{kl}^{(1)}/z + O(1/z^2) \qquad (16.17)$$

where $M^{(0)}$ as discussed in the context of (13.14,15,16) is given by

$$M_{kl}^{(0)} = i\beta\langle \dot{q}_k; q_l \rangle = \langle [q_k, q_l] \rangle / \hbar \ . \qquad (16.18)$$

Collecting all this together, one concludes a spectral representation for the memory function of the form

$$M_{kl}(z) = M_{kl}^{(0)} + \frac{1}{\pi} \int \frac{M''_{kl}(\omega)}{z - \omega} d\omega \qquad (16.19)$$

with a non-negative spectral function $M''_{kl}(\omega)$.

16.3 Low-Frequency Response

For approximately quasi-static processes (16.3) is valid for arbitrary successive times

$$S(t + \delta t) \geq S(t) \quad , \quad \delta t \geq 0 \quad . \tag{16.20}$$

Furthermore, memory effects in the kinetic equation (15.5) can be neglected. This leads to

$$\dot{S}(t) = \sum f_k(t) \mu_{kl} f_l(t) / T \geq 0 \quad . \tag{16.21}$$

If one chooses Onsager's notation, one has for a system with $dE/dt = 0$

$$\dot{S}(t) = \sum \frac{\partial S}{\partial Q_k} \dot{Q}_k = \sum \frac{\partial S}{\partial Q_k} L_{kl} \frac{\partial S}{\partial Q_l} \geq 0 \quad . \tag{16.22}$$

For the sake of completeness, it may be pointed out that the condition of constant energy, which is of course always fulfilled for closed systems, can also hold for open systems in steady state "flow equilibria". A typical example is steady state heat conduction.

Finally, we mention that, for *symmetric* matrices L_{kl} (for instance if all variables q_k have the same time reversal parity), a general statement can be made concerning the *second* derivative of the entropy. Let us for conciseness denote partial derivatives of S with respect to Q_k by a subscript k, than one finds by differentiation of (16.22)

$$\begin{aligned}\ddot{S}(t) &= \sum (\dot{Q}_k \dot{S}_k + S_k \ddot{Q}_k) \\ &= \sum \left(\dot{Q}_k S_{kl} \dot{Q}_l + S_k \sum L_{kl} \dot{S}_l \right) = 2 \sum \dot{Q}_k S_{kl} \dot{Q}_l \quad .\end{aligned} \tag{16.23}$$

The matrix S_{kl} of second derivatives of the entropy now is the negative static susceptibility divided by T, and as such a negative (semidefinite) matrix. Thus the r.h.s. of (16.23) is negative semidefinite.

In the general case this statement is only correct for the first term on the r.h.s. of the first equation in (16.23). One then has the weaker but more general statement

$$[\ddot{S}(t)]_f = \sum \dot{Q}_k \dot{S}_k = \sum \dot{Q}_k S_{kl} \dot{Q}_l \leq 0 \quad , \tag{16.24}$$

denoted as the "general evolution criterion" by Glansdorf and *Prigogine* [16.2].

PROBLEMS

16.1 Using (5.17,18) and (16.21) determine $dS(t)/dt$ for the Brownian oscillator. Discuss $d^2 S(t)/dt^2$.

16.2 Using (16.16) determine the symmetry properties of $M_{kl}(z)$ and $M_{kl}(t)$.

17. The Increase of Entropy: A Critical Discussion**

The irreversibility of processes, expressed mathematically by the positivity of kinetic coefficients and the corresponding increase of entropy in time, is a fundamental fact of our world. Life as we know it, starting with birth and ending with death, would be unthinkable without irreversibility. Measurements, the collection of data, "common experience" generally speaking would be impossible without it. The asymmetry of time is so obvious to us that it is hard to consider it in a detached way.

A physicist, nevertheless, should be surprised by it. The basic laws of mechanics, quantum mechanics, electrodynamics and elementary particle physics (with the exception of K^0 decay into a pion, electron and neutrino, or muon and neutrino) are all invariant under time reversal: a movie of a process obeying these laws run backwards is again a movie of a process obeying these laws. On the other hand, a movie of everyday life run backwards is not a movie of everyday life.

So, one wonders how it is possible to account for the difference between past and future, if the basic laws of physics are symmetric in time; how it is possible to derive irreversible thermodynamics from reversible mechanics. One wonders how and where the asymmetry is introduced in such a derivation. It may be mentioned first of all, in a negative sense, that it has not been introduced by the use of statistical considerations alone: von Neumann's equation of statistical mechanics is just as much time reversal invariant as Schrödinger's equation of quantum mechanics. The concepts of neither order nor disorder have ingredients that distinguish past and future.

Ever since the early days of statistical physics, therefore, doubts have been raised as to the validity of the derivation of irreversible kinetic equations from reversible basic laws. In the following sections we are going to discuss several objections raised against the law of increase of entropy in order to learn about its foundations.

17.1 Maxwell's Demon

Maxwell proposed an arrangement aimed at decreasing the entropy in a closed system. A vessel filled with gas is divided into two by a wall with a small hole in it. The hole can be opened and closed by a shutter, attended by "Maxwell's demon". His task is to allow only fast particles from one side and slow particles from the other side through the hole. Obviously this would lead to a buildup of a temperature difference between the two parts of the vessel, going against the laws of thermodynamics. Now why would the demon be unable to fulfill his task? Well, one has to apply the laws of thermodynamics to the demon: either he is already in equilibrium with the gas, then, because of thermal fluctuations his "fingers will tremble" too much to do his job, or else he is not (yet) in equilibrium, then he might be able to increase the entropy of the gas for a while. During this time his entropy, however, will increase in such a way that the entropy of the closed system gas + demon also increases.

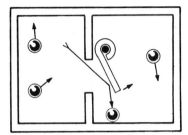

Fig. 17.1. A model of the Maxwellian demon

A simple example is shown in Fig. 17.1. A hole is covered by a lid suspended at one side. One expects, at first sight, that sufficiently fast particles from the left will push the lid open, whereas particles from the right will close it. However, in order to respond to collisions with the gas particles, its mass has to be comparable to their mass. Then in thermal equilibrium it will undergo random motions, opening and closing the hole in a random manner, in contrast to its task.

17.2 Gibbs' Ink Parable

As a consequence of von Neumann's equation and the cyclic invariance of the trace, one has

$$\frac{d\langle \ln \varrho \rangle}{dt} = \frac{-i\,\text{Tr}\{[H, \varrho \ln \varrho]\}}{\hbar} = 0 \quad . \tag{17.1}$$

Thus the so-called information entropy $S = -k\langle \ln \varrho \rangle$ is constant in time. This was mentioned in [Ref. 17.1, Sect. 10], when the thermodynamical entropy was introduced in addition to the information entropy. The thermodynamic entropy is defined as the maximum of the information entropy under the condition that the expectation values of certain observables have certain given values. The corresponding statistical operator is then said to describe quasi-static equilibrium. In

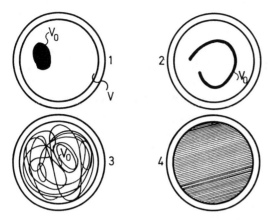

Fig. 17.2. Gibbs' illustration of the increase of entropy by "coarse graining". N ink "molecules" with the initial volume V_0 are distributed in water by stirring. N remains constant in time, and for incompressible flow the total volume occupied by the ink particles does too. Consequently also the entropy "S" $= -N\ln(N/V_0)$ remains unchanged. After stirring for a sufficiently long time, and viewed with the naked eye, the ink has "dissolved" in the water and seems to have spread out over the whole volume V occupied originally by the total system ink + water. The "coarse grained" entropy corresponding to an average density N/V of the ink, i.e. "S" $= -N\ln(N/V)$ is larger than the initial entropy

the previous chapter we demonstrated that the thermodynamic entropy increases when time moves away from the initial time. The increase of thermodynamic entropy corresponds to a continuous loss of information as compared to the initial information.

The evolution of the thermodynamic entropy can be calculated from the evolution of a limited number of observables. But the knowledge of the values of these observables is not sufficient to reconstruct the initial conditions completely.

The loss of information in going from the exact statistical operator to the coarse-grained operator of quasi-static equilibrium is illustrated by the "ink parable" of Gibbs, see Fig. 17.2. In accordance with this illustration one calls the density operator $\varrho(t)$ of quasi-static equilibrium "coarse grained" in contrast to the exact density operator ϱ.

For the sake of clarity, it should be mentioned that the increase of entropy introduced by coarse graining does not lead to a distinction between past and future. The thermodynamic entropy increases starting from any improbable state in both directions of time, past as well as future.

17.3 Zermélo's Recurrence Paradox

Poincaré demonstrated that finite mechanical systems behave quasi-periodically in time: almost any initial state is reached again (for increasing time with increasing accuracy). This recurrence theorem of Poincaré, when applied to statistical mechanics, leads to the so-called recurrence paradox. Thermodynamic systems are finite mechanical systems and should obey Poincaré's recurrence theorem. Therefore, quantities such as the entropy, even if they first increase in time (starting from an improbable initial situation) should behave quasi-periodically and thus decrease again later on. This is Zermélo's objection to general proofs of irreversibility.

Let us consider a simple plausibility argument for Poincaré's recurrence theorem in quantum mechanics. Start from an expression for the statistical operator in terms of the eigenstates $|\mu\rangle$ of H

$$\varrho(t) = \sum \varrho_{\mu\nu}(0)\exp[-i(E_\mu - E_\nu)t/\hbar]|\mu\rangle\langle\nu| \qquad (17.2)$$

and represent the energy difference $E_\mu - E_\nu$ as a decimal number with an accuracy of n decimal places in units of an average difference $1/\Omega(E)$

$$E_\mu - E_\nu = \frac{1}{\Omega(E)}\left(N_{\mu\nu} + \frac{n_{\mu\nu}}{10^n}\right) \qquad (17.3)$$

with integers $N_{\mu\nu}, n_{\mu\nu}$ and n obeying

$$0 \leq n_{\mu\nu} < 10^n \quad . \qquad (17.4)$$

Let us not discuss the (nontrivial) problem of the neglect of further decimal places. Then, for a time

$$\Delta t = 2\pi\hbar 10^n \Omega(E) \qquad (17.5)$$

one has

$$\varrho(t + \Delta t) = \varrho(t) \quad . \qquad (17.6)$$

This is a sloppy "proof" of a quantum version of Poincaré's recurrence theorem.

Introducing an energy interval ΔE of the order of the energy fluctuation of a canonical ensemble for the system, one can write, taking into account $S = k \ln[\Omega(E)\Delta E]$,

$$\Delta t = \frac{2\pi\hbar}{\Delta E} 10^n e^{S/k} \quad . \qquad (17.7)$$

This result demonstrates that the recurrence objection, though correct in principle, is not relevant from a practical point of view: since S/k is of the order of $N = 10^{23}$ the actual recurrence times are many, many ages of the universe long.

One may argue against this that a recurrence of the full $\varrho(t)$ in general may not be required, since one is only interested in the observation of a small set of physically relevant quantities q_k. Now the recurrence times of such quantities are, indeed, smaller than for the full $\varrho(t)$, but as soon as one considers values of these quantities sufficiently far away from average thermal deviations from their equilibrium values the recurrence times are still large compared to the age of the universe. Table 17.1 contains estimates of recurrence times for density

Table 17.1. Waiting times for the occurrence of density fluctuations in a subvolume of 1cm of a dilute gas; see Problem 17.1

Relative deviation $\Delta n/\sqrt{\langle n \rangle}$	Average waiting time
2×10^{-10}	4×10^{-3} s
3×10^{-10}	1 s
4×10^{-10}	1.3×10^3 s = 21 min
5×10^{-10}	1.3×10^3 s = 5 months
6×10^{-10}	1×10^{12} s = 3×10^4 years
7×10^{-10}	5×10^{17} s = 2×10^{10} years

fluctuations in a subvolume of a dilute gas. As one can see, the recurrence times vary from a fraction of a second at twice the average thermal deviation from equilibrium to about the age of the universe at seven times this deviation. Note that for the first time in our discussion a new cosmological time scale occurs: the age of the universe.

Large deviations of a quantity from its equilibrium value can, in principle, always occur in two ways: either by spontaneous fluctuation, or by external perturbation. If, however, the recurrence time of a spontaneous fluctuation is very large compared to the age of the universe, it is unlikely to occur (in our present universe). Thus one can conclude: if a large deviation from thermal equilibrium is observed, one can be pretty sure that it was not produced by spontaneous fluctuation, but by an external perturbation.

17.4 Loschmidt's Reversibility Objection

All considerations so far do not distinguish past and future. Indeed, for a closed system there is no such distinction. This fact is described precisely by the so-called reversibility objection: for every process there is a corresponding one obtained by the operation of time reversal. If the entropy increases for a certain process, it obviously decreases for the time reversed process. From purely statistical considerations there is no reason why the time reversed states should be less probable than the original ones.

On the other hand, if one goes back to Chap. 7 one sees that time reversal invariance led to symmetry properties of the correlation functions, but not to a decrease of entropy with increasing time. The reason is that time reversal was combined with time translation invariance, in order to stay within the framework of retarded correlation functions. So why does one always want to consider retarded instead of advanced functions and initial value problems instead of of final value problems?

To answer this question let us consider a certain closed system with a large deviation of certain variables from their equilibrium values. According to the discussion of Sect. 17.3 one can be pretty sure that at some time there was an external perturbation, which produced this deviation. Obviously, since the system can be considered closed only after this time, one is led to the solution of an inital value problem. Before this time the system cannot be considered closed, but, one can enlarge the original system by including those devices (and persons if necessary) such that the enlarged system can be considered closed for even earlier times. Let us continue this procedure to larger and larger systems, going back further and further in time. If one starts from originally closed systems on earth one will notice that, pretty soon, one reaches systems of a size of the order of our solar system, which may be considered approximately closed for a time of the order of its age, which may be only a little smaller than the age of the universe. The next stage then may be our galactic system and finally the whole universe. Many other, originally different smaller systems end up the same way and may be considered as different "branches" of the same larger system. So finally our question has been reduced to the problem of the evolution in time of our universe.

Although many problems in this context are unsolved as yet, one can be pretty sure that the universe is finite, started with the "big bang" and is presently expanding. The answer to the question, why the entropy of closed smaller subsystems increases with increasing time thus is: Because they all are different branches of the same expanding universe, which imprints the arrow of time on them. This simple answer may have to be modified by the influence of *gravitational* forces, which become more and more important with increasing size of the systems. The potential energy of these forces is negative and has no lower limit. The statistical mechanics of gravitating systems, therefore, is quite different from the one of nongravitating systems. The formation of stars, star systems, galaxies and black holes demonstrates this quite clearly. Gravitational forces somehow have a tendency to produce "improbable" situations, and thus act similarly to external perturbations.

Although at present there are still many unsolved problems in describing and understanding the evolution of our universe, and in estimating the entropy balance of the world, one can be pretty sure that the irreversibility of terrestrial processes is ultimately of cosmological origin.

PROBLEMS

17.1 Expanding the entropy S in $\Delta t = t_0 \exp(S/k)$ near a maximum Q_0 as $S = S(Q_0) - (\Delta q)^2/(2\langle \Delta q^2 \rangle) + \cdots$, estimate waiting times as occurring in Table 17.1. See [17.2].

Part II

Irreversible Thermodynamics

Irreversible thermodynamics deals with the solution of irreversible equations (relaxation equations, kinetic equations, transport equations, etc.) as well as the determination of the thermal and quantum mechanical fluctuations of certain specified variables. Sometimes these variables are "macroscopic" and/or "slow" but sometimes one is also interested in "microscopic" parameters (such as the position and momentum of a Brownian particle), and/or the high-frequency behaviour of quantities (such as electronic charges and currents).

The kinetic coefficients in the kinetic equations are considered as adjustable phenomenological parameters – in contrast to the third part of this volume, where they are explicitly calculated from microscopic theory.

The chapters of this part are ordered more or less according to the number of variables occurring in the various problems. The first chapter deals with just a single variable: the charge $Q = CV$ or the current $J = V/R$ in an RC circuit. We then proceed to two or more discrete variables, to continuous variables in coordinate space (treating diffusion, hydrodynamics, electrodynamics), continuous variables in momentum space, and finally in phase space.

18. The Nyquist Formula

In this chapter we derive an expression for the spectrum of thermal agitations of electrons in an electric RC circuit [18.1] (the spectral function of the current–current correlation function) as a measure of the "noise" in the system. These results were first derived, including quantum effects, by *Nyquist* [18.2]. The system is another version of the Brownian relaxator considered earlier Chap. 5. Let us use linear response theory and consider the RC circuit under the influence of an external voltage V^e (Fig. 18.1). The total current j in the circuit then consists of the average Ohmic current $J = V/R$ and the noise component Δj, where V is the voltage drop at the resistor with resistance R, which is in thermal contact with a heat bath of temperature T. For the average charge Q on the capacitor of capacitance C one has

$$Q = C(V + V^e) \quad , \tag{18.1}$$

and thus, taking into account the conservation of charge $dQ/dt = -J$,

$$\dot{Q}(t) + Q(t)/RC = V^e(t)/R \quad . \tag{18.2}$$

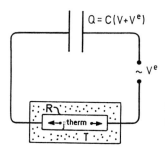

Fig.18.1. Schematic of an RC circuit with an external voltage and a heat bath

After Fourier transformation

$$[-i\omega + 1/(RC)]Q(\omega) = V^e(\omega)/R \quad , \tag{18.3}$$

the solution of this equation can be written as

$$Q(\omega) = \chi(\omega)V^e(\omega) \tag{18.4}$$

with the dynamic susceptibility

$$\boxed{\chi(\omega) = \frac{i/R}{\omega + i/RC}} \qquad (18.5)$$

The static susceptibility in our system is equal to the isothermal one (since the resistor is always in thermal contact with the bath) and given by

$$\chi(0) = \chi^T = \langle q; q \rangle / kT = C \qquad (18.6)$$

in agreement with $Q(\omega = 0) = CV^e(\omega = 0)$.

For the relaxation function $\Phi(\omega) = [C - \chi(\omega)]/\omega$ one finds after continuation into the complex plane

$$\boxed{\Phi_q(z) = \frac{C}{z \pm i/RC}} \qquad (18.7)$$

The dissipative part of Φ thus is given by

$$\Phi_q'' = \frac{1/R}{\omega^2 + 1/(RC)^2} \qquad (18.8)$$

and from this, according to the fluctuation–dissipation theorem (Chap. 8), one finds the spectrum of the charge fluctuations $q(\omega)$ by using (8.5)

$$[s_q(\omega) + s_q(-\omega)]/2 = \varepsilon(\omega, T) \frac{1/R}{\omega^2 + 1/(RC)^2} \qquad (18.9)$$

Remember that (Problem 8.2.), if $q(\omega)$ is the Fourier transform of $q(t)$ in a large but finite time interval, then the l.h.s. of (18.9) is identical to $\langle |q(\omega)|^2 \rangle$.

Instead of (18.2) one may also consider the corresponding Langevin equation

$$\dot{q}(t) + q(t)/RC = v^{\text{therm}}/R = -j^{\text{therm}} \qquad (18.10)$$

The two r.h.s. are the fluctuating "forces" in this case and may be interpreted in the absence of external voltages as the noise voltage or current, respectively. The minus sign in front of the thermal current j^{therm} has been introduced because of $dq/dt = -j$. The second term on the l.h.s. subtracts the "systematic part" from the thermally induced current Δj.

From (18.10) one obtains

$$q(\omega) = \frac{j^{\text{therm}}}{\omega + i/RC} \qquad (18.11)$$

and after averaging the absolute square of both sides

$$\boxed{\langle |q(\omega)|^2\rangle = \frac{\langle |j^{\text{therm}}(\omega)|^2\rangle}{\omega^2 + 1/(RC)^2}} \quad . \tag{18.12}$$

Comparison with (18.9) leads to

$$\boxed{\langle |j^{\text{therm}}(\omega)|^2\rangle = \varepsilon(\omega, T)/R} \quad , \tag{18.13}$$

which is, of course, nothing but Langevin's equation for a relaxator, generalized to the quantum mechanical case. In the classical limit the Planck function approaches kT.

Another, more formal, way to express the foregoing results is to remember the connections between fluctuating forces and memory functions, see Chaps. 12 and 13. We can identify the memory function $N(z)$ by comparing (18.8) with the general definition of $N(z)$ from (12.1) as

$$N_q(z) = \mp i/R \quad \text{for Im}\{z\} \gtrless 0 \quad . \tag{18.14}$$

Then (18.13) takes the simple form

$$N_q''(\omega) = \frac{1}{R} \quad . \tag{18.15}$$

Equation (18.14) can be taken as a starting point for a simple demonstration of what we said about separation of time scales and also about continued fraction expressions for correlation functions. Step 1 in the hierarchy of time scales (or continued fractions) is (18.7). The charge is expressed in terms of the conductance $1/R$, step 2: R is frequency dependent. The simplest frequency dependence would be a Drude behaviour $R(z) = (1 \mp iz\tau)R$. Step 3: there are cases (for instance, if electron localization in strong random potentials occurs, see Chap. 43), where the collision time τ is also frequency dependent. In this case one writes $\mp i\tau = 1/M(z)$ with another memory kernel M. Combing all three steps one has a representation

$$\Phi_q(z) = \frac{C}{z - \{iM(z)/RC[z - M(z)]\}} \quad . \tag{18.16}$$

The three time, or frequency, scales involved in the first three steps of a continued fraction are: the relaxation frequency of the RC circuit $\gamma = 1/(RC)$, the electron collision frequency $1/\tau$ and the characteristic frequency in the memory kernel $M(z)$.

PROBLEMS

18.1 Integrate (18.12), using (18.13), in the limit $\hbar/RC \ll kT$ to determine $\langle q; q\rangle$, and check the static sum rule (11.12) and (18.6).

19. Thermomechanical Effects

Irreversible equations with more than one variable contain off-diagonal terms describing couplings between diagonal effects (cross effects). In this chapter we are going to consider a system with two variables: the particle number and energy of a subsystem of a larger system. The diagonal processes are the flow of particles driven by a pressure difference between the subsystem and the rest, and a flow of heat driven by a temperature difference. The cross processes are obtained when in the foregoing sentence the words pressure and temperature are exchanged.

The discussion of this chapter with two discrete variables is a simplified treatment of continuous situations with the continuous variables particle density $n(\mathbf{r})$ and energy density $\varepsilon(\mathbf{r})$, the diagonal processes particle and energy diffusion, and the cross processes thermodiffusion (Soret effect) and its inverse "diffusive heating" (Dufour effect).

For more than one variable the symmetry of the kinetic coefficients, in particular the Onsager reciprocity relations [see Table 7.1 and (12.1)] following from time reversal symmetry, become important. In order to incorporate these relations it is useful to use Onsager's method of deriving kinetic equations as described in Sect. 15.2. In order to fix our attention we consider a system as depicted in Fig. 19.1.

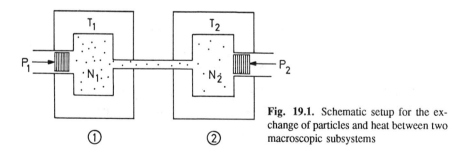

Fig. 19.1. Schematic setup for the exchange of particles and heat between two macroscopic subsystems

Two systems 1, 2 are connected by a capillary, porous wall, or permeable membrane. They have pressures $P_{1,2}$ and temperatures $T_{1,2}$. Processes can be induced by varying these parameters. In order to have only *two* independent variables, we keep the volumes of the subsystems fixed.

The variables which we choose first will be the two particle numbers $N_{1,2}$ and energies $E_{1,2}$ of the subsystems. They are not independent, because we consider the total system as closed. Hence $N_1 + N_2 = N_{\text{tot}} = \text{const}$, and similarly for E, leading to

$$\dot{N}_1 = -\dot{N}_2 = \dot{N} \tag{19.1}$$

and

$$\dot{E}_1 = -\dot{E}_2 = \dot{E} \ . \tag{19.2}$$

So, the only two independent variables in our choice are $N_1 = N$ and $E_1 = E$. Onsager, in his version of the kinetic equations, uses instead of the forces f_k directly the entropy derivatives for the closed system, see (15.19). In our case this means

$$\begin{aligned}\dot{N} &= L_{11}(\partial S/\partial N) + L_{12}(\partial S/\partial E) \ , \\ \dot{E} &= L_{21}(\partial S/\partial N) + L_{22}(\partial S/\partial E) \ .\end{aligned} \tag{19.3}$$

Now the total entropy of the closed system can be written as

$$S(E, N) = S_1(E, N) + S_2(E_{\text{tot}} - E, N_{\text{tot}} - N) \tag{19.4}$$

and a differential change of S takes the form

$$dS = -\Delta(\mu/T)dN + \Delta(1/T)dE \tag{19.5}$$

where ΔA stands for $A_1 - A_2$.

Differentiating μ/T one finds, using (remember $s = S/N, v = V/N$, etc.)

$$\Delta\mu = -s\Delta T + v\Delta P \quad \text{(to first order, assuming } \Delta \text{ to be small)} \tag{19.6}$$

and ($i = e + Pv$, the enthalpy per particle)

$$\mu = e + Pv - Ts, \quad \text{and} \quad dE = \delta Q + i dN = \delta Q + (e + Pv)dN, \tag{19.7}$$

$$\boxed{T\, dS = -(\Delta T/T)\delta Q - \Delta P(v\, dN)} \ . \tag{19.8}$$

This corresponds to a new choice of independent variables (δQ and $v dN$) instead of dE and dN. The corresponding kinetic equation then takes the form

$$\begin{aligned} v\dot{N} &= -\mu_{11}\Delta P - \mu_{12}\Delta T/T \ , \\ \dot{Q} &= -\mu_{21}\Delta P - \mu_{22}\Delta T/T \ .\end{aligned} \tag{19.9}$$

Here $\dot{Q}$ is not the derivative of a function $Q(t)$ (which does not exist, as expressed in the second law of thermodynamics) but only an abbreviation for

$$\dot{Q} = \dot{E} - (e + Pv)\dot{N} \ . \tag{19.10}$$

The connection of the kinetic coefficients μ occurring in these equations with the original L's can be determined easily (Problem 19.1.) but at the moment we only need to know that the μ's, just as the L's, obey Onsager's symmetry relations

$$L_{12} = L_{21} \quad \text{and} \quad \mu_{12} = \mu_{21} \ . \tag{19.11}$$

In order to measure (and interpret) the μ_{ik} let us consider two special situations.

19.1 Diffusion and the Mechanocaloric Effect ($\Delta T = 0$)

For vanishing temperature difference one has according to (19.9) a flow (diffusion) of particles through the membrane according to

$$\boxed{v\dot{N} = -\mu_{11}\Delta P \ ,} \tag{19.12}$$

where μ_{11} is some kind of *diffusion coefficient*. At the same time one has a heat flow according to

$$\dot{Q} = -\mu_{12}\Delta P \ , \tag{19.13}$$

which because of (19.12) is proportional to $\dot{N}$

$$\boxed{\dot{Q} = (\mu_{21}/\mu_{11})v\dot{N} = q^*\dot{N} \ .} \tag{19.14}$$

The fact that a particle flow at constant temperature is always coupled with heat flow is called the mechanocaloric effect and $q^* = v(\mu_{12}/\mu_{11})$ is called the *transfer heat*.

19.2 Heat Diffusion and Thermomechanical Pressure Difference ($\dot{N} = 0$)

If the temperature difference is nonzero, one can maintain a stationary state with respect to particle flow if an appropriate pressure difference is introduced. This is called the thermomechanic effect. The relation between pressure difference and temperature difference according to (19.9) is given by

$$\Delta P = -(\mu_{12}/\mu_{11})\Delta T/T \ . \tag{19.15}$$

Here we can make use of the Onsager symmetry (19.11) and express the r.h.s. in terms of q^*

$$\boxed{\Delta P = -q^* \Delta T / vT} \quad . \tag{19.16}$$

At the same time one has a heat flow, which by (19.9, 15) is given by

$$\boxed{\dot{Q} = (\mu_{12}^2/\mu_{11} - \mu_{22})\Delta T/T} \quad . \tag{19.17}$$

The combination

$$\lambda = \mu_{22} - \mu_{12}^2/\mu_{11} \tag{19.18}$$

is a kind of "heat diffusion coefficient" or *heat transfer coefficient*.

Equation (19.16) has a seductive similarity to the Clausius-Clapeyron equation of *equilibrium* thermodynamics. It may be interesting to mention in this context that (19.16) and similar relations in other nonequilibrium processes had been "derived" before the Onsager relations by invoking the stationarity condition of the entropy valid in equilibrium thermodynamics.

The result, however, cannot be right, since the process is definitely irreversible: there is heat conduction with the corresponding entropy production. The validity of the Onsager symmetry relations is an important ingredient of the result. If they are invalid, for instance in the presence of magnetic fields, the result is no longer valid. For more details see [19.1].

In order to make use of the results (19.14,16) one needs the transport heat q^*. We consider three different examples.

a) *A large hole.* According to (19.15), q^* is the heat transferred per particle, thus, taking into account (19.10): $q^* = e^* - (e + Pv)$. Now if one additional particle is introduced into subsystem 1 it first adds its energy e to it, but it also does work $P(dV)_{1\text{particle}} = Pv$. Thus $q* = 0$, in agreement with the idea that for a macroscopic hole (i.e. a hole large compared to the mean free path) for vanishing particle flow one cannot maintain a pressure difference between the subsystems.

b) *A small hole.* If the hole is small compared to the mean free path, the situation is totally different. In this case the particle does not do work, but e^*, nevertheless, is different from e because more fast particles fly through the hole than slow ones. The (kinetic) energy transported on the average through the hole then has to be calculated from the Boltzmann distribution

$$e^* = \frac{\int\limits_{v_x \geq 0} (mv^2/2) v_x e^{-\beta mv^2/2} d^3v}{\int\limits_{v_x \geq 0} v_x e^{-\beta mv^2/2} d^3v} \quad . \tag{19.19}$$

The easiest way to do the integrals is to introduce the energies

$$E_x = mv_x^2/2 \quad \text{and} \quad E_{\parallel} = m(v_y^2 + v_z^2)/2 \tag{19.20}$$

and write (19.20) as

$$e^* = \frac{\int_0^\infty (E_x + E_{\parallel})e^{-(E_x+E_{\parallel})}d^2E}{\int_0^\infty e^{-(E_x+E_{\parallel})}d^2E} = 2kT \quad . \tag{19.21}$$

Thus, although the contribution $Pv(=kT)$ of the work does not occur for particles passing through a narrow hole, the transport energy is larger than $e = 3kT/2$ but only by $kT/2$. Thus we have

$$q^* = e^* - e - Pv = -kT/2 \tag{19.22}$$

and hence because of (19.12) a pressure difference

$$\Delta P/P = \Delta T/2T \quad . \tag{19.23}$$

c) *Fountain effect in superfluid* He. If one connects two vessels filled with superfluid He at equal temperature by a ultrathin capillary, only the superfluid component can pass through it. According to the two fluid model [19.2] of He II the superfluid component has no entropy and zero temperature. So if particles of the superfluid component are transferred into system 1 the corresponding heat bath has to deliver amount $T(dS)_{1\text{particle}} = Ts$ to the liquid in order to avoid it cooling down. Thus the transport heat is given by

$$q^* = -Ts \quad .$$

For a small temperature difference ΔT because of (19.16) one expects a corresponding pressure difference

$$\boxed{\Delta P = s\Delta T/v \quad .} \tag{19.24}$$

In this equation all quantities are independently measurable. Experimental confirmation of (19.24) thus contributes to a confirmation of the two fluid model. The existence of a fountain pressure can also nicely be demonstrated in a "circus experiment" [19.3], see also Fig. 19.2.

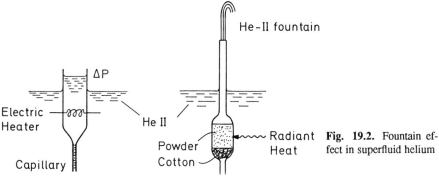

Fig. 19.2. Fountain effect in superfluid helium

PROBLEMS

19.1 Determine the relation between the L_{kl} and μ_{kl}.

19.2 Using the results of Problem 19.1, check the validity of the Onsager symmetry (19.11).

20. Diffusion and Thermodiffusion

In this chapter we discuss a spatially continuous version of the thermomechanic effects. This is somewhat unsystematic in the sense of the ordering of chapters in Part II according to the number of variables, but since our discussion is very closely related to the preceding chapter we think it is justified.

We restrict ourselves to diffusion of a single substance through an inert material. Then the variables will be only a single-particle density $n(r,t)$ and energy density $\varepsilon(r,t)$. They are functions of the continuous coordinate r and hence amount to infinitely many variables (one for each space point).

We take it for granted that the reader is familiar with the usual derivation of the diffusion law via Fick's law and the continuity equation and refrain from repeating it. Instead we start from the general Onsager form (15.19) and use all relevant symmetries of the problem. This may look artificial and complicated for such a simple problem. Nevertheless we choose this path, not only to give an example of the application of the very general Onsager formula in a simple case, but also to demonstrate that the usual derivation contains an approximation, which is mostly not mentioned explicitly, and is not generally valid.

In our case (15.19) becomes an integrodifferential equation

$$\dot{n}(r,t) = -\int L(r,r')\mu(r',t)d^3r'/T \quad . \tag{20.1}$$

So far, T is considered constant. The form of the matrix (or better integral kernel) L of kinetic coefficients is restricted by symmetries and conservation laws. Let us consider first the

a) *Homogeneity* of the medium. This means: a translation by a does not change its properties

$$L(r+a, r'+a) = L(r,r') \quad \text{for all} \quad a \tag{20.2}$$

and, in particular, for $a = -r'$

$$L(r,r') = L(r-r', 0) = L(r-r') \quad . \tag{20.3}$$

The r.h.s. of (20.1) then becomes a convolution which in Fourier space becomes a product. We therefore use the Fourier transform

$$A(k) = \int A(r)\exp(-ikr)d^3r \tag{20.4}$$

with its inverse

$$A(r) = \frac{1}{(2\pi)^3} \int A(k)\exp(ikr)d^3r \quad . \qquad (20.5)$$

Then (20.1) takes the form

$$n(k, t) = -L(k)\mu(k, t)/T \quad . \qquad (20.6)$$

Homogeneity allows no couplings between different Fourier components of the density. Next we consider the

b) *Isotropy* of the medium. Then L cannot depend on the direction of the vector k but only on its magnitude $|k| = k$ or the square of it

$$L = L(k^2) = L(k^2) \quad . \qquad (20.7)$$

Finally we consider
c) *Particle Number Conservation* which in Fourier space yields

$$\dot{N}(t) = \dot{n}(k = 0, t) = 0 \quad \text{(for all } \mu\text{)} \quad , \qquad (20.8)$$

which is only possible if $L(0) = 0$.

This ends our general discussion. The diffusion equation, in Fourier space, takes the form (20.6) where the function L depends only on k and vanishes for $k = 0$. If one assumes in addition that L can be expanded for small k in k (which may be expected for short-range forces) L has to start out as

$$L(k) = \ell k^2 + \ldots \qquad (20.9)$$

In fact this is all one can say in general. Figure 20.1 exhibits a typical shape of $L(k)$. If the wavelength $1/k$ becomes comparable to interatomic spacings a, $L(k)$ develops a "wiggle" known from phonons in solids or rotons in liquid helium. The corresponding momentum transfers k can be reached by inelastic neutron scattering.

The usual procedure for deriving a diffusion equation goes via the continuity equation

$$\dot{n}(r, t) = -\text{div} j_n(r, t) \qquad (20.10)$$

and Fick's law

$$\boxed{j_n = -nB\,\text{grad}\,\mu = D\,\text{grad}\,n} \qquad (20.11)$$

with the Einstein relation

$$\boxed{D = Bn(\partial\mu/\partial n)} \qquad (20.12)$$

and leads to

$$\dot{n}(r, t) = -\Delta D n(r, t) \qquad (20.13)$$

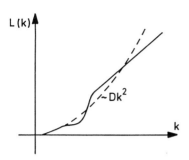

Fig. 20.1. Dependence of diffusion coefficient on wave number

or, in Fourier space,

$$\dot{n}(\boldsymbol{k},t) = -nBk^2\mu(\boldsymbol{k},t) \quad . \tag{20.14}$$

In the usual approach L is proportional to k^2 : Fick's law is not the most general linear ansatz even in a homogeneous medium, but only a long-wavelength approximation.

Let us, from now on, stay with this approximation. We want to consider the entropy production in diffusion and thermodiffusion. The idea underlying Onsager's approach is to treat the system as in quasi-static equilibrium at every instant with the variables $Q_k(t)$. In our case, with variables depending on position $\boldsymbol{r}$ in space, it means quasi-static equilibrium between volume elements that are small enough compared to macroscopic variations to be treated as infinitesimal in the formalism, but large enough compared to mean free paths that equilibrium can be achieved (often called "local equilibrium"): besides a separation of time scales one has also one of length scales (in brief, this is the background of the long-wavelength approximation).

The entropy then can be written as an integral over an entropy density $\sigma = sn$:

$$S(t) = \int \sigma(\boldsymbol{r},t)d^3r \quad . \tag{20.15}$$

For local equilibrium one can do standard thermodynamics within d^3r. Since the volume elements are fixed (as in Euler's description of hydrodynamics) there is no term like Pdv, but particle exchange exists; there is always μdn

$$T\,d\sigma = d\varepsilon - \mu\,dn \tag{20.16}$$

where d can mean ∂_x or ∂_t. As before, $\varepsilon(\boldsymbol{r},t) = en$ is the energy density.

We first disregard thermal effects and consider only diffusion. Then one obtains from (20.16) and the continuity equation (20.11)

$$\dot{\sigma} = -\mu\dot{n}/T = \text{div}\,(\mu n/T) - (\boldsymbol{j}_n\text{grad}\,\mu)/T \quad . \tag{20.17}$$

After inserting this into (20.15) the div term contributes zero for a closed system, and the rest, after using Fick's law (20.11), yields for the entropy increase the (positive) expression

$$\dot{S} = (Bn/T) \int (\operatorname{grad} \mu)^2 d^3 r \quad . \tag{20.18}$$

In the general case one has to take into account energy conservation

$$\dot{\varepsilon} = -\operatorname{div} \boldsymbol{j}_\varepsilon \quad . \tag{20.19}$$

The generalization of (20.17) then reads

$$\dot{\sigma} + \operatorname{div} \boldsymbol{j}_\sigma = -\boldsymbol{j}_n \cdot \operatorname{grad}(\mu/T) + \boldsymbol{j}_\varepsilon \cdot \operatorname{grad}(1/T) \tag{20.20}$$

with the entropy current density, see (20.16),

$$\boldsymbol{j}_\sigma = (\boldsymbol{j}_\varepsilon - \mu \boldsymbol{j}_n)/T \quad . \tag{20.21}$$

The linear relation (20.11) can now be generalized to

$$\boldsymbol{j}_k = \sum \ell_{kl} \operatorname{grad}(\partial \sigma / \partial \ell) \tag{20.22}$$

with

$$k = 1, 2 = n, \varepsilon \quad . \tag{20.23}$$

The driving forces thus are the gradients of $-\mu/T$ and $1/T$.

If one differentiates (μ/T) and goes through the steps in analogy to Chap. 19 again, one can write the result in the form

$$\begin{aligned} \boldsymbol{u} &= -\mu_{11} \operatorname{grad} P - \mu_{12}(\operatorname{grad} T)/T \quad , \\ \boldsymbol{j}_q &= -\mu_{21} \operatorname{grad} P - \mu_{22}(\operatorname{grad} T)/T \quad , \end{aligned} \tag{20.24}$$

where

$$\boldsymbol{u} = v \boldsymbol{j}_n \quad , \tag{20.25}$$

because $vn = 1$, is nothing but the drift velocity of the diffusion and

$$\boldsymbol{j}_q = T(\boldsymbol{j}_\sigma - s \boldsymbol{j}_n) = \boldsymbol{j}_\varepsilon - (e + Pv) \boldsymbol{j}_n \tag{20.26}$$

[compare the analogous equation (19.10)] is the heat current density.

If (20.24) is inserted into (20.15) it is easy to see that the positivity of the entropy increase is equivalent to the positivity of the matrix μ_{kl} (Problem 20.4.). Various processes can be considered in more or less complete analogy to this chapter. We leave this excercise to the reader.

An important cross process is *thermodiffusion*: a temperature gradient sets up a diffusion current. The mass dependence of μ_{12} is the physical basis of the application of thermodiffusion in technical and industrial isotope separation.

PROBLEMS

20.1 What would the homogeneity discussion look like in momentum space if (20.1) had been written from the beginning in momentum space?

20.2 Determine $L(k)$ for a discrete version $\partial n_m/\partial t = -\sum \mu_{mn} n_n$ of (20.1), where n and m label points on a cubic lattice.

20.3 Einstein relation: evaluate $n(\partial \mu/\partial n)$ for the ideal Boltzmann gas at T and for the ideal Fermi gas at $T = 0$.

20.4 Write the entropy production (20.15) in terms of the ℓ_{kl}, see the discussion below (20.26), and investigate its positivity.

21. Thermoelectric Effects

Thermoelectric effects are closely related to thermomechanic ones, hence it seems reasonable to treat both of them more or less together. If a thermocouple has different temperatures T at its ends, see Fig. 21.1, it develops (for zero current) a thermovoltage between its ends. This is known as the Seebeck effect (T.J. Seebeck, 1821)

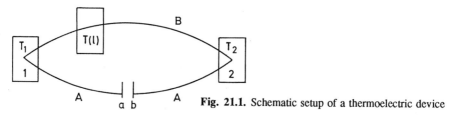

Fig. 21.1. Schematic setup of a thermoelectric device

If, on the other hand, the current J is nonzero but the temperature difference T is zero heat is produced at the contact points of the two metals, which in contrast to the usual Joule heat is not proportional to J^2 but to $\pm J$. This is called the Peltier effect (J.C.A. Peltier, 1834). The analogy of these effects with the thermomechanic effects treated in the two previous chapters is obvious. If one replaces the variables $\Delta P, v$ from the foregoing chapters by $\Delta \Phi$, e_0 (e_0 being the electric charge per particle), then practically all results from there may be carried over to here. Let us introduce in addition the electric charge $Q_e = N e_0$ then the analogue of the formulae (19.13,14) for the mechano-caloric effect can now be written

$$\dot{Q} = \Pi \dot{Q}_e = \Pi J \quad . \tag{21.1}$$

It describes the proportionality beween Peltier heat $\dot{Q}$ and current J. The Peltier coefficient Π corresponds in the mechanocaloric effect to q^*/v. In analogy to (19.16) for the thermomechanical pressure difference one now finds

$$\Delta \Phi = \Pi \Delta T / T \tag{21.2}$$

for the thermopower in the Seebeck effect.

The simplified treatment does not allow one to treat the Thomson heat [21.3], for which a continuous description along the wires of the thermocouple is needed. A suitable choice of variables in this case is to work with energy derivatives instead of entropy derivatives, start out from

$$d\varepsilon = T d\sigma + \mu dn \tag{21.3}$$

and use correspondingly

$$\begin{aligned} \boldsymbol{j}_n &= -\mu_{11} \operatorname{grad}\mu - \mu_{12} \operatorname{grad}T \ , \\ \boldsymbol{j}_\sigma &= -\mu_{21} \operatorname{grad}\mu - \mu_{22} \operatorname{grad}T \ . \end{aligned} \tag{21.4}$$

[This may also be achieved by starting out from entropy derivatives as in the foregoing section; write $\operatorname{grad}(\mu/T) = (\operatorname{grad}\mu)/T - \mu(\operatorname{grad}T)/T^2$. We leave it to the reader to find the relations between the old and new kinetic coefficients and check the symmetry relations.]

As in Chap. 19 we now consider two special cases:

a) $\operatorname{grad}T = 0$: The first line of (21.4) describes again diffusion or, since we are dealing with charged particles, conduction. The chemical potential (for charged particles often called the electrochemical potential) contains one part μ_0 which depends on density n (and on $\boldsymbol{r}$ only via n), and one depending on the electrostatic potential $e_0\Phi$. Consequently the current is the sum of a concentration-driven part and a field-driven part. The proportionality constant in both cases is the electrical conductivity $e_0^2\mu_{11}$

$$\boldsymbol{j}_{\mathrm{el}} = -\sigma_{\mathrm{el}} \operatorname{grad}(\mu_0/e_0 + \Phi) \tag{21.5}$$

(remember $E = -\operatorname{grad}\Phi$).

In analogy to (19.13,14), connected with the electric current there is an entropy current

$$\boldsymbol{j}_\sigma = s^*\boldsymbol{j}_n \quad \text{with} \quad s* = \frac{\mu_{21}}{\mu_{11}} = \frac{\mu_{12}}{\mu_{11}} \ ; \tag{21.6}$$

s^* may now be called transport entropy.

b) $\operatorname{grad}T \neq 0$: In the general case one can eliminate $\operatorname{grad}\mu$ from the second equation by inserting its value from the first one in (21.4). If one uses the notation introduced in a) and furthermore the heat conductivity

$$\lambda = T(\mu_{22} - \mu_{12}^2/\mu_{11}) \ , \tag{21.7}$$

the two equations (21.4) take the form (remember $\boldsymbol{j}_\sigma = \boldsymbol{j}_\varepsilon - \mu\boldsymbol{j}_n$)

$$\begin{aligned} \boldsymbol{j}_{\mathrm{el}} &= -\sigma_{\mathrm{el}}[\operatorname{grad}(\mu/e_0) + s^*\operatorname{grad}(T/e_0)] \ , \\ \boldsymbol{j}_q &= [\boldsymbol{j}_\varepsilon - (\mu + s^*T)\boldsymbol{j}_{\mathrm{el}}/e_0)] = -\lambda \operatorname{grad}T \ . \end{aligned} \tag{21.8}$$

The first one is a generalized Ohm's law, containing a current, driven by a temperature gradient; the second one is the law of heat conduction.

Our aim is to calculate the heat produced along the wire. We restrict ourselves to the stationary case $\text{div} \boldsymbol{j}_\varepsilon = \text{div} \boldsymbol{j}_{el} = 0$. In order to do so we integrate the divergence of the heat current over length elements $d\ell$ of the wire. When taking the divergence of the r.h.s. of (21.8) it is now essential to take into account the temperature (and material) dependence of s^*, since this leads also to an r dependence. Ohm's law then may be written as

$$\boldsymbol{j}_{el} = -\sigma_{el} \,\text{grad}[(\mu + Ts^*)/e_0] - T\,\text{grad} s^*/e_0 \qquad (21.9)$$

and the divergence of the heat conduction law in the stationary case reads

$$\text{div} \boldsymbol{j}_q = -\{\text{grad}[(\mu + Ts^*)/e_0]\} \boldsymbol{j}_{el} \quad . \qquad (21.10)$$

Combining these two equations one finds

$$\text{div} \boldsymbol{j}_q = j_{el}^2/\sigma_{el} - T(\text{grad} s^*) \boldsymbol{j}_{el}/e_0 \quad . \qquad (21.11)$$

Here the first term on the r.h.s. is obviously the Joule heat. The second will turn out to yield the Peltier heat (and Peltier coefficient) but also a second effect (the Thomson heat), a kind of continuous Peltier heat along the wire which has its origin in the continuous change of s^* with temperature instead of the abrupt change with material.

Let us first rederive the *Peltier effect* from (21.11). For this we only have to integrate it over a small volume element of the wire enclosing a contact between the two conductors, say a and b. Integrating from a to b, one obtains from (21.11) (in agreement with (21.1))

$$\dot{Q} = \Pi_{ab} J \quad \text{with} \quad \Pi_{ab} = T(s_a^* - s_b^*)/e_0 \quad , \qquad (21.12)$$

where Π is now expressed in terms of transport *entropies*.

In order to derive the *Thomson effect* one integrates again over a small volume element along the wire but this time outside the contacts. Let the length coordinate along the wire be ℓ. Then one obtains a formula completely analogous but with $s_a^* - s_b^*$ being replaced by $\Delta s^* = s^*(\ell) - s*(\ell + d\ell)$. If one writes (taking into account that s^* depends on ℓ only via T)

$$\dot{Q} = \dot{q} d\ell \quad \text{and} \quad \Delta s^* = -\frac{\partial s^*}{\partial T} \frac{\partial T}{\partial \ell} d\ell \quad , \qquad (21.13)$$

then the result takes the form

$$\begin{aligned} \dot{q} d\ell &= \tau \frac{\partial T}{\partial \ell} J d\ell \quad , \\ \tau_{a,b} &= -\partial s_{a,b}^*/\partial T \quad . \end{aligned} \qquad (21.14)$$

Comparison with (21.12) leads immediately to the *Thomson relation*

$$\boxed{\frac{d(\Pi_{ab}/T)}{dT} = -\frac{\tau_a - \tau_b}{T}} \quad . \tag{21.15}$$

Finally we come to the derivation of the *Seebeck effect* in our continuous model. We start out from Ohm's law (21.8a) and integrate it in the current-free state from the first contact to the second contact over both conductors a and b, and add up. The result is

$$\Phi_a - \Phi_b = \int (s_a^* - s_b^*)dT = \int \Pi_{ab}/T \, dT$$

which could, in principle, also be obtained by direct integration of (21.2).

PROBLEMS

21.1 Derive (21.4) from entropy derivatives, as suggested in the text below (21.4).

22. Chemical Reactions

We now come back to discrete variables und study chemical reactions. Chemical reaction kinetics deals with the time evolution of chemical reactions and the fluctuations of the number of reaction partners in equilibrium. We restrict ourselves to homogeneous, closed systems. Then the relevant variables are the particle numbers $N_k(t)$ of the various substances that participate in the reaction.

The thermodynamically conjugate variables are the chemical potentials μ_k. Consequently we expect kinetic equations of the form

$$\dot{N}_k = -\sum L_{kl}\Delta\mu_l/T \ . \tag{22.1}$$

Here $\Delta\mu_l = \mu_l - \mu_{0l}$ represent the deviations of the chemical potentials from their equilibrium values μ_{0l}. While for many transport and relaxation phenomena the linear approximation is good enough, one often finds large deviations from it for chemical reactions. Within a simple model and a simple example we want to determine which kind of nonlinearities and functions $L_{kl}(\mu_1,...,\mu_n)$ one has to expect. Let us consider the NO oxidation

$$2\mathrm{NO} + \mathrm{O}_2 \rightleftharpoons 2\mathrm{NO}_2 \ . \tag{22.2}$$

The chemical symbols are usually called A_k, the numbers in front of them (2,1,2 in our example) are known as the stoichiometric coefficients, and called ν_k. Usually one introduces negative ν_k's and writes reaction equations in the form

$$\boxed{\sum \nu_k A_k \rightleftharpoons 0 \ .} \tag{22.3}$$

The values of the ν and A for our example can be found in Table 22.1

The model for the reaction we consider will be single collisions in gas space. The collision probabilities will be incorporated in a rate equation. The collisions are assumed to proceed just as described by the reaction equation (22.2). A word

Table 22.1. Stoichiometric coefficients of the NO oxidation

k	1	2	3
A_k	NO	O_2	NO_2
ν_k	2	1	-2

of caution is appropriate, however, at this point. In general the actual path of a reaction is by no means described by gross reaction equations such as (22.2). In most cases there are several reactions involving more complicated intermediates which lead to the same end products. Even in our simple example, for instance, there may be intermediate steps such as $NO + NO \Rightarrow N_2O_2$, $N_2O_2 + O_2 \Rightarrow 2NO_2$, $NO + O_2 \Rightarrow NO_3$, $NO_3 + NO \Rightarrow 2NO_2$, etc.

Further complications occur if intermediate steps lead to an increase in the number of atoms or radicals in the form of chain reactions. A simple example is an intermediate step in the detonating gas reaction. Then the steps $H + O_2 \Rightarrow OH + O$ and $O + H_2 \Rightarrow OH + H$ both produce two radicals from one. If there are no reactions leading to a cutoff of such chains, they evolve into an explosion. In our model we pretend that the gross reaction (22.2) describes the reaction path.

The forward reaction then is a consequence of the collision of two NO molecules and one O_2 molecule. The rate of change of number of O_2 molecules, for instance, then will contain a term proportional to the probability of finding these molecules at the same time in the "reaction volume" multiplied by the collision cross section for going from the channel on the l.h.s. of (22.2) to the r.h.s. This reduces, for example, the oxygen concentration. Then there is an equivalent term for the "reverse" reaction, which increases oxygen,

$$\dot{N}_2 = -k' c_1^2 c_2 + k c_3^2 \quad . \tag{22.4}$$

Here $c_k = N_k/N$ ($N = \sum N_k$) are the concentrations, k and k' are concentration-independent rate constants proportional to the cross sections mentioned in the text. The two rate constants are not independent, they have to fulfill the "equilibrium condition" that in equilibrium $c_k = c_{0k}$; the rate of change of all N_k's has to vanish. For (22.4) this means

$$k c_{03}^2 = k' c_{01}^2 c_{02} \quad . \tag{22.5}$$

This allows us to express, for instance, k in terms of k'. Equation (22.4) then takes the form

$$\dot{N}_2 = k c_3^2 \left[1 - \left(\frac{c_1}{c_{01}}\right)^2 \left(\frac{c_2}{c_{02}}\right) \left(\frac{c_{03}}{c_3}\right)^2 \right] \quad . \tag{22.6}$$

Now for dilute systems, see for instance [Ref. 22.1, Eq. (27.11)],

$$\mu = g(P, T) + kT \ln c \tag{22.7}$$

so that (22.6) may also be written

$$\boxed{\dot{N}_2 = k c_3^2 \left[1 - \exp \sum \beta(\nu_k \Delta \mu_k) \right] = k c_3^2 (1 - e^{\beta \Delta \alpha})} \tag{22.8}$$

with the so-called "affinity"

$$\alpha = \sum \nu_k \mu_k \quad . \tag{22.9}$$

Since in chemical equilibrium because of the law of mass action (see e.g. [Ref. 22.1, Eq. (20.16)]) $\sum \nu_k \mu_{0k} = 0$, one may just as well omit the Δ in (22.8). In the vicinity of equilibrium, i.e. for $\beta\alpha \ll 1$ one may expand the exponential in (22.8), obtaining $dN_2/dT = -kc_{03}\beta\alpha$: the reaction rates are proportional to the affinity. Now in the arbitrary case one may "force" (22.8) into a form (22.1) by introducing the rate "constant"

$$\lambda = kc_3^2 T(1 - e^{\beta\alpha}/\alpha) \quad . \tag{22.10}$$

This form may look rather artificial. We will see later on, however, (Chap. 52) that this form is directly obtained from the microscopic theory (15.6,7). Taking the stoichiometric conditions $dN_1 = 2dN_2 = -dN_3$ into account the final result can indeed be written in the form

$$L_{kl} = -\nu_k \lambda \nu_l \quad . \tag{22.11}$$

As one sees, the L_{kl} are not independent. There is only one independent rate λ. The physical reason behind the simple factorization (22.11) is that there is only one reaction in the game. For more reactions one obtains sums of terms like the r.h.s. of (22.11).

As is known from equilibrium theory, in the case of chemical reactions one can avoid redundances if instead of the N_k and μ_k one transforms to so-called progress variables x_r and affinities α_r. In our case with only one reaction we have

$$dN_k = \nu_k dx \quad . \tag{22.12}$$

Then (22.1), taking into account (22.11,12), can be simplified to

$$\dot{x} = \lambda \alpha \quad . \tag{22.13}$$

For the sake of completeness we consider the general case of R reactions according to equations

$$\sum A_k \nu_{kr} \rightleftharpoons 0, \quad r = 1, ..., R \quad . \tag{22.14}$$

The generalization of (22.12) then reads

$$dN_k = \sum_{r=1}^{R} \nu_{kr} dx_r \quad . \tag{22.15}$$

Finally one has to define the affinities

$$\boxed{\alpha_r = \sum \mu_k \nu_{kr}} \quad , \tag{22.16}$$

which again vanish in equilibrium. The entropy change then is

$$\sum \Delta \mu_k dN_k = \sum \Delta_k \nu_{kr} dx_r = \sum \Delta \alpha_r dx_r \tag{22.17}$$

and the kinetic equations take the form

$$\boxed{\dot{x}_r = \sum \lambda_{rs} \Delta \alpha_s} \tag{22.18}$$

with

$$L_{kl} = -\sum \nu_{kr} \lambda_{rs} \nu_{ls} \quad . \tag{22.19}$$

In contrast to our simple example above, where the symmetry relation for L (22.11) could be derived [essentially from the equilibrium condition (22.5)], one now needs the Onsager symmetry for the λ

$$\lambda_{rs} = \lambda_{sr} \tag{22.20}$$

as additional conditions.

PROBLEMS

22.1 From (22.1) derive linear rate equations $dN_k/dt = \sum r_{kl} \Delta N_l$ for the devations ΔN_k from their equilibirum values. Using (22.1,7) determine the r_{kl}.

23. Typical Time Evolutions of Simple Chemical Reactions

The nonlinearities of chemical rate equations lead to deviations in the time evolutions of chemical reactions, as compared to the linear rate equations with their exponential decay. In this chapter we consider some typical examples of such time evolutions for a single reaction.

We use again the progress variables $x(t)$ so that the $N_k(t)$'s evolve like

$$N_k(t) = N_k(t_0) + \nu_k x(t) \tag{23.1}$$

and the evolution of $x(t)$ is governed by the equation

$$\dot{x}(t) = -\lambda(\alpha)\alpha(x(t)) \ . \tag{23.2}$$

The linear regime is always reached near equilibrium, where $\alpha = 0$. We then take λ at equilibrium and expand $\alpha(x)$ around $x = x_0$

$$\alpha = (\partial\alpha/\partial x)_0 x + \cdots \tag{23.3}$$

to obtain the usual exponential decay equation

$$\Delta\dot{x} = -\lambda(0)(\partial\alpha/\partial x)_0 \Delta x = -\Delta x/\tau \ . \tag{23.4}$$

In a certain sense the opposite limit is given "far away from equilibrium", where one has only the "forward" or the "backward" reaction. In this case the concentrations occur as a product of powers. The sum of the power exponents in this case is called the *order of the reaction*. The x dependence of the r.h.s. of the kinetic equation for a reaction of an order m then is a polynomial of mth order in x. Let us consider a few simple examples.

23.1 Zero-Order Reactions

For instance: the "adsorption" or "desorption" at the end of an infinitely long chain according to $A_\infty \rightleftharpoons A_\infty + A$. In this case one has

$$\boxed{\dot{x} = k} \tag{23.5}$$

with a solution $x(t) = (t - t_0)k$ linear in t.

23.2 First-Order Reactions

There are cases, when the exponential decay law is valid for all times, not only near equilibrium. For instance monomolecular transformations between isomers or other excited states of a molecule according to $A^* \rightleftharpoons A$. In this case one has $dx/dt = -kx$ and the corresponding exponential time dependence for all times.

23.3 Second-Order Reactions

An example of a reaction of second order is the back reaction of the NO oxidation. Generally every reaction of the type $A + A \rightleftharpoons B$ is of second order. Let us work with concentrations generally instead of particle numbers, in order to have intensive quantities, and call $c_A = a$. Then one has a differential equation of the form ($a(t) = a_0 - 2x$)

$$\dot{x} = k(a_0 - 2x)^2 \quad . \tag{23.6}$$

The solution is

$$a(t) = a_0/(1 + 2a_0 k t) \quad . \tag{23.7}$$

The concentration decay of the source substance $a(t)$ is thus much slower than the exponential decay for a reaction of first order.

Somewhat more general is the reaction $A + B \rightleftharpoons C$. The x dependence of the concentrations is given by

$$a = a_0 - x, \quad b = b_0 - x, \quad c = x \tag{23.8}$$

and the differential equation reads

$$\dot{x} = k(a_0 - x)(b_0 - x) \quad . \tag{23.9}$$

The solution can now be obtained by partial fraction decomposition of $(a_0 - x)^{-1}(b_0 - x)^{-1}$ [in fact this is the general method for a polynomial, (23.6) being a degenerate case with coinciding zeros of the polynomial] and has the form

$$a(t) = \frac{a_0(a_0 - b_0)}{a_0 - b_0 \exp[(b_0 - a_0)kt]} \quad . \tag{23.10}$$

The solution for $b(t)$ can be obtained, of course, by interchanging a and b in (23.10). For $a = b$ (23.10) should go over into (23.7) (Problem 23.1). The reaction $A + B \rightleftharpoons 2B + C$, which can be written $A - B \rightleftharpoons C$, is mathematically quite similar. This is a simple example of "autocatalysis". The kinetic equation and its solution can be obtained from (23.8,9,10) by changing the sign in front of b (which is the stoichiometric coefficient) in all formulae. See Problem 23.2.

23.4 Third-Order Reactions

The simplest reaction of third order obviously would be $A + A + A \rightleftharpoons B$. In analogy to the second-order reaction above, the differential equation is

$$\dot{x} = k(a_0 - 3x)^3 \tag{23.11}$$

with the solution

$$a(t) = a_0\sqrt{1 + 6a_0^2 kt} \quad . \tag{23.12}$$

More general cases, for instance $2A + B \rightleftharpoons C$ or $A + B + C \rightleftharpoons D$ can be treated in an analogous manner.

Figure 23.1 displays the results for the reactions $mA \rightleftharpoons B$ for $m = 0, 1, 2, 3$.

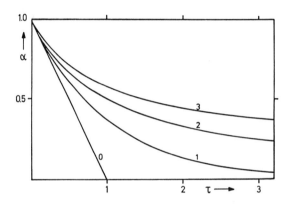

Fig. 23.1. Concentration of source substance $a(t)$ for different orders of chemical reactions, normalized to equal initial slope

In conclusion, one may say that the order of a chemical reaction makes itself felt in the time dependence of the concentrations. Measurement of this time dependence gives information about the order of the reaction. A lack of definite order in the time evolution of a reation usually indicates the superposition of more than one reaction (with different orders).

On the other hand, one cannot, in general, deduce the reaction mechanism or reaction path from the order of a reaction. Here one has to take into account the time scales of intermediate steps. If there are intermediate steps which are fast on the time scale of the total reaction, then the intermediate products are approximately in equilibrium on a time scale of the total reaction but not in general.

In order to illustrate what happens in such a case let us take again the example of the NO oxidation of Chap. 22. Here the two intermediate reactions $N_2O_2 + O_2 \rightleftharpoons 2NO_2$ and $NO_3 + NO \rightleftharpoons 2NO_2$ mentioned in the text are fast. Now, far away from equilibrium, the decrease of O_2-concentration will be proportional to the corresponding concentration products occurring on the l.h.s. of the above chemical reaction equations. But since these reactions are fast, the l.h.s. are approximately in equilibrium with the r.h.s., i.e. the concentration products will be proportional to the corresponding concentration products of the r.h.s. In our case this is in both cases $c(NO_2)^2$. The equation for $c(O_2, t)$ then will be in both cases approximately $dc(O_2)/dt = -kc(NO_2)^2$. This however is just the same as for the gross reaction (22.4).

In such cases one needs further information about the time dependence of the concentration of intermediates. One can monitor such time dependences spectroscopically. For instance, in our example the existence of the peroxide NO_3 could be spectroscopically detected during the reaction.

PROBLEMS

23.1 Consider the transition from (23.10) to (23.7). Note that, except for the expansion of the exponential, you have to take into account that for $b \to a$ the concentration of a doubles at $b = a$.

23.2 Using (22.10) complete Fig. 22.1 by including the autocatalytic reaction (22.9) with $b \to -b$.

24. Coupled Nonlinear Reactions

The solutions of coupled nonlinear equations exhibit a number of interesting phenomena that do not occur in the linear regime. Chemical reactions provide many examples of such coupled nonlinear equations and hence may be used to demonstrate such phenomena. In this chapter we discuss three of them briefly.

24.1 Nonequilibrium Phase Transitions

We consider a model defined by the reaction scheme [24.1]

$$A + X \rightleftharpoons 2X \;,$$
$$B + X \rightleftharpoons C \;. \tag{24.1}$$

Introducing the corresponding concentrations a, b, c, n the rates r_1 and r_2 for these reactions may be written as

$$r_1 = k_1 a n - k_1' n^2 \;,$$
$$r_2 = -k_2 b n + k_2' c \;. \tag{24.2}$$

The species X participates in both reactions and the rate of change of n is the sum of r_1 and r_2: $dn/dt = r_1 + r_2$. We now assume that the concentrations a, b and c are kept constant by external supply. Then this is the only equation one has to consider. Furthermore we choose the units of x and time t such that two of the rate constants are absorbed: $k_1 a = k_1' = 1$. We then call $k_1 a - k_2 b = \varepsilon$ and $k_2' c = \gamma$ so that the rate equation takes the form

$$\boxed{\dot{n} = \varepsilon n - n^2 + \gamma \;.} \tag{24.3}$$

An interesting analogy to phase transitions occurs in the discussion of the steady-state solution $dn/dt = 0$, leading to

$$\gamma = n^2 - \varepsilon n \;. \tag{24.4}$$

This equation obviously has a one-to-one correspondence to the molecular field theory of a ferromagnetic phase transition, leading to

$$B = M^2 - \left(1 - \frac{T}{T_c}\right) M \qquad (24.5)$$

for the magnetic field B, the magnetization M and the critical temperature T_c. The case $\gamma = 0$ also has an application as a simplified model for laser action. In this case ε is the difference of gain due to stimulated emission and loss through the end faces of the laser. The quadratic term describes the decrease of the gain term arising from the reduction of the number of excited atoms due to the presence of photons. For positive ε (i.e. if the gain terms exceed the loss terms) there is a steady-state solution $n_s \neq 0$ far away from the equilibrium solution $n_s = 0$.

The solution of (24.3) can be written down in analogy to (23.9,10). It takes the form

$$n(t) = \frac{\varepsilon}{2} - \frac{\lambda}{2} \frac{c\exp(-\lambda t) - 1}{c\exp(-\lambda t) + 1} \qquad (24.6)$$

with

$$c = \frac{|\varepsilon| + \varepsilon - 2n_0}{|\varepsilon| + \varepsilon + 2n_0} \qquad (24.7)$$

and

$$\lambda = \sqrt{\varepsilon^2 + 4\gamma} \quad . \qquad (24.8)$$

For large t this solution approaches the steady-state solution

$$n(\infty) = (\varepsilon + \lambda)/2 = n_s \quad . \qquad (24.9)$$

Only in the case of negative ε does this lead to the equilibrium state $n_s = 0$ in the limit of $\gamma = 0$. Otherwise the system approaches a steady state which is held far away from equilibrium by means of the external supply or the pumping processes.

In another related model [24.2] the first reaction in (24.1) is replaced by $A + 2X \rightleftharpoons 3X$. The steady-state condition in this case has a one-to-one correspondence to the van der Waals equation of equilibrium phase transition theory (Problem 24.1).

24.2 Kinetic Oscillations

Oscillatory solutions in space as well as in time are another feature which is often encountered in the nonlinear regime. A simple model describing temporal oscillations is defined by the reaction scheme

$$\begin{aligned} A + X &\rightarrow 2X \quad , \\ X + Y &\rightarrow 2Y \quad , \\ Y &\rightarrow 2E \quad . \end{aligned} \qquad (24.10)$$

The concentrations a and e of the initial and final reactant are maintained externally at constant values in such a way that the reverse reactions can be neglected. Then for the concentrations m, n of the species X, Y one obtains

$$\dot{m} = k_1 a m - k_2 m n ,$$
$$\dot{n} = k_2 m n - k_3 n .$$
(24.11)

This system is isomorphic to the Lotka–Volterra model [24.3] of predator–prey interactions: the species X (concentration m) feeds on an unlimited supply "a" of plants, whereas species Y (concentration n) feeds on X and dies with a death rate k_3.

Equation (24.11) has two steady-state solutions. The trivial one, $m_s = n_s = 0$, which is the equilibrium solution, and another one far from equilibrium

$$m_s = \frac{k_3}{k_2} , \quad n_s = \frac{k_1}{k_2} a .$$
(24.12)

Linearization around the steady-state solutions according to

$$m = m_s + \delta m, \quad n = n_s + \delta n$$
(24.13)

(keeping only linear terms in the small quantities $\delta m, \delta n$) leads to

$$\delta \dot{m} = k_1 a \delta m, \quad \delta \dot{n} = -k_3 \delta n$$
(24.14)

in the vicinity of the equilibrium solution. This corresponds to an exponential growth of δm and an exponential decrease of δn.

In the vicinity of (24.12) one has

$$\delta \dot{m} = -k_3 \delta n , \quad \delta \dot{n} = k_1 a \delta m .$$
(24.15)

In this case the time dependence is oscillatory: $\delta m, \delta n \propto \exp(i\omega t)$ with a frequency ω given by

$$\omega^2 = k_1 k_3 a .$$
(24.16)

In contrast to case of Sect. 24.1, none of the two steady-state solutions is approached in the limit of large times. One therefore says the steady-state solutions are not "asymptotically" (for large t) stable. The oscillatory solution (24.15,16), however, is stable. The general reason for all this is the existence of a special conservation law which can be derived for (24.11). For this one writes (24.11) as

$$d \ln \frac{m}{dt} = k_1 a - k_2 n , \quad d \ln \frac{n}{dt} = k_2 m - k_3 .$$
(24.17)

Then one multiplies the first equation by k_3, the second one by $k_1 a$ and adds them together. One combines the result with the sum of (24.11a,b) and uses

(24.12) to obtain the conservation law

$$m_s\left(\frac{m}{m_s} - \ln\frac{m}{m_s}\right) + n_s\left(\frac{n}{n_s} - \ln\frac{n}{n_s}\right) = \text{const} \quad . \tag{24.18}$$

This conservation law defines the trajectories of the solutions in the (m,n)-plane. In the vicinity of the nontrivial steady state they are ellipses around the steady-state point m_s, n_s, and for larger deviations from this point they become deformed (Fig. 24.1) but remain closed curves. This means the solutions describe stable oscillations in time, similar to the linear regime (Fig. 24.2).

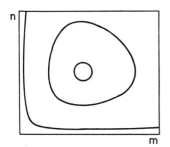
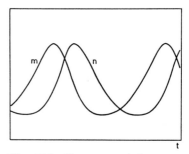

Fig. 24.1. Three trajectories of the Lotka–Volterra equation in the (m,n)-plane: two in the vicinity of the steady states and an arbitrary one

Fig. 24.2. The two populations m, n of the Lotka–Volterra model as functions of time t

The existence of a conservation law for nonlinear kinetic equations is an exception. Nevertheless kinetic oscillations occur frequently. Often these oscillations occur in connection with so-called limit cycles. A reaction scheme with limit cycle behaviour is, for instance,

$$\begin{aligned} A &\rightleftharpoons X \quad , \\ 2X + Y &\rightleftharpoons 3X \quad , \\ B + X &\rightleftharpoons Y + B \quad , \\ X &\rightleftharpoons E \quad . \end{aligned} \tag{24.19}$$

The concentrations of A, B, D, E are again maintained from outside. One has again only two variables, say, m and n. The system has been analyzed in great detail [24.4]. One of the interesting results is that below a certain critical affinity for the overall reaction $A+B \rightleftharpoons D+E$, the system has a single nontrivial steady-state point. In contrast to the Lotka–Volterra model, however, this steady state is asymptotically stable: the solutions approach the steady state in the limit of large times. Above the critical affinity the model asymptotically (24.19) exhibits kinetic oscillations. In contrast to the Lotka–Volterra model, however, the orbits in the (m,n)-plane all spiral asymptotically towards a closed curve, the so-called limit cycle, and finally oscillate along this curve with a frequency independent of the initial conditions.

So far we have treated the chemical reactions as homogeneous in space. If one allows inhomogeneities and their relaxation by diffusion one may well expect oscillatory solution in space, or in space and time. Experimental examples of kinetic oscillations in space and time are the Belousov–Zhabotinskii reaction [24.5] and the catalytic oxidation of CO on Pt [24.6]. The occurrence of spatially periodic and other structures, more generally "kinetic pattern formation", is a fascinating field of current research [24.7]. It is not restricted to chemical reactions. There are early examples in hydrodynamics [24.8], for instance, and many other fields. It may well be that kinetic nonlinearities are able to explain the self-organization of biological systems and the evolution of biological macromolecules [24.9].

24.3 "Chaos"

Another very interesting feature of nonlinear equations is the occurrence of quite irregular ("chaotic") solutions of quite simple and regular equations. This phenomenon is often called "deterministic chaos" or simply "chaos". The simplest equation leading to chaotic solutions seems to be a discretized version of the laser equation, i.e. (24.3) with $\gamma = 0$. With a small time interval τ, (24.3) may be approximated as $n_{\nu+1} = (1 + \varepsilon\tau)n_\nu - \tau n_\nu^2$. Introducing $\tau n_\nu/(1 + \varepsilon\tau) = x_\nu$ and $(1 + \varepsilon\tau) = \alpha$ it takes the form

$$x_{\nu+1} = \alpha x_\nu(1 - x_\nu) = f(x_\nu) \quad . \tag{24.20}$$

If x is kept between 0 and 1 and α between 0 and 4, (24.20) describes a mapping of the interval $0 \leq x \leq 1$ onto itself. This mapping is nowadays called the "logistic map" and (24.20) the "logistic equation". Sequences $x_1, \cdots, x_\nu, \cdots$ obeying (24.20) can easily be calculated using a pocket calculator. The qualitative behaviour of the solutions depends strongly on the parameter α: below a certain critical value α_1 the x_ν converge towards a stable steady-state fixed point. With increasing α the sequences x_ν after a certain "transition time" $\nu_0\tau$ start to oscillate back and forth between two values $\bar{x}$ and $\underline{x}$ with a period one. [Then, of course, (24.20) is no longer an approximation for the differential equation (24.3), which would require a small τ]. With further increasing α the distance between these two points increases until at a second value α_2 another "bifurcation" occurs: the solution asymptotically jumps back and forth between four different x values with a period four. This scenario continues. At a sequence α_l of parameters α, period doubling of the asymptotic solution occurs. Finally above a finite critical value α_c the solutions x_ν become chaotic. They jump back and forth in a random manner, except for certain *window intervals* of α in which periodic solutions reappear. For the "bifurcation points" α_l of the logistic equation (24.20) a relation

$$\lim_{l\to\infty} \frac{\alpha_{l+1} - \alpha_l}{\alpha_{l+2} - \alpha_{l+1}} = 4.6692016\ldots \qquad (24.21)$$

was found [24.10]. It turned out [24.11] that this law is more universal and holds for a whole class of functions $f(x)$ in the mapping (24.20).

Continuous nonlinear differential equations can lead to nonperiodic chaotic solutions behaving qualitatively similarly to the simplified model above: the representative point in the space of variables after a transition time is attracted to some finite region in that space (called the "attractor") in which it remains and executes some kind of circling motion. Often the attractor has different sections in the form of lobes. The representative point circles around in one lobe for some time and then suddenly jumps to the other lobe where it starts a circling motion again. Often the route to such random jumps is via period doublings of periodic motions in limit cycles. A frequently studied system is a model for the onset of turbulence in hydrodynamics [24.12]. The simplest model exhibiting chaos seems to be one with three variables and only one nonlinear term [24.13]. It is plausible that at least three dimensions are required for the chaos described above: the random jumps between different lobes in two dimensions would lead to intersecting trajectories which would violate the uniqueness of the solutions of the differential equations. Projections of three-dimensional orbits into a plane, however, are allowed to intersect. The model proposed is defined by

$$\begin{aligned}\dot{x} &= -(y+z) \ , \\ \dot{y} &= x + ay \ , \\ \dot{z} &= b + xz - cz \ .\end{aligned} \qquad (24.22)$$

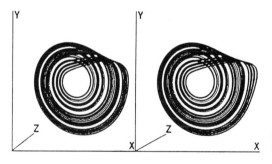

Fig. 24.3. Stereo plot of trajectories of (24.22) [24.13]

Figure 24.3 indicates the attractor for this system. The investigations of chaos in hydrodynamics, chemical reactions and many other examples have opened up a wide field of current research [24.14].

PROBLEMS

24.1 Replace (24.1a) by $A + 2X \rightleftharpoons 3X$ and discuss the corresponding system of equations. In particular, determine the steady-state solutions and note the analogy to the van der Waals equation.

24.2 Write down the equations for the concentrations m, n of the species X, Y in the reaction scheme (24.19) if the reverse reactions can be neglected. Linearize the equations in the vicinity of the steady state(s).

25. Chemical Fluctuations

The fluctuations of the progress variables x_r around their averages may be called "chemical fluctuations". A measure of such fluctuations are the correlations $\langle \Delta x_r \Delta x_s \rangle$, which are given in terms of the susceptibilities by

$$\langle \Delta x_r \Delta x_s \rangle = kT \frac{\partial x_r}{\partial \alpha_s} \quad . \tag{25.1}$$

Because

$$dN_i = \sum \nu_{ir} dx_r \quad , \tag{25.2}$$

one can calculate the fluctuations of the N_i from those of the x_r (25.1)

$$\langle \Delta N_i \Delta N_j \rangle = \sum \nu_{ir} \nu_{rs} \langle \Delta x_r \Delta x_s \rangle \quad . \tag{25.3}$$

For the calculation of susceptibilities we assume that the reactions proceed at constant pressure and temperature (as in the derivation of the law of mass action in [Ref. 25.1, Eq. (27.11)]). Furthermore, we restrict ourselves to dilute systems (gases or solutions). Then one can make use of $\mu_i = kT \ln c_i + g(P, T)$, and hence

$$d(\beta \alpha_r) = \sum \nu_{ir} d \ln c_i = \sum \nu_{ir} \frac{dc_i}{c_i} \quad . \tag{25.4}$$

So we have to differentiate $c_i = N_i/N = N_i/\sum N_k$

$$dc_i = d\left(N_i/\sum N_k\right) = dN_i/N - c_i \sum dN_k/N \quad . \tag{25.5}$$

Finally we use (25.2) to express dN_i in terms of dx_r

$$dc_i = \sum_r \left(\nu_{ir} - c_i \sum_k \nu_{kr}\right) dx_r/N \quad . \tag{25.6}$$

Then one finds for the inverse susceptibility $N\beta(\partial \alpha_r/\partial x_s)_{P,T} = \alpha_{r,s}$

$$\alpha_{rs} = \sum_{i,j} \nu_{ir} \nu_{js}(\delta_{ij}/c_i - 1) \quad . \tag{25.7}$$

The dynamics of fluctuations follows as usual from the relaxation functions using the fluctuation dissipation theorem (8.5). In writing the equations of motion we chose the zeros of the progress variables such that their averages $\langle x_r \rangle$ vanish.

We make the further simplifying assumption that the λ_{rs} are diagonal. This condition is fulfilled, for instance, for the "Stosszahlansatz" (22.4) and similar ones. The equations of motion $dx_r/dt = \lambda_r a_r$ now read after linearization

$$\dot{x}_r = \lambda_r \sum_s (\partial a_r/\partial x_s) x_s = -\kappa_r \sum_s \alpha_{rs} x_s \quad . \tag{25.8}$$

Here the new rate constants $\kappa_r = N\beta\lambda_r$ are now (just like the matrix elements α_{rs}) intensive quantities in the sense of thermodynamics. Note that, although the kinetic coefficients are diagonal, there exists a coupling between the reactions via the off-diagonal elements of the susceptibility.

For a simple illustration we consider a single reaction with only two participating substances "1" and "2" and the stoichiometric coefficients $\nu_1 = -\nu_2 = 1$. (For example a monomolecular transformation, or the association or repulsion of water by salt in an electrolyte.) Then

$$\alpha_{11} = \frac{1}{c_1} + \frac{1}{c_2} = 1/c_1(1-c_1) \quad , \tag{25.9}$$

and for the relaxation function one obtains

$$\Phi(\omega) = \frac{N\beta}{\alpha_{11}} \frac{1}{(\kappa_1 \alpha_{11} + i\omega)} \quad . \tag{25.10}$$

The fluctuations of the particle numbers can now be calculated from this using the fluctuation dissipation theorem. For the Fourier components ($i = 1, 2$) (Problem 8.2.)

$$\Delta N_i(\omega) = \Delta x(\omega)/\nu_i = (2t_0)^{-1/2} \int_{-t_0}^{+t_0} \Delta N(t) e^{i\omega t} dt \tag{25.11}$$

of the particle number fluctuations one finds

$$\langle |\Delta N_i(\omega)|^2 \rangle = s(\omega) = \frac{N}{\alpha_{11}} \frac{2\kappa_1 \alpha_{11}}{[(\kappa_1 \alpha_{11})^2 + \omega^2]} \tag{25.12}$$

in the classical approximation.

Next we consider two coupled reactions. For instance if one takes into account the dissociation of the salt in the electrolyte. Generally we may take two reactions like

$$A \rightleftharpoons BC \rightleftharpoons B + C \quad . \tag{25.13}$$

For the sake of completeness we write down the generalization of (25.10) for the case of two coupled reactions. One then has the relaxation matrix

$$\Phi(\omega) = \frac{\begin{pmatrix} (\kappa_2\alpha_{22} + i\omega)\alpha_{22} + \kappa_1\alpha_{12}^2 & -(\kappa_2\alpha_{22} + i\omega)\alpha_{12} - \kappa_1\alpha_{11}\alpha_{12} \\ -(\kappa_1\alpha_{11} + i\omega)\alpha_{12} - \kappa_2\alpha_{22}\alpha_{12} & (\kappa_1\alpha_{11} + i\omega)\alpha_{11} + \kappa_2\alpha_{12}^2 \end{pmatrix}}{(\alpha_{11}\alpha_{22} - \alpha_{12}^2)D(\omega)}$$

where (25.14)

$$D(\omega) = (\kappa_1\alpha_{11} + i\omega)(\kappa_2\alpha_{22} + i\omega) - \kappa_1\kappa_2\alpha_{12}^2 \qquad (25.15)$$

is the determinant of the matrix $(\kappa\alpha + i\omega)_{rs}$.

In most cases one has a simplification: Because of the Arrhenius law for most reaction rates $[\kappa_r \propto \exp(-\beta E_2)]$, they vary widely. In our example the

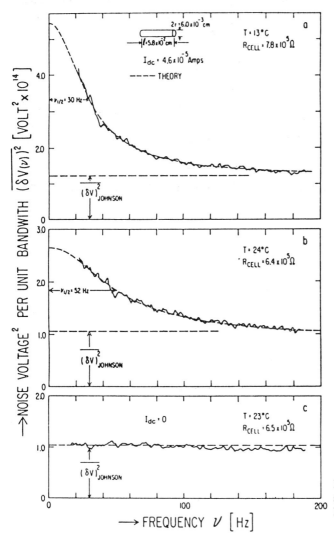

Fig. 25.1. Chemical noise spectrum of a BeSO$_4$ solution [25.2]

dissociation rate of the salt is fast compared to the association of the water. Let's say $\kappa_1 \ll \kappa_2$. In this case in the coupled system the fast rate is simply given by $\gamma_2 \approx \kappa_2 \alpha_{22}$. The rate of the slow reaction is, of course, not simply $\kappa_1 \alpha_{11}$ but gets "renormalized" by the coupling to the fast reaction. By an approximate determination of the zeros of $D(\omega)$ one finds

$$\gamma_1 \approx \kappa_1(\alpha_{11} - \alpha_{12}^2/\alpha_{22}) \qquad (25.16)$$

and the corresponding approximate factorization of D:

$$D(\omega) \approx [\kappa_1(\alpha_{11} - \alpha_{12}^2/\alpha_{22}) + i\omega](\kappa_1 \alpha_{11} + i\omega) \quad . \qquad (25.17)$$

An experimental observation of the chemical fluctuation spectrum in electrolytes is possible, since the electric resistance R of the electrolyte is proportional to the number $N_p = N_n$ of ions. If a current J is passed through the electrolyte one finds voltage fluctuations ΔV with a spectrum

$$\Delta V(\omega) = 2J(\partial R/\partial N_p)\Delta N_p(\omega) \qquad (25.18)$$

which are superimposed on the usual resistance noise (Chap. 18).

Figure 25.1 exhibits a fluctuation spectrum that was observed this way. It shows only the slow component represented by the corresponding Lorentz curve for three different temperatures. The constant background represents the thermal resistance noise (Johnson–Nyquist noise).

PROBLEMS

25.1 Prove the approximate factorization (25.17).

26. Sticking, Desorption, Condensation and Evaporation

In this chapter we consider the simplest surface reactions: either a particle is in a dilute gas and reacts with the surface of a solid or liquid phase, or it is in a dilute solution and reacts with a solid surface. If, after the reaction, the particle stays on the surface for a time τ_{res} (the residence time) long compared to microscopic time scales, one calls this process sticking or adsorption if surface and particle are different substances, and condensation if the substances are the same. In the first case one deals with a chemical reaction, in the second with a phase transition. The chemical reaction, however, is of a special kind. For instance, the stoichiometric coefficients have no precise meaning. Furthermore, there are gradual transitions between both processes in molecular beam epitaxy, when gradually more and more layers of one substance are grown on another. Since the physical mechanism of all these processes is the same, one should treat them on the same footing. This is what we are going to do. The corresponding inverse processes are adsorption and sublimation or evaporation.

Let us consider the case where the dilute phase "f" (free) is at first in equilibrium with the top layer "a" (adsorbed) of the dense phase. Then the system is brought out of equilibrium, for instance by a sudden pressure or temperature change in the dilute phase. Then the above-mentioned processes will start.

Now let N_a be the number of particles in the top monolayer of the "a" phase and n_f the density of the "f" phase. Then one will expect a rate equation of the form

$$\dot{N}_a = k' n_f - k_d N_a \quad . \tag{26.1}$$

Instead of the rate constant k' one often introduces the so-called sticking coefficient s. In order to do so one first defines the average thermal velocity v_T so that $n_f v_T$ is the current density of thermally impinging "f" particles on the "a" surface. If A is the surface of the "a" phase, then s is given by

$$s v_T A = k' \quad . \tag{26.2}$$

Thus s is the fraction of particles of the "f" phase that stick when hitting the "a" surface. The rate constant k_d again is not independent of k' or s; it has to obey the equilibrium condition

$$k_d(N_a)_0 = k'(n_f)_0 = sv_T A(n_f)_0 \quad . \tag{26.3}$$

Since the number $(N_a)_0$ is proportional to the surface area A, k_d is independent of A, as it should be.

We now assume that not only does the "f" phase form a dilute (3-dimensional) gas but the "a" phase also forms a dilute (2-dimensional) gas on the surface of the underlying bulk phase. Then the equilibrium thermodynamics can be determined from the first term of the virial expansion (see e.g. [Ref. 26.1, Sect. 30]), i.e.

$$(n_f)_0 = z_f e^{\beta \mu}, \qquad (N_a)_0 = z_v z_i' e^{\beta V_0} e^{\beta \mu} A/d^2 \quad ; \tag{26.4}$$

$z_f = z_i/(\lambda_T)^3$ is the partition function per unit volume of the "f" phase, z_i the partition function of the internal excitations of the "f" particles and $\lambda_T = h/(2\pi m k T)^{1/2}$ their thermal de Broglie wavelength. d^2 is the area of the "atomic cell" of the "a" particles on the surface, V_0 the potential well depth which binds the particles to the bulk phase and z_v the partition function of a single "a" particle bound in that potential well. If one approximates this well by an isotropic (3-dimensional) oscillator one obtains

$$z_v = z_i' \frac{e^{-3\beta\hbar\omega/2}}{(1 - e^{-\beta\hbar\omega})^3} = z_i' \begin{cases} e^{-3\beta\hbar\omega/2}, & kT \ll \hbar\omega \\ (kT/\hbar\omega)^3, & kT \gg \hbar\omega \end{cases} , \tag{26.5}$$

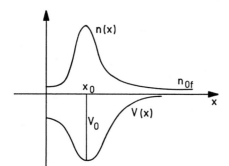

Fig. 26.1. Potential $V(x)$ and density $n(x)$ for particles in the vicinity of a surface, x being the coordinate perpendicular to the surface. The number of "a" particles is the sum of particles in the bound states of the well [approximately the integral of the bump on top of $(n_f)_0$]

If one finally introduces the average thermal "impinging" velocity defined in (26.2)

$$v_T = (m/2\pi kT)^{1/2} \int v e^{-\beta m v^2/2} dv = \sqrt{kT/2\pi m} \quad , \tag{26.6}$$

then one obtains with (26.8,3) the equations

$$(N_a)_0 d^2/A = (c_a)_0 = (n_f)_0 \lambda_T^3 (z_v/z_i) e^{-V_0/kT} \tag{26.7}$$

and

$$k_d = s(d/\lambda_T)^2 (z_i/z_v)(kT/h) e^{-V_0/kT} \quad . \tag{26.8}$$

So for the desorption rate k_d one finds an Arrhenius law with an activation energy $E_a = V_0 - 3\hbar\omega/2$, see (26.5).

This law becomes particularly simple if the sticking coefficient s becomes of order unity. Then (26.8) becomes a special case of the so-called "absolute" rate theory proposed by *Eyring* [26.2] in which rate constants are expressed solely in terms of equilibrium quantities.

At this point we allow ourselves a little excursion into the field of adsorption. One calls the chemical binding of particles to solid surfaces adsorption. In particular one speaks of *chemisorption* for ionic, covalent and metallic binding, and of *physisorption* for van der Waals binding. In quantitative terms the boundary between them lies at a binding energy of about 1eV per particle.

Chemisorption plays an important role in heterogeneous catalysis (increase of reaction rates at solid surfaces, which in many cases makes reactions technically feasible that would otherwise be practically impossible), in epitaxy, in lubrification and as a first step in reactions of solids with gases or liquids.

Particularly clear are the situations for monolayers on solids. The occupation properties of such monolayers are measured by the "coverage" $\Theta = N_a/N$, where N_a is the number of particles in the monolayer and N the maximum number of particles in a monolayer. So Θ is a number between 0 and 1. From (26.7) one realizes that in equilibrium such situations can only be produced by going to ultrahigh vacuum (Problem 26.1) of the order of 10^{-10} torr and lower.

For coverages near $\Theta = 1$, (26.7) has to be corrected, since the assumption of low two-dimensional density is no longer fulfilled. Langmuir has proposed a simple way to do this. He considers the adsorption equilibrium as a chemical equilibrium for the reaction between an empty place "e" on the surface and a free particle "f", forming the surface molecule "a". This simple prescription takes into account (via stoichiometric coefficients) that each surface place "e" can be occupied only once. The reaction equation looks like

$$e + f \rightleftharpoons a \quad .$$

Now (26.7) reads in the form of a law of mass action

$$(c_a/n_f)_0 = K(T) \quad [\propto \exp(-V_0/kT)] \quad , \tag{26.9}$$

whereas the stoichiometric conditions expressed in (26.9) require

$$(c_a/n_f c_e)_0 = K(T) \quad . \tag{26.10}$$

The connection with the Θ defined above is $(c_a)_0 = \Theta$ and $(c_e)_0 = 1 - \Theta$.

Solving (26.10) for Θ then leads to the Langmuir isotherm

$$\Theta(n_f, T) = n_f K(T)/[1 + n_f K(T)] \quad , \tag{26.11}$$

which approaches (26.7) for small Θ but saturates at the maximum $\Theta = 1$. The rate equation (26.1) does not change. Equation (26.3) also remains valid. If here, instead of (26.7), (26.10) is inserted, one finds instead of (26.8) an equation in which s is replaced by $s/(1-\Theta)$. On the other hand, one must expect that the stoichiometric conditions which are expressed in (26.9) will somehow also change the temperature dependence of $s(T)$: sites which are already occupied will have a reduced sticking coefficient. In the simplest way this may be taken into account by the ansatz $s = s_0(1-\Theta)$ with a Θ-independent s_0. For the desorption rate then one finds again a formula similar to (26.8) with s replaced by s_0. In general one will have, however, more complicated laws.

PROBLEMS

26.1 Using (26.7) or (26.11) and $P = (n_\mathrm{f})_0 kT$ plot a few curves $\Theta(P,T)$ for T around room temperature and below, and Θ below 1.

27. Nucleation

Solid surfaces (walls of vessels, dust particles) in contact with saturated vapour are usually covered with a rather thick adsorption layer of the liquid phase ("wet"). An oversaturation of the vapour by a decrease of temperature (undercooling) or increase of pressure is therefore impossible, since the wet surfaces act as condensation nuclei which start to grow if one tries to oversaturate the vapour [27.1].

The overheating of liquids is also usually not possible. Evaporation from a liquid follows an Arrhenius law (see the previous chapter) and may be slow, but is always possible.

An oversaturation of saturated vapours, however, may be possible by means of a sufficiently fast pressure jump (as in a cloud chamber) in the absence of dust particles. An overheating of solids and liquids can occur if they are heated inside while their surfaces are kept below the phase transition temperatures. Such deviations from phase equilibrium are unstable. In the long run they have to proceed to a stable state with phase equilibrium along a phase separation surface. Such equilibration processes occur via the formation of small embryos of the other phase in the homogeneous original phase. Such a process is called "nucleation". In this chapter we want to consider just one simple example of such processes: the kinetics of nucleation in an oversaturated vapour.

The embryos in this case will be droplets of the liquid phase. They form during thermal fluctuations. What we have to look for is the kinetics of the formation and decay of such droplets in the vapour. In order to do so we call N_l the number of droplets with l atoms (or molecules) of the liquid. Hence $N = \sum N_l$ is the total number of doplets (the case $l = 0$ being included) and the pressure is given by $PV = NkT$.

Using the grand canonical ensemble we can write the chemical potential of all l-droplets as

$$N_l = V z_{il} e^{\mu_l/kT} / \lambda_l^3 \quad . \tag{27.1}$$

Here

$$z_{il} = e^{-f_{il}/kT} \tag{27.2}$$

is the partition function of the internal excitations of a droplet l. Solving for μ yields

$$\mu_l = f_{il} + kT \ln(\lambda_l^3 P/kT) + kT \ln c_l = g_l(P, T) + kT \ln c_l \quad . \tag{27.3}$$

Between the droplets there are chemical reactions according to

$$A_l + A_m \rightleftharpoons A_{l+m} \quad . \tag{27.4}$$

The concentration of droplets in equilibrium can be determined from the law of mass action [Ref. 27.2, Eq. (20.16)] for the reactions (26.4), i.e.

$$\mu_{l,0} + \mu_{m,0} = \mu_{l+m,0} \tag{27.5}$$

with the solution

$$\mu_{m,0} = m\mu_{1,0} \quad . \tag{27.6}$$

Using (27.3) one finds

$$c_{l,0} = (c_{1,0})^l e^{-\Delta g_l/kT} \tag{27.7}$$

with

$$\Delta g_l = g_l - lg_1 = g_l - kT \ln(\lambda_l^3/kT) \quad . \tag{27.8}$$

Now f_{il} in g_l contains something like the surface energy of the droplets which will increase strongly with l. Thus $c_{l,0}$ will always be very small compared to $c_{1,0}$, the concentration of molecules in the vapour. To get a rough idea of the l dependence of g_l one may try an expansion

$$g_l = l\mu_{\text{liq}} + 4\pi r_l^2 \sigma + \cdots \quad . \tag{27.9}$$

Here μ_{liq} is the chemical potential of the (infinitely extended) liquid and σ is the surface tension between liquid and vapour. The implicit assumption that this quantity is independent of l is, of course, only a rough approximation. $4\pi r_l^2$ is the surface area of the droplet. The implcit assumption of a spherical surface is no longer valid if the surface tension is small, for instance near critical points.

Now, since liquid and vapour are in equilibrium at the saturation pressure P_s,

$$\mu_{\text{liq}} = \mu_{1,0} = kT \ln(\lambda_1 P_s/kT) \quad , \tag{27.10}$$

and one can write for (27.8)

$$g_l = -lkT \ln(P/P_s) + 4\pi r_l^2 \sigma + \cdots \quad . \tag{27.11}$$

Taking into account that the number of molecules l in the droplet is approximately proportional to the volume, and thus to r_l^3, one finds for Δg_l a set of curves of the form shown in Fig. 27.1.

For nonzero oversaturation $\ln(P/P_s)$, Δg_l has a maximum at, say, $l = k$, given by

$$\frac{\partial \Delta g_l}{\partial l} = \frac{\partial g_l}{\partial l} - \mu_1^0 = \tilde{\mu}_l - \mu_1^0 \quad . \tag{27.12}$$

Taking into account that the volume of the droplet and the volume per particle v_{liq} in the liquid are connected by

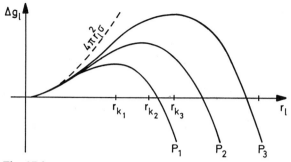

Fig. 27.1. Free enthalpy Δg_l for the formation of a droplet from the vapour phase as a function of its radius r_l and the pressure ratio P/P_s

$$4\pi r_l^3/3 = lv_f \quad, \tag{27.13}$$

the differentiation (27.12) yields

$$\ln(P/P_s) = 2\sigma v_f/kTr_k \tag{27.14}$$

and hence

$$\Delta g_k = 16\pi\sigma^3 v_f^2 [kT\ln(P/P_s)]^2/3 \quad. \tag{27.15}$$

In (27.12) $\tilde{\mu}_l$ is the chemical potential of a molecule in an l-droplet.

Hence (27.12) can be interpreted as the condition for phase equilibrium between k-droplets and vapour of pressure P. In other words, droplets with k molecules have the vapour pressure P, which is larger than the vapour pressure of the infinitely extended liquid (at least for small droplets). Physically this is a consequence of surface tension: it costs too much surface energy to form a small droplet out of the bulk liquid.

The maximum at $l = k$ is unstable: droplets below k would shrink, and above k would grow. In order to describe the time evolution of such processes we set up kinetic equations describing the corresponding reactions (27.4). The corresponding progress variables may be called $x_{l,m}$. Then one expects rate equations of the form

$$\dot{x}_{l,m} = k_{l,m}c_l c_m - k'_{l,m}c_{l+m} \quad. \tag{27.16}$$

If one furthermore neglects in the term quadratic in the concentrations contributions with m, l *both* larger than one (because they are quadratic in the small c_m's) then one is left with

$$\dot{x}_{m,l} = k_m c_m - k'_m c_{m+1} \quad. \tag{27.17}$$

The contribution of these reactions to the change of the droplet number N_l is then given by

$$\dot{N}_l = -\dot{x}_{l,1} + \dot{x}_{l-1,1} \quad. \tag{27.18}$$

If finally the equilibrium conditions

$$k_m c_m^0 = k'_m c_{m+1}^0 \tag{27.19}$$

are taken into account, the resulting equation can be cast into a form which has an amazing similarity with the diffusion equation for a particle in an external potential with the coordinate l. One only has to introduce the concentration $c_l = N_l/N$ and the "current density" $j_l = \dot{x}_{l,1}/N$. The differences on the r.h.s. of (27.18) are approximated by differential quotients. Then (27.18) becomes the continuity equation

$$\dot{c}_l + \frac{\partial j_l}{\partial l} = 0 \ . \tag{27.20}$$

Using the same notations and approximations and introducing a "position dependent" diffusion "constant" $D_l = k_l/N_l$, we may write the rate equations (27.17) in the form of a generalized Fick's law for the diffusion current in a potential g_l

$$j_l = -D_l c_l^0 \frac{\partial (c_l/c_l^0)}{\partial l} = -D_l \left(\frac{\partial c_l}{\partial l} - c_l \frac{\partial \Delta g_l/kT}{\partial l} \right) \ . \tag{27.21}$$

We now look for a solution of these equations for a situation in which one starts out from equilibrium in saturated vapour that is perturbed by a fast pressure increase from P_s to P. The simplest solution is a stationary one. This corresponds to a situation in which a current j independent of l flows from $l = 1$ to infinity (i.e. the bulk liquid). This, of course, would be continuously filled up.

From the continuity equation (27.20) one obtains

$$j = -D_l c_l^0 \frac{d(c_l/c_l^0)}{dl} \ . \tag{27.22}$$

This then has to be integrated with the boundary conditions

$$\frac{c_l}{c_l^0} \to \begin{cases} 1, & l = 1 \ , \\ 0, & l \to \infty \ . \end{cases} \tag{27.23}$$

One therefore obtains the stationary nucleation rate

$$j = \left(\int_0^\infty \frac{dl}{D_l c_l^0} \right)^{-1} \ . \tag{27.24}$$

The integral can be evaluated approximately, taking into account that $1/c_{l,0}$, because of (27.7,11) (see Fig. 27.1), has a sharp maximum at $l = k$. In the roughest approximation one just replaces the integral by its value at the maximum. An improvement can be obtained by approximating the integrand by a Gauss function [expanding the exponent up to second order in $(l - k)$]. Evaluation of the integral over the Gauss function then yields (k_B is the Boltzmann constant)

$$j = D_k(\Delta g_k/3\pi k_B T k^2)^{-1/2} e^{-\Delta g_k/k_B T} \qquad (27.25)$$

where Δg_k is given by (27.15). Again the nucleation rate obeys an Arrhenius law. An estimate of the preexponential factor in (27.25) can be obtained if in (27.17) a sticking coefficient s is introduced, such that

$$D_k = snv_T 4\pi r_k^2 \qquad (27.26)$$

where $n = N_1/V$ is the density of vapour molecules. For s of the order of 1 it turns out that the whole prefactor for the experimentally relevant values of k is of the order of a gas kinetic collision frequency [10^{10} s^{-1} for atmospheric pressure, 10^6 s^{-1} for the saturation pressure (1/100 atm) of water vapour at room temperature]. A precise determination of the prefactor is difficult. But since the exponential depends very strongly on P/P_s, this is not so important for a first orientation.

Experimentally easily accessible are the values of k. They follow from the condition that the number n_j of the nuclei generated per second and cubic centimetre is somewhat greater than 1, so that the nuclei can be detected as fog in a cloud chamber. Table 27.1 exhibits several (historic) results for P_s/P determined from the condition of "visible clouds".

Recent computer experiments [27.3] indicate that the restriction to spherical droplets, inherent in (27.11), and the restriction to droplet reactions with only single particles (27.17) lead to errors. These errors become particularly large near critical points T_c of phase transitions. If the correlation length $\xi(T)$, see

Table 27.1. "Critical" visible-cloud values of P/P_s for various substances [27.4]. K: number of molecules in critical nucleus

Vapour	K	Radius r_k [Å]	P/P_s theory	P/P_s exp.
Water, 275.2 K	80.0	8.9	4.2	4.2 ± 0.1
Water, 261.0 K	72.0	8.0	5.0	5.0
Methanol, 270.0 K	32.0	7.9	1.8	3.0
Ethanol, 273.0 K	128.0	14.2	2.3	2.3
n-Propanol, 270.0 K	115.0	15.0	3.2	3.0
Isopropyl alcohol, 265.0 K	119.0	15.2	2.9	2.8
n-Butyl alcohol, 270.0 K	72.0	13.6	4.5	4.6
Nitromethane, 252.0 K	66.0	11.0	6.2	6.0
Ethyl acetate, 242.0 K	40.0	11.4	10.4	8.6 to 12.3

Chap. 48, becomes larger than the droplet radius, scaling arguments can be used [27.5] and lead to good agreement with the results of computer experiments [27.3].

PROBLEMS

27.1 Electric charges on condensation nuclei are known to have large effects on their behaviour in cloud chambers. Which term has to be added on the r.h.s. of (27.11) if the droplets have an electric charge q and what is the effect of such a term on the curves of Fig. 27.1?

28. The Oscillator with Mechanical and Thermal Attenuation *

As in the case of thermomechanical and thermoelectric effects (Chaps. 19,20) we start our consideration of continuous hydrodynamics with a simple discrete model, see Fig. 28.1. A piston with mass m and cross section A can move between two equal parts (1) and (2) of a closed system (for instance a gas). It is impermeable to particles but permeable to heat.

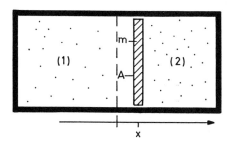

Fig. 28.1. Model for the dynamics of pressure and temperature fluctuations

If the piston has velocity v and the mass of the gas is neglected, the total energy of the system is given by

$$E = E_0 + e_1 + e_2 + mv^2/2 \quad , \tag{28.1}$$

where $e_{1,2}$ are the deviations of the internal energies of the two parts from their average $E_0/2$. For the total entropy one has similarly

$$S = S_0 + s(e_1, V_1) + s(e_2, V_2) \quad . \tag{28.2}$$

Here

$$V_1 = V_0 + Ax, \quad V_2 = V_0 - Ax \tag{28.3}$$

are the volumes of the parts, and x is the displacement of the piston from its equilibrium position. The two subsystems exert forces f_i on the piston, which are given by

$$de_1 = T_1 ds_1 + f_1 dx \quad , \quad de_2 = T_2 ds_2 - f_2 dx \quad , \tag{28.4}$$

where T_i are the temperatures of the subsystems.

For a differential change of the total entropy one obtains

$$dS = de_1/T_1 + de_2/T_2 + \Delta(f/T)dx \quad . \tag{28.5}$$

We are using the notation $\Delta A = A_2 - A_1$.

We now introduce new variables. First, using energy conservation $dE/dT = 0$:

$$de_1 + de_2 = -vdp \quad, \tag{28.6}$$

introducing $p = mv$. Instead of $f_{1,2}$ we introduce f_0 and f by

$$f_1 = f_0 + f/2 \quad, \quad f_2 = f_0 - f/2 \quad, \tag{28.7}$$

and finally we call the heat transfer from (1) to (2)

$$\delta q = \Delta(Tds)/2 = \Delta de/2 - f_0 dx \quad. \tag{28.8}$$

Collecting everything together, the differential entropy change is given by

$$\boxed{TdS = -[fdx + vdp + \Delta T(\delta q/T)]} \quad, \tag{28.9}$$

correct to terms linear in ΔT.

In the same linear approximation the inexact differential δq can be written as an exact differential dq of the function

$$q = (s_2 - s_1)T/2 \quad. \tag{28.10}$$

Onsager's prescription (15.19) then leads to the kinetic equations

$$\begin{pmatrix} \dot{x} \\ \dot{p} \\ \dot{q}/T \end{pmatrix} = -\mu \begin{pmatrix} f \\ v \\ \Delta T \end{pmatrix} \quad. \tag{28.11}$$

For the matrix μ of the "mobilities" we write

$$\mu = \begin{pmatrix} 0 & -1 & 0 \\ 1 & m\gamma & 0 \\ 0 & 0 & \lambda/T \end{pmatrix} \quad. \tag{28.12}$$

In this matrix the first column is dictated by the trivial relation $dx/dt = v$. The first row then follows from symmetry (Chap. 7) in particular time reversal symmetry. The term $m\gamma$ then defines the friction coefficient. The second irreversible term is the heat transfer coefficient λ. We have not been able to prove that the remaining two terms are zero, but we put them to zero anyway, because we want a close analogy to hydrodynamics, and there the corresponding terms vanish because of symmetry. Both irreversible coefficients γ and λ must be positive, so that the entropy production

$$\dot{S} = [\gamma mv^2 + \lambda(\Delta T)^2/T]/T \tag{28.13}$$

is positive.

For the determination of the relaxation function we need the susceptibilities χ and their reciprocals. Because of (28.10), instead of q we may also use the entropies directly and write

$$\begin{pmatrix} (\partial x/\partial f)_T & (\partial x/\partial T)_f \\ (\partial s/\partial f)_T & (\partial s/\partial T)_f \end{pmatrix} = \begin{pmatrix} \chi_T & \alpha \\ \alpha & C_f/T \end{pmatrix} \qquad (28.14)$$

with the inversion

$$\begin{pmatrix} (\partial f/\partial x)_s & (\partial f/\partial s)_x \\ (\partial T/\partial x)_s & (\partial T/\partial s)_x \end{pmatrix} = \begin{pmatrix} 1/\chi_s & a \\ a & T/C_x \end{pmatrix} . \qquad (28.15)$$

The off-diagonal elements of this matrix can be expressed in terms of the thermal expansion coefficients α, either by the usual "thermodynamic differential gymnastics" [Ref. 28.1, Sect. 17] or in this case more simply by direct inversion of (28.14):

$$\begin{pmatrix} 1/\chi_s & a \\ a & T/C_x \end{pmatrix} = \begin{pmatrix} C_f/T & -\alpha \\ -\alpha & \chi_T \end{pmatrix} /(\chi_T C_f/T - \alpha^2) \qquad (28.16)$$

yielding

$$\chi_T - \alpha^2 = \chi_s C_f/T = \chi_T C_x/T . \qquad (28.17)$$

This leads to the well-known thermodynamics relations

$$\chi_T/\chi_s = C_f/C_x \qquad (28.18)$$

and

$$C_x = C_f - T\alpha^2/\chi_T, \quad \chi_s = \chi_T - T\alpha^2/C_f , \qquad (28.19)$$

and also the desired connection between a and α

$$a = -\alpha T/\chi_s C_f = -(C_f/C_x - 1)/\alpha = -(\chi_T/\chi_s - 1)/\alpha \qquad (28.20)$$

If one, finally, takes into account $\partial p/\partial v = m$, one finds for χ

$$\chi = \begin{pmatrix} \chi_T & 0 & a \\ 0 & m & 0 \\ \alpha & 0 & C_f/T \end{pmatrix} \qquad (28.21)$$

and the inverse matrix

$$\chi^{-1} = \begin{pmatrix} 1/\chi_s & 0 & a \\ 0 & 1/m & 0 \\ a & 0 & T/C_x \end{pmatrix} . \qquad (28.22)$$

Furthermore, one obtains for the frequency matrix

$$\nu = \mu\chi^{-1} = \begin{pmatrix} 0 & -1/m & 0 \\ 1/\chi_s & \gamma & a \\ \lambda a/T & 0 & \lambda/C_x \end{pmatrix} . \qquad (28.23)$$

For the relaxation matrix $\Phi(\omega)$ one has to calculate the inverse of $(\nu - i\omega)$. Using (28.23) one obtains

$$(\nu - i\omega)^{-1} = D^{-1}$$
$$\times \begin{pmatrix} (\gamma - i\omega)(\lambda/C_x - i\omega) & (\lambda/C_x - i\omega)/m & -a/m \\ \lambda a^2/T - m\omega_s^2(\lambda/C_x - i\omega) & -(\lambda/C_x - i\omega)i\omega & i\omega a \\ -\lambda a(\gamma - i\omega)T & \lambda a/(mT) & \omega_s^2 - \omega^2 - i\omega\gamma \end{pmatrix} .$$
(28.24)

Here $\omega_s^2 = 1/m\chi_s$ is the square of the adiabatic vibrational frequency of the oscillator and

$$D = (\lambda/C_x - i\omega)(\omega_s^2 - \omega^2 - i\gamma\omega) - \omega_s^2\lambda(1/C_x - 1/C_f) . \quad (28.25)$$

The zeros of this determinant are the poles of $\Phi(\omega)$ and hence the excitation energies of the system.

Looking at (28.25 and 19), one sees that if $\lambda\alpha$ were zero the eigenfrequencies would be given by $\omega_{1,2}^2 \approx \omega_s^2 - i\omega_s\gamma$ and $\omega_3 = -i\lambda/C_x$. They correspond to a pure vibrational mode and a pure heat transfer mode.

For nonvanishing $\lambda\alpha$ there exists a coupling of these modes: during an increase of ΔT, α leads to an expansion and hence a change in x, and conversely, during an adiabatic shift of x there will be a ΔT. Due to these effects there will be a shift in the frequencies and a mixing of the zero-order modes. To first order the shifts and mixing coefficients will be proportional to $\lambda\alpha$. One finds for the shifted frequencies (Problem 28.1) an approximate factorization of D as

$$D = (\gamma_T - i\omega)(\omega_s^2 - \omega^2 - i\omega\gamma_s) + O(\lambda^2\alpha^2) \quad (28.26)$$

with the thermal attenuation coefficient

$$\gamma_T = \lambda/C_f \quad (28.27)$$

and the mechanical one

$$\gamma_s = \gamma + \gamma_T(C_f/C_x - 1) . \quad (28.28)$$

In order to obtain the full relaxation matrix one has to multiply (28.24) by the susceptibility matrix (28.21). We consider here only the diagonal matrix element $\Phi_{xx}(\omega)$. For this one finds in the same approximation as in (28.26–28).

$$\Phi_{xx} = \chi_T[(\gamma - i\omega)(\lambda/C_x - i\omega) + \omega_s^2(1 - \chi_s/\chi_T)]/D$$
$$= -i\chi_T\left[\frac{C_x}{C_f}\frac{[\gamma_s + \gamma_T(C_f/C_x - 1) - i\omega]}{(\omega_s^2 - \omega^2 - i\omega\gamma_s)} + \frac{1 - C_x/C_f}{\gamma_T - i\omega}\right], \quad (28.29)$$

see Problem 28.2.

As one can see, the coupling of the two modes shows up directly in the vibrational relaxation function: besides the vibrational poles there is a heat transfer pole near $\omega = 0$ with a width γ_T. The corresponding peak is often called the Landau-Placzek peak [28.2]. The strength of this peak is a factor $1 - C_x/C_f$ smaller than the two vibrational peaks.

It is interesting to consider the limit where the two attenuation constants (28.27,28) approach zero. In this case the three peaks of the spectrum approach delta functions at $\omega = \pm\omega_s$ and $\omega = 0$. The central peak then does not contribute any longer to $\chi''_{xx}(\omega) = \omega\Phi''_{xx}(\omega)$. The dynamic susceptibility $\chi(\omega)$, see (28.18), thus approaches the *adiabatic* susceptibility at $\omega = 0$.

On the other hand, the relaxation function at $t = 0$ can be read off from the strength of $\Phi(\omega)$ for ω approaching infinity. A glance at (28.29) shows that this strength is still given by the *isothermal* susceptibility as required by the general rules (compare the discussion in Chap. 3).

PROBLEMS

28.1 Determine the zeros of the cubic equation $D(\omega) = 0$ to first order in $\lambda\alpha$ and the corresponding factorization of D.

28.2 In the same approximation determine Φ_{xx} and verify (28.29). Plot $\Phi''(\omega)$ and $\Phi'(\omega)$.

29. Hydrodynamics

In this chapter we consider the hydrodynamic motions in normal fluids at long wavelengths. Superfluidity, liquid crystals, multicomponent systems, in particular, and also charged systems ("plasmas"), are not considered here.

Our guideline for deriving the equations of motion will again be Onsager's relation between forces and fluxes. We start out from the energy e per particle, consisting of the internal energy e_{in} and the kinetic energy e_{kin}:

$$e = e_{\text{in}} + e_{\text{kin}} \; . \tag{29.1}$$

The first law of thermodynamics then reads

$$de = Tds - Pdv + \sum v_k dp_k \quad (k = 1, 2, 3) \; . \tag{29.2}$$

Here the additional quantities per particle (or better per fluid element) s: entropy, v: volume, p_k: momentum are introduced. We have also introduced the components v_k of the velocity as thermodynamically conjugate "forces" to the p_k. Instead of quantities per fluid element one often uses quantities per (fixed) volume element or simply "densities" in the so-called Eulerian description [29.1]. A few relations and notations for such densities are

$$\begin{aligned} n &= 1/v & &\text{particle number density} \; , \\ g &= np_k & &\text{momentum density} \; , \\ \varepsilon &= ne & &\text{energy density} \; , \\ \sigma &= ns & &\text{entropy density} \; . \end{aligned}$$

If one multiplies (29.2) by n one finds after a Legendre transformation

$$d\varepsilon = T\,d\sigma + \mu\,dn + \sum v_k dg_k \tag{29.3}$$

with the chemical potential

$$\mu = e + Pv - Ts - \sum v_k p_k \; , \tag{29.4}$$

which now, in generalization of (19.7), has a differential change

$$d\mu = -s\,dT + v\,dP - \sum p_k\,dv_k \; . \tag{29.5}$$

As in the case of diffusion (Chap. 20) we can make use of the fact that the three "coordinates" ε, n, g_k obey differential conservation laws. Onsager's kinetic equations can then be directly formulated in terms of the corresponding current densities, instead of time derivatives. These current densities can be read off from the conservation laws ($\partial_k = \partial/\partial x_k$)

$$\dot{n} + \sum \partial_k(nv_k) = 0 \quad , \tag{29.6}$$

$$\dot{g}_i + \sum \partial_k \Pi_{ik} = 0 \quad , \tag{29.7}$$

$$\dot{\varepsilon} + \sum \partial_k j_{\varepsilon k} = 0 \tag{29.8}$$

with the components Π_{ik} of the "momentum flux" or stress tensor. We disregard possible external (electromagnetic, gravitational) forces, the density f_i of which would otherwise occur on the r.h.s. of (29.7). We now interpret the "d" in (29.3) as a "∂_t" and insert the continuity equations (29.6–8). We solve for $\partial_t \sigma$ and then on the r.h.s. of the resulting equations take the force terms under the divergences, leading to an expression for the entropy change:

$$\begin{aligned}\dot{\sigma} + \partial_k[(j_{\varepsilon k} - \sum v_i \Pi_{ik} - \mu n v_k)/T] \\ = \sum [j_{\varepsilon k} \partial_k(1/T) - \Pi_{ik} \partial_k(v_i/T) - n v_k \partial_k(\mu/T)] \quad .\end{aligned} \tag{29.9}$$

As in the case of diffusion we introduce new variables after differentiating the gradient terms of the forces and making use of (29.5). This leads to

$$\boxed{\dot{\sigma} + \operatorname{div} j_\sigma = -\sum [(j_{qk}/T)\partial T + \Pi'_{ik}\partial_k v_i]/T} \tag{29.10}$$

with the entropy current density

$$j_{\sigma k} = \left(j_{\varepsilon k} - \sum v_i \Pi_{ik} - \mu n v_k + P v_k\right)/T \tag{29.11}$$

the heat current density

$$\boldsymbol{j}_q = T(\boldsymbol{j}_\sigma - sn\boldsymbol{v}) \tag{29.12}$$

and the "viscous" part of the stress tensor

$$\Pi'_{ik} = \Pi_{ik} - g_i v_k - P\delta_{ik} \quad . \tag{29.13}$$

Since we have introduced the velocity as nothing but the thermodynamical conjugate to the momentum density, the relation between g and v is not obvious. We expect that momentum density is mass density times velocity, but there is a little more to say about that.

If we restrict ourselves to small velocities we expect a linear relation, which for a homogeneous and isotropic system (after Fourier transformation) reads

$$\boldsymbol{g}(\boldsymbol{k}) = \varrho^l(\boldsymbol{k})\boldsymbol{v}^l(\boldsymbol{k}) + \varrho^t(\boldsymbol{k})\boldsymbol{v}^t(\boldsymbol{k}) \tag{29.14}$$

with the longitudinal and transverse mass densities ϱ^l and ϱ^t. The mass density defined this way is a static susceptibility, and hence a static correlation function of the current operator with itself. Let us first consider the long-wavelength limit of these correlation functions. They are particularly simple if the current density operator has a well defined $k = 0$ limit, since this has to be $g(0) = P/V^{1/2}$ (with P the total momentum and V the volume of the periodicity box). For the correlation functions one then obtains

$$\chi_{ij} = \frac{1}{V}\beta\langle P_i; P_j\rangle = \frac{2}{V}\sum \frac{\langle\mu|P_i|\nu\rangle\langle\nu|P_j|\mu\rangle}{\varepsilon_\nu - \varepsilon_\mu} \exp[\beta(F - \varepsilon_\mu)] \ . \qquad (29.15)$$

The r.h.s. of this equation is well known from the dipole sum rule. For a system of N points of mass m one has $\langle\mu|P_i|\nu\rangle = (i/\hbar)Nm(\varepsilon_\mu - \varepsilon_\nu) \times \langle\mu|X_i|\nu\rangle$. Then using the commutation relations between P_i and X_j one finds

$$\chi_{ij} = nm\delta_{ij} \ . \qquad (29.16)$$

Hence in the long-wavelength limit one has

$$g = mn v \ . \qquad (29.17)$$

In the long-wavelength limit there is no difference between longitudinal and transverse mass density. This may be expected, since at $k = 0$ there is no preferred direction.

We close our consideration of the mass with a few remarks concerning more general situations. (i) First of all, the dipole sum rule is derived using only operator relations and is valid independent of the state, in particular the velocity of this state. Hence (29.17) is valid for arbitrary velocities. (ii) For higher k values in the longitudinal case the dipole sum rule can be generalized to arbitrary k (see Chap. 31 and Problem 29.1): the longitudinal mass is given by m for arbitrary k. In the transverse case there is no such generalization. The transverse current correlation function indeed decreases with increasing k from its $k = 0$ limit. This decrease which is proportional to k^2 is calculated at the end of Chap. 34. It is the origin of the Landau diamagnetism for electrons. Furthermore, if the particles are molecules, even if they are treated as points for their translational motion, they may have an intrinsic angular momentum, say, S. The angular momentum density $S(r)$ then contributes a term curl $S(r)/2$ to the momentum density[1]. The corresponding contribution to the transverse current

---

[1] We adopt the point of view, that $g = g^0 + \text{curl } S/2$ is the *total* momentum density and Π is the total stress tensor. *Total* angular momentum conservation then is a consequence of momentum conservation (29.7) and no new conservation law, which might introduce new hydrodynamic modes. (See Martin, P. C., Parodi, O., Pershan, P. S.: Phys. Rev. A6, 2401 (1972), Appendix A. There may be circumstances, however, (for instance in external magnetic fields, in rotating vessels, or in dilute gases at low temperatures) where the internal angular momentum is an additional slow variable, varying on a time scale large compared to intermolecular collision times.

correlation function also leads to a "paramagnetic" correction to the transverse mass, which is also proportional to k^2. Both "magnetic" corrections are quite small, however, and have not been detected so far for ordinary fluids. The final conclusion is that (going back to coordinate space)

$$g(r, t) = mn(r, t)v(r, t) \tag{29.18}$$

is a very good approximation for practically all purposes.

The third remark (iii) concerns superfluids. In this case the longitudinal and transverse masses are different even at $k = 0$. Equation (29.16) is no longer valid because the correlation functions of $g(k)$ for small k depend on the direction of k in a discontinuous way. One has [29.5] $\varrho^l = \varrho = mn$, in accord with the longitudinal f-sum rule. On the other hand, $\varrho^t = \varrho_n$ the mass density of the so-called normal component of the two fluid model of superfluidity. The superfluid component then has mass density $\varrho_s = \varrho - \varrho_n$.

Let us now continue the discussion of (29.10–13). First of all they can be used to derive a relation between heat current and energy current density

$$j_{\varepsilon k} = (\varepsilon + P)v_k + v_i \Pi'_{ik} + j_{qk} \quad . \tag{29.19}$$

Note again the occurrence of the enthalpy density $\varepsilon + P$ instead of simply the energy density ε in the energy current density.

Looking at (29.10) we can now write down the Onsager relations between current densities and force gradients. We start with the definition of the heat conductivity κ

$$j_{qk} = -\kappa \frac{\partial T}{\partial x_k} \quad . \tag{29.20}$$

In writing down the relations for the stress tensor Π' one has to take into account the symmetry of Π. Hence it can only be related to the two symmetric parts $(\partial v_i/\partial x_k + \partial v_k/\partial x_i)$ and $\sum(\partial v_l/\partial x_l)\delta_{ik}$. Thus one can define two viscosities η and ζ by [29.6,7]

$$\Pi'_{ik} = -\eta \left(\frac{\partial v_i}{\partial x_k} + \frac{\partial v_k}{\partial x_i} \right) - \left(\sum \frac{\partial v_l}{\partial x_l} \right) \delta_{ik} \left(\frac{\zeta - 2\eta}{3} \right) \quad . \tag{29.21}$$

So far we have three kinetic coefficients. How about cross terms between j_{qk} and $\partial_k v_i$? This would mean a linear isotropic relation between a vector and a

---

The r.h.s. of (29.3) then contains the terms $\sum(v_k dg_k^0 + \omega_k dS_k) = \sum v_k dg_k + (\omega - \text{curl } v/2)dS$ (plus a divergence) and the additional term leads to a relaxation eq. $dS/dt = -\eta_r(\omega - \text{curl } v/2)$. See de Groot, S. R., Mazur, P.: *Non-equilibrium Thermodynamics*, Chap. XII, North Holland (1969), and Kreuzer, H. J.: *Nonequilibrium Thermodynamics and its Statistical Foundations*, Chap. 2, Clarendon Press, Oxford, (1981)

tensor. The only vector one can form out of a tensor gradient is the antisymmetric combination $\partial_i v_k - \partial_k v_i$. This would imply a relation $j_q = \lambda \nabla \times v$ which would be compatible with isotropy but not with reflection symmetry. (This is, by the way, the argument that could not be applied in the foregoing section.) More generally, since the stress tensor is symmetric, only the symmetric part of $\partial_k v_i$ occurs in (29.10). Furthermore, the antisymmetric part $\partial_i v_k - \partial_k v_i$ would remain for a uniform rotation of the fluid, which should not cause a heat current.

The symmetry of the stress tensor follows from the fact (see footnote 1 of this chapter) that for short-ranged intermolecular forces the torque of these forces on an arbitrary volume element $V \int (x_i \sum \partial_l \Pi_{kl} - x_k \sum \partial_l \Pi_{il}) d^3 x$ must be equal to the torque of the forces on the surface of this volume $\int \sum (x_i \Pi_{kl} - x_k \Pi_{il}) n_l d^2 a$, where n_l are the components of the unit vector along the surface normal. Using Gauss's integral theorem one then obtains

$$x_i \sum \partial_l \Pi_{kl} - x_k \sum \partial_l \Pi_{il} = \sum \partial_l (x_i \Pi_{kl} - x_k \Pi_{il}) \quad . \tag{29.22}$$

Differentiating the r.h.s. one finds indeed $\Pi_{ik} = \Pi_{ki}$.

The conservation laws (29.6–8) together with the constitutive equations (29.18, 20, 21) form a closed system of differential equations that, in principle, suffice to determine the functions $n(r,t)$, $\varepsilon(r,t)$ and $g(r,t)$. Our derivation has provided us with the full nonlinear set of hydrodynamic equations. In the following part of this chapter we want to confine ourselves to the linear regime, however. We therefore consider small deviations of all quantities from their equilibrium values n_0, ε_0, etc. In linearizing the equations one has to take into account that the equilibrium values of v and g vanish. Hence products of such quantities are already of second order.

In order not to have too many indices, we omit the labels "0" from the equilibrium quantities and indicate the first-order deviations with a "δ" if necessary. Then the linearized momentum conservation together with (29.13,16) yields

$$\dot{g} = -\text{grad}\, P + \eta \Delta v + (\zeta + \eta/3)\text{grad div}\, v \quad . \tag{29.23}$$

In isotropic systems it is always useful to decompose vectors into their longitudinal and transverse parts according to

$$a = a^l + a^t \quad \text{with} \quad \text{div}\, a^t = 0 \quad \text{and} \quad \text{curl}\, a^l = 0 \quad . \tag{29.24}$$

In Fourier space this decomposition becomes particularly simple: $a^l(k)$ is the part of $a(k)$ *parallel* to k, $a^t(k)$ the one *perpendicular* to k. Now the transverse part of (29.23) is

$$\dot{g}^t(k,t) = -\eta k^2 g^t(k,t)/mn = -D^t k^2 g^t(k,t) \quad . \tag{29.25}$$

This is nothing but a simple diffusion equation for the transverse components of the momentum density with a diffusion constant

$$D^t = \eta/mn \quad . \tag{29.26}$$

The longitudinal part of (29.23) we combine with the two other longitudinal equations (29.6,8). Introducing the notation $a^l = a^l \mathbf{k}/|\mathbf{k}|$ we obtain

$$\dot{n}(\mathbf{k},t) = -iknv^l(\mathbf{k},t), \quad (k = |\mathbf{k}|) \quad , \tag{29.27}$$

$$\dot{g}^l(\mathbf{k},t) = -ik\delta P(\mathbf{k},t) - k^2(\zeta + 4\eta/3)v^l(\mathbf{k},t) \quad , \tag{29.28}$$

$$\dot{q}(\mathbf{k},t) = -\kappa k^2 \delta T(\mathbf{k},t) \; [= T\dot{\sigma}(\mathbf{k},t)] \quad . \tag{29.29}$$

For the last equation see Problem 29.2.

From now on we can proceed in analogy to the previous chapter. We first write the longitudinal equations in matrix form in analogy to (28.11):

$$\begin{pmatrix} \dot{n}/n \\ \dot{g}^l \\ \dot{q}/T \end{pmatrix} = -\mu \begin{pmatrix} \delta P \\ v^l \\ \delta T \end{pmatrix} \tag{29.30}$$

where μ, the matrix of mobilities, is now given by

$$\mu = \begin{pmatrix} 0 & -ik & 0 \\ -ik & mn\gamma & 0 \\ 0 & 0 & \lambda/T \end{pmatrix} \tag{29.31}$$

with

$$\gamma = (\zeta + 4\eta/3)k^2/mn = D^l k^2 \tag{29.32}$$

and

$$\lambda = \kappa k^2 \quad . \tag{29.33}$$

The susceptibilities corresponding to (28.14) are now given by

$$\begin{pmatrix} (\partial n/\partial P)_T/n & (\partial n/\partial T)_P/n \\ (\partial s/\partial P)_T & (\partial s/\partial T)_P \end{pmatrix} = \begin{pmatrix} \kappa_T & \alpha \\ \alpha & c_P/T \end{pmatrix} \tag{29.34}$$

and the full susceptibility matrix analogous to (28.21) is

$$\chi = \begin{pmatrix} \kappa_T & 0 & \alpha \\ 0 & mn & 0 \\ \alpha & 0 & c_P/T \end{pmatrix} \tag{29.35}$$

and its inverse

$$\chi^{-1} = \begin{pmatrix} 1/\kappa_S & 0 & a \\ 0 & 1/mn & 0 \\ a & 0 & T/c_V \end{pmatrix} \quad . \tag{29.36}$$

The matrix $\nu = \mu/\chi$ then results in

$$\nu = \begin{pmatrix} 0 & -ik/mn & 0 \\ -ik/\kappa_S & \gamma & a \\ \lambda a & 0 & \lambda/c_V \end{pmatrix} \quad . \tag{29.37}$$

From here on the calculations go more or less parallel to the one the previous section. We leave the details to the reader and quote only a few important results:

First the vibrational frequencies are approximately given by

$$\omega_s^2 = k^2/mn\kappa_S = (c_s k)^2, \quad \text{with the attenuation}$$
$$\gamma_s = D^l k^2 + D_T k^2(c_P/c_V - 1) \tag{29.38}$$

and the thermal attenuation frequency

$$\omega_T = -i\gamma_T = -iD_T k^2, \quad \text{with} \quad D_T = \kappa/c_P \quad . \tag{29.39}$$

The diagonal relaxation function of Chap. 28 corresponds to $\Phi_{nn}(\omega)$ in our case now. It can be obtained from (28.29) by the trivial replacements $\chi_T \to \kappa_T$, $C_f \to c_P$ and $C_x \to c_V$.

Again one finds two adiabatic sound wave peaks ("Rayleigh" peaks), which are propagating with real frequencies $\omega_s = c_s k$ but slightly attenuated $\propto -ik^2$, and the Landau–Placzek peak (the thermal diffusion component of the density waves), which has a strength $(c_P/c_V - 1)$. Usually the Landau–Placzek ratio $(c_P/c_V - 1) = (\kappa_T/\kappa_S - 1)$ is small compared to one, except near the critical point of the fluid, when κ_T diverges. Then the density–density relaxation function is completely dominated by the thermal diffusion peak, see Chap. 46.

The fact that the frequencies of all three modes tend to zero when k tends to zero has the same reason as already discussed in the context of diffusion (Chap. 20): it is the existence of three differential conservation laws for particle number, momentum and energy. In diffusion and thermodiffusion one has no momentum conservation, and thus only two slow modes for $k \to 0$.

PROBLEMS

29.1 Using the continuity equation (31.20) and the commutation relations (31.22) derive (29.18) in the longitudinal case.

29.2 Generalizing the procedure of Chap. 20 for thermodiffusion derive an expression for the change δq in heat and the heat current density.

29.3 Derive the expression for the entropy production

$$\dot{S} = \int (\delta P \dot{n}/n + v^l \dot{g}^l + \delta T \dot{q}/T) d^3x / T$$

correct to second order in the small quantities of linear theory.

30. Hydrodynamic Long-Time Tails

In recent years a new kind of data has become available through the use of fast computers. Basically the idea underlying these computer "experiments" is quite simple: one solves Newton's equations for N particles in a box (N of the order of 10^3) by a finite difference numerical approximation. Hard-sphere interactions are particularly simple since their dynamics consists of straight line motions interrupted by sudden deflections obeying simple geometrical laws. Various results have been obtained this way, in particular, velocity autocorrelation functions of single particles have been determined [30.1]. These correlation functions for short times (of the order of the relaxation time) showed an exponential decrease as expected from the simple theory of Brownian motion, see (4.10). It turned out [30.1], however, that after a few relaxation times, when the correlations had decreased to a few percent of their initial value $\langle v^2 \rangle$, the decrease slowed down to a power law, which in three-dimensional space was $\propto t^{-3/2}$, see Fig. 30.1. One sees that after sufficiently many collisions the correlation functions decay away very slowly according to a power law.

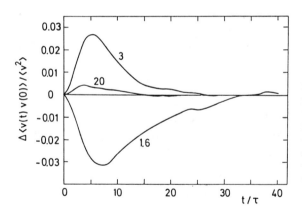

Fig. 30.1. Plot of the deviation of the velocity autocorrelation function of a hard sphere gas from exponential decay as a function of time t/τ, for three different average volumes V per particle ($V/V_0 = 1.6, 3, 20$), V_0: the volume for close packing [30.1]

We now know that this result could have been foreseen from hydrodynamics [30.2]. During the first few collision times the particle shares its initial velocity with a few particles by means of binary collisions. After this, many-body effects come into play. The simplest way to take them into account is by hydrodynamics: in the long-time limit the particle sets up hydrodynamic flow in its vicinity. Since the longitudinal modes propagate with the velocity of sound, they are fast processes which can be neglected. The remaining transverse modes propagate according to a diffusion equation (29.25) and spread over a characteristic distance $\delta = (\eta t/mn)^{1/2}$ after a time t. The number of particles sharing the initial velocity then will be of the order of $n\delta^d$ in d dimensions and the decrease of the velocity will be expected to be of the order of

$$\langle v(t)v(0)\rangle \approx \langle v(0)^2\rangle \frac{1}{n(\eta t/nm)^{d/2}} . \qquad (30.1)$$

Another hydrodynamic way of looking at the results is suggested by a problem in *Landau* and *Lifshitz*'s *Fluid Mechanics* [30.3]. They calculate the frequency-dependent corrections to Stokes' formula $F = 6\pi\eta R$ for the drag F of a hard sphere of radius R in a fluid of viscosity η. They find (apart from a term corresponding to a mass renormalization)

$$F_\omega = 6\pi\eta R[1 + (1-i)R/\delta]v_\omega \quad \text{with} \quad \delta = \sqrt{2\eta/nm\omega} . \qquad (30.2)$$

The physics behind the correction term is, of course, again the diffusion equation (29.25), whose solution at nonzero ω contains a factor $\exp(ikr)$ [with $k = (1-i)/\delta$] besides the zero-frequency dipole field, responsible for the Stokes result for $d = 3$.

The Fourier transform of (30.2) leads to [30.3]

$$F(t) = 6\pi\eta R\left(v(t) + \frac{R}{\sqrt{\pi nm\eta}}\int_{-\infty}^{t} \frac{dv}{dt'}\frac{dt'}{\sqrt{t-t'}}\right) . \qquad (30.3)$$

The equation of motion for $v(t)$ then has the form $mdv(t)/dt = F(t)$. (If the mass renormalization term were taken into account, m *in this equation* would have to be replaced by $m^* \approx 1.5m$.) Without the correction term the solution would be $v(t) = v(0)\exp(-t/\tau)$ with the relaxation time

$$\tau = m/6\pi\eta R . \qquad (30.4)$$

The correction term for large t can be handled by power expansion in t'.

Since the $v(t')$ in the velocity relaxation function [which obeys the same equation of motion as $v(t)$] jumps discontinuously from 0 to $v(0)$ at $t' = 0$, one has

$$\int \frac{dv}{dt'}dt' = 0 \quad \text{and} \quad \int t'\frac{dv}{dt'}dt' = -v(0)\tau . \qquad (30.5)$$

Both integrals for large t can be extended from minus to plus infinity and for the evaluation of the second integral we have neglected the long-time tail. Inserting everything into (30.3) leads to

$$F(t) = 6\pi\eta R\left(v(t) - \frac{Rv(0)\tau}{\sqrt{(\pi mn\eta)}}t^{-3/2}\right) \qquad (30.6)$$

for large t and $d = 3$. Since for large t the acceleration term is $\propto t^{-5/2}$ it can be neglected and $v(t)$ is determined by $F(t) = 0$. Using (30.6,4) we find

$$\langle v(t)v(0)\rangle = \frac{2\langle v(0)^2\rangle}{3n(4\pi\eta t/nm)^{3/2}} \quad ; \quad d = 3 \ . \tag{30.7}$$

This fixes the undetermined factors of order 1 in (30.1).

If (30.2) is expressed in terms of a frequency-dependent diffusion coefficient $D(\omega)$, one obtains a nonanalytic behaviour

$$D(\omega) - D(0) \propto \sqrt{i\omega} \tag{30.8}$$

after Fourier transformation this frequency dependence corresponds to memory effects, which are expressed explicitly in (30.3). The memory effects indicate that the velocity of the particle is not the only slow variable in the game: one has to take into account the hydrodynamic transverse modes as additional slow variables. If one considers other hydrodynamic correlation functions (for instance the energy density and momentum flux) one finds similar low-frequency anomalies [30.4].

Another interesting result is the density dependence of the long-time tails (30.1), which is also nonanalytic. It would be missed in a simple expansion of transport coefficients in powers of the density n. Since n introduces a characteristic length (the inter-particle distance $n^{-1/d}$) it is plausible that also the k dependence of generalized transport coefficients, for instance $D(k,\omega)$, contains nonanalytic terms.

A serious difficulty arises for $d = 2$. Equation (30.1) indicates long-time tails proportional to $1/t$. This has the consequence that the time integral of the velocity correlation function (which is the zero-frequency self-diffusion constant) diverges logarithmically: $D(\omega)$ is proportional to $\ln \omega$ for small ω. Since similar results hold for other hydrodynamic diffusion coefficients one has the paradox that the assumption of the existence of a diffusive mode leads to the result that the diffusion coefficient does not exist. This may be called *hydrodynamic suicide*.

PROBLEMS

30.1 The solution of the diffusion equation (29.25) in k-space reads

$$v(\boldsymbol{k}, t) = v(\boldsymbol{k}, 0)\exp(-Dk^2 t) \ .$$

If the initial situation describes a velocity field localized in a small region in space near the origin $r = 0$, $v(\boldsymbol{k}, 0)$ is only weakly dependent on $\boldsymbol{k}$. Determine the resulting velocity field $v(\boldsymbol{r} = 0, t)$ for an arbitrary dimension d by Fourier back transformation.

30.2 Read [Ref. 3, Sect. 24], in particular Problems 5 and 6, to check (30.2,3).

31. Matter in Electromagnetic Fields

Electromagnetic fields and currents are coupled systems. The electromagnetic field influences the motion of charged particles, and they produce the electromagnetic fields. The structure of the system is this: the Hamiltonian may be (neglecting all details and constants) symbolically written as $H = H_{00} - j \cdot A$. Here H_{00} is the "bare" unperturbed Hamiltonian (without electromagnetic interactions between the particles), and j and A are the four-vectors of current and vector potential.

A possible method of solution (avoiding the radiative correction problems of quantum field theory) is to start out from the solution of the wave equation for A: $A = A^e + Gj$. Here A^e, the solution of the homogeneous equation, is some "external" field. G is the Green's function of the wave equation. Inserting this into H one may write $H = H_0 - j \cdot A^e$, where H_0 is now the "dressed" unperturbed Hamiltonian (including the electromagnetic interactions between the particles, in particular the Coulomb interaction). Then one (i) solves for j to first order in A^e: $j = j_0 + \chi A^e + \cdots$. However, one usually wants to know j in terms of A. Thus (ii) one inserts the j into the solution of the wave equation and solves for A^e in terms of A.

In this chapter we want to consider the last two steps of the procedure. We begin with the decomposition of the four-vector $A = (A_\mu) = (\mathbf{A}, -\varphi)$,

$$A = A^e + A^j = A^e + A_0 + A_1 \quad . \tag{31.1}$$

Here A is the total field, A^e the external one, and A^j the internal field, produced by the current density j with the equilibrium contribution A_0 and the contribution A_1 induced by the external field to first order.

As customary in nonrelativistic quantum mechanics, we use the Coulomb gauge

$$\text{div } \mathbf{A} = 0 \tag{31.2}$$

[valid also for all contributions in (31.1) separately]. The field equations are particularly simple if the fields are decomposed into their longitudinal and tranverse parts, for instance $\mathbf{B} = \mathbf{B}^t$, $\mathbf{A} = \mathbf{A}^t$, $\mathbf{E} = \mathbf{E}^l + \mathbf{E}^t$, $\mathbf{j} = \mathbf{j}^l + \mathbf{j}^t$. Maxwell's equations then take the form

$$\left(\Delta - \frac{1}{c^2}\frac{\partial^2}{\partial t^2}\right) \mathbf{A}^j = -\frac{4\pi e}{c}\mathbf{j}^t \quad , \quad \text{and} \quad \Delta \varphi^j = -4\pi e \varrho \tag{31.3}$$

where
$$j = (j_\mu) = (\boldsymbol{j}, c\varrho) \tag{31.4}$$

are the components of the *particle* current density and *particle* density.
The external field is not produced by j [it is the homogeneous part of the solution of (31.3)]. Often one considers, however, external sources j^e which are the sources of the external field.

The electromagnetic fields E and B result from the potentials A in the usual way

$$\boldsymbol{B} = \operatorname{curl} \boldsymbol{A} \quad , \quad \boldsymbol{E}^l = -\operatorname{grad} \varphi \quad , \quad \boldsymbol{E}^t = -\dot{\boldsymbol{A}}/c \quad . \tag{31.5}$$

The nonrelativistic expressions for j are

$$\varrho(r) = \sum \delta(\boldsymbol{r} - \boldsymbol{r}_i) \quad , \tag{31.6}$$

$$\boldsymbol{j}(\boldsymbol{r}) = \frac{1}{2} \sum [\boldsymbol{v}_i \delta(\boldsymbol{r} - \boldsymbol{r}_i) + \delta(\boldsymbol{r} - \boldsymbol{r}_i)\boldsymbol{v}_i] + \frac{g}{m} \operatorname{curl} \boldsymbol{s}(\boldsymbol{r}) \tag{31.7}$$

and the velocity $\boldsymbol{v}$ is connected with the canonical momentum $\boldsymbol{p}$ by

$$m\boldsymbol{v}_i = \boldsymbol{p}_i - e\boldsymbol{A}^j(\boldsymbol{r}_i)/c \quad . \tag{31.8}$$

Here $\boldsymbol{s}(\boldsymbol{r})$ is the spin density

$$\boldsymbol{s}(\boldsymbol{r}) = \sum \boldsymbol{s}_i \delta(\boldsymbol{r} - \boldsymbol{r}_i) \quad . \tag{31.9}$$

We have written down only the contributions of the electrons. In general there are also those of the nuclei. For electrons the Landé factor g has in very good approximation the value $g = 2$.

For the average of j we write

$$\langle j \rangle = (\langle \boldsymbol{j} \rangle, cn) \quad . \tag{31.10}$$

Since in (31.8) we have only taken into account the internal part of the vector potential, the average *induced* current is given to first order in A^e by

$$J_1 = (\langle \boldsymbol{j} \rangle_1 - en_0 \boldsymbol{A}^e/mc, n_1) \quad . \tag{31.11}$$

For the Hamiltonian of matter in an external field we now write to first order

$$H = H_0 - \frac{e}{c} \int \sum j_\mu A_\mu^e d^3r \quad . \tag{31.12}$$

In second order it also contains the diamagnetic term $(e^2/2mc^2) \int \varrho(A^e)^2 d^3r$, which in linear response theory can be neglected. The term H_0 is given by

$$H_0 = T + V \quad . \tag{31.13}$$

Here V includes all internal interactions, including relativistic corrections, non-electromagnetic and field-mediated interactions, magnetic and nonmagnetic interactions, in particular the electron–electron Coulomb interaction

$$V_{ee} = \frac{1}{2} \sum_{i \neq k} \frac{e^2}{r_{ik}} \quad . \tag{31.14}$$

Within linear response theory one finds the linear relation (3.5)

$$e\langle j_\mu(x)\rangle_1 = \int \chi_{\mu\nu}(x,x') A_\nu^e(x') d^4x' \tag{31.15}$$

where $x = (\boldsymbol{r}, t)$ and

$$\chi_{\mu\nu}(x,x') = i \frac{e^2}{\hbar c} \langle [j_\mu(x), j_\nu(x')] \rangle \Theta(t - t') \tag{31.16}$$

with the Fourier transformation

$$\chi_{\mu\nu}(\boldsymbol{r}, \boldsymbol{r}', \omega) = i \frac{e^2}{\hbar c} \int_0^\infty \langle [j_\mu(\boldsymbol{r}, t), j_\nu(\boldsymbol{r}', 0)] \rangle e^{i\omega t} dt \quad . \tag{31.17}$$

We now pick up only the spatial part, $\mu = m$ ($m = 1, 2, 3$),

$$e\langle j_m \rangle_1 = \int \chi_{mn}(x, x') A_n^e(x') d^4x' - \int \chi_m(x, x') \Phi^e(x') d^4x' \tag{31.18}$$

with

$$\chi_m(x, x') = i \frac{e^2}{\hbar} \langle [j_m(x), \varrho(x')] \rangle \Theta(t - t') \quad . \tag{31.19}$$

We now use the differential conservation law of charge

$$\sum \partial_n j_n(x') + \partial_{t'} \varrho(x') = 0 \quad , \tag{31.20}$$

which, applied to (31.19), yields

$$-\partial_{t'} \chi_m = c \partial_n \chi_{mn} + i \frac{e^2}{\hbar} \langle [j_m(x), \varrho(x')] \rangle \delta(t - t') \quad . \tag{31.21}$$

Now from (31.6,7) and the commutation relations between coordinate and momentum one finds the commutation relation

$$[j_m(r), \varrho(r')] = (\hbar/im) \varrho(r) \partial_m \delta(r - r') \quad . \tag{31.22}$$

Using this, the commutator on the r.h.s. of (31.21) can be evaluated: because of the delta function it also only involves equal times $t = t'$.

Finally one collects everything together and adds the diamagnetic term according to (31.11) to $\boldsymbol{j}$. Then, after Fourier transformation, using (31.4) one finds after several intermediate steps the amazingly simple result

$$\boxed{eJ_{1m}(\boldsymbol{r},\omega) = \int \sum \sigma^{e}_{mn}(\boldsymbol{r},\boldsymbol{r}',\omega)E^{e}_{n}(\boldsymbol{r}')d^{3}r'} \quad , \tag{31.23}$$

which is a generalized Ohm's law with a frequency-dependent (nonlocal) conductivity

$$\boxed{\begin{aligned}\sigma^{e}_{mn} &= \frac{e^{2}}{\hbar\omega}\int_{0}^{\infty}\langle[j_{m}(r,t),j_{n}(r',0)]\rangle e^{i\omega t}dt \\ &\quad + \frac{ie^{2}}{m\omega}n_{0}(r)\delta_{mn}\delta(r-r')\end{aligned}} \quad . \tag{31.24}$$

This is indeed the *full* content of (31.15); it is simpler (it depends only on E^{e}) and has the advantage of being independent of the choice of gauge. The fourth component of (31.15) is, indeed, also contained in (31.23). One obtains it by making use of the continuity equation (which is also valid for J_{1} separately) in Fourier space

$$n_{1}(\boldsymbol{r},\omega) = \sum \partial_{m}J_{1m}(\boldsymbol{r},\omega)/(i\omega) \quad . \tag{31.25}$$

Now we come to the second step of our whole procedure: As mentioned already at the beginning of this chapter, in electrodynamics one usually wants relations between currents and total fields instead of the external field, so one has to use Maxwell's equations to solve for the total fields. The solution is somewhat complicated in general. We therefore restrict ourselves to the homogeneous, isotropic case. Then everything can be decomposed again into transverse and longitudinal parts:

$$eJ_{1}^{t} = \sigma^{te}E^{te} \quad \text{and} \quad eJ_{1}^{l} = \sigma^{le}E^{le} \quad . \tag{31.26}$$

We now define a conductivity σ by $\sigma E = \sigma^{e}E^{e}$, and correspondingly

$$eJ_{1}^{t} = \sigma^{t}E^{t} \quad \text{and} \quad eJ_{1}^{l} = \sigma^{l}E^{l} \quad . \tag{31.27}$$

Maxwell's equations become simpler when one uses dielectric functions ε instead of conductivities. The usual relation between them is

$$\varepsilon = 1 + 4\pi i\sigma/\omega \quad \text{(valid for both ``}l\text{'' and ``}t\text{'')} \quad . \tag{31.28}$$

Then in Fourier space the Maxwell equations take the form

$$[\varepsilon^{t}(k,\omega) - (ck/\omega)^{2}]E^{t}(\boldsymbol{k},\omega) = [1 - (ck/\omega)^{2}]E^{te} \tag{31.29}$$

and

$$\varepsilon^{l}(k,\omega)E^{l}(\boldsymbol{k},\omega) = E^{le}(\boldsymbol{k},\omega) \quad . \tag{31.30}$$

In the last equation we have already divided out the factor k^2.

Going back to (31.26,27) one finds the relations

$$\sigma^{te}[\varepsilon^t - (ck/\omega)^2] = \sigma^t[1 - (ck/\omega)^2] \tag{31.31}$$

and

$$\sigma^{le}\varepsilon^l = \sigma^l \quad. \tag{31.32}$$

Taking into account (31.28) one obtains the solution

$$\sigma^t = \sigma^{te}\left(1 - \frac{4\pi i \sigma^{te}/\omega}{\omega^2 - c^2 k^2}\right)^{-1} \tag{31.33}$$

and

$$\sigma^l = \frac{\sigma^{le}}{1 - 4\pi i \sigma^{le}/\omega} \quad. \tag{31.34}$$

From Maxwell's equations (31.29,30) one can easily read off the conditions for eigenvibrations of the system ($E \neq 0$ but $E^e = 0$), namely $\varepsilon^l = 0$ and $\varepsilon = (ck/\omega)^2$. Or, in more detail, the eigenfrequencies $\omega(k)$ obey the equations

$$[\omega^t(k)]^2 \varepsilon^t(k, \omega^t(k)) = c^2 k^2 \tag{31.35}$$

and

$$\varepsilon^l(k, \omega^l(k)) = 0 \quad. \tag{31.36}$$

The total average field E is the sum of the external and the average internal field. The average internal field is nothing but the Hartree field in the Hartree approximation for the electrons. Thus in the Hartree approximation, σ describes the response of the bare Hamiltonian H_{00} (without any Coulomb interaction) to the *total* field E. In the Hartree–Fock approximation, H_{00} would include the (short-ranged) exchange interactions. Naturally it is easier to calculate the reponse σ of a bare Hamiltonian to the total field than σ^e of the dressed Hamiltonian to the external field and afterwards solve for σ.

The calculation of dielectric functions (conductivities, susceptibilities, etc.) is one of the main tasks of solid-state theory. Here we content ourselves with a few examples of simple models. They are intended to demonstrate that, indeed, not only dielectric, but *all* electromagnetic properties of matter can be expressed in terms of dielectric functions.

We start with a dielectric substance with a polarizibility χ_e. Then one has

$$\varepsilon = 1 + 4\pi \chi_e \quad \text{(dielectric substance)} \quad. \tag{31.37}$$

On the other hand, for a conductor with conductivity σ one has

$$\varepsilon = 1 + 4\pi i \sigma/\omega \quad \text{(Ohmic conductor)} \quad. \tag{31.38}$$

Magnetic properties, too, can be described in terms of ε. Suppose one has a magnetically polarizable substance with the susceptibility χ_m, then in a field

B it has a magnetization
$$M = \chi_{\mathrm{m}} B \quad . \tag{31.39}$$
The magnetization current corresponding to this magnetization is
$$e\boldsymbol{J} = c\,\mathrm{curl}\,\boldsymbol{M} = c\chi_{\mathrm{m}}\mathrm{curl}\,\boldsymbol{B} \quad . \tag{31.40}$$
Now B can be expressed in terms of the transverse electric field by means of Faraday's law of induction: $\boldsymbol{B} = -ic\,\mathrm{curl}\,\boldsymbol{E}^t/\omega$. Inserting all this, one obtains a dielectric function
$$\varepsilon^t = 1 + 4\pi(ck/\omega)^2\chi_{\mathrm{m}} \quad \text{(magnetizable substance)} \quad . \tag{31.41}$$

Another interesting example is a superconductor obeying London's equation $e\boldsymbol{J} = n_s e^2 \boldsymbol{A}/mc$, where n_s is the density of superconducting electrons. If A is expressed in terms of the transverse E, via (31.5), one finds
$$\varepsilon^t = 1 - 4\pi e^2 n_s/m\omega^2 \quad \text{(London superconductor)} \quad . \tag{31.42}$$
This may be interpreted, via (31.38), as $\sigma = e^2 n_s/\omega$ (perfect conduction) or, via (31.41), as $\chi_{\mathrm{m}} = -e^2 n_s/mc^2k^2$ (perfect diamagnetism).

In a dielectric the dielectric function approaches a constant for ω and k approaching zero, in a conductor it diverges as $1/\omega$, in a magnetizable substance $\varepsilon - 1$ behaves as $(k/\omega)^2$. Currents varying in time may be interpreted as either polarization currents or conduction currents; if they vary in space and time they may also be interpreted as magnetization currents.

The constants χ_e, σ, χ_{m}, n_s, occurring in (31.37,38,41,42) will in general depend on k and ω. They will, however, usually approach nonzero constants for zero k and ω.

If in Maxwell's equations (31.3) the current j is replaced by the average current J_1 of linear response theory, a closed equation for an average A results. According to Onsager's regression theorem a corresponding equation is valid for the correlation functions of the fields A. In principle, one can base the whole procedure on such equations for the field correlation functions [31.1,2].

PROBLEMS

31.1 Write the relations (31.33,34) in terms of the functions ε and χ instead of σ.

31.2 Determine, either by using the appropriate formulae from above or directly from linear response theory, the response of the density $n_1(x)$ to a change in the chemical potential $\delta\mu(x') = -e\varphi^e(x')$. Prove:
   a) The susceptibility is given by $\chi^e(x, x') = i\langle[\varrho(x), \varrho(x')]\rangle\Theta(t - t')/\hbar$.
   b) The connection between χ^e and σ^e is given by $e^2\chi^e = ik^2\sigma^e/\omega$.
   c) If one introduces by $e^2\chi = ik^2\sigma/\omega$ the response to the total field φ, then χ and χ^e are related by
$$\chi^e(k,\omega) = \chi(k,\omega)/[1 + 4\pi e^2\chi(k,\omega)/k^2] \quad .$$

32. Rate Equations (Master Equation, Stosszahlansatz)

In this chapter we discuss several kinds of rate equations: equations which govern the time evolution of particle numbers. We have already encountered such equations in the context of chemical reactions. In that case we had a few equations for different sorts of particles. Here we want to consider only one sort of particle but concentrate on the occupation numbers of (in principle infinitely many) different quantum states of this particle. We call the quantum numbers of these states p because later on we want to identify them with the momenta (in some periodicity volume). But for the time being the meaning of p may be arbitrary.

The variables which we are going to deal with may be called n_p. As in equilibrium theory, such occupation numbers play an important role in the description of elementary excitations in condensed matter at low temperatures (electrons, phonons, magnons, etc., in solids, as well as others in liquid ^{3}He and liquid He II) and in dilute gases and plasmas at higher temperatures. In equilibrium these occupation numbers have their well-known form for fermions and bosons [Ref. 32.1, Eqs. (37.9,17)]. Deviations from equilibrium may be described by keeping these forms but shifting the energies ε_p by an amount, say $-\varphi_p$. Then one has

$$n_p = \frac{1}{e^{\beta(\varepsilon_p - \mu - \varphi_p)} \pm 1} \quad ; \quad \begin{cases} + : & \text{fermions}, \\ - : & \text{bosons} \end{cases} \quad (32.1)$$

Instead of shifting the energy, one may just as well say one has a p-dependent chemical potential $\mu_p = \mu + \varphi_p$. [However, one should not mix this up with the k dependence of μ in hydrodynamics, for example, where one deals with the Fourier transform of $\mu(r)$.]

For the change in energy one has

$$dE = TdS + \sum (\varepsilon_p - \mu - \varphi_p) dn_p \quad . \quad (32.2)$$

For a closed system with $dE = dN = \sum dn_p = 0$ one finds

$$TdS = -\sum \varphi_p dn_p \quad . \quad (32.3)$$

Hence Onsager's prescription would lead to (calling $\gamma_{pq} = L_{pq}/T$)

$$\dot{n}_p = -\sum \gamma_{pq} \varphi_q = r_p(\ldots, \varphi_q, \ldots) \quad (32.4)$$

as the differential equation for the time evolution of $n_p(t)$.

We now want to discuss several examples for the "rates" r_p (dimension 1/time). We had some examples in Chap. 22 on chemical reactions. They were of the form of so-called collision rates. Historically the first of such forms was Boltzmann's "Stosszahlansatz" [32.2]. We plan to start somewhat more generally from (32.4), but, in agreement with Boltzmann, write the rate as a function of the occupation numbers n_p instead of the generalized "forces" φ_p.

$$\dot{n}_p = r_p(\cdots, n_q, \cdots) \ . \tag{32.5}$$

The first condition which the rates have to fulfill [looking at (32.4)] is that they vanish at equilibrium $\varphi_p = 0$. Since (32.5) does not obey this condition automatically, one has to impose it on the rates:

$$r_p(\cdots, n_q^0, \cdots) = 0 \ . \tag{32.6}$$

Next we consider small deviations from equilibrium and expand in the small quantities $\delta n_q = n_q - n_q^0$:

$$r_p = \sum (\partial r_p / \partial n_q)_0 \delta n_q \ . \tag{32.7}$$

The connection to the φ description again goes via the susceptibility

$$\delta n_q = \chi_q \varphi_q \tag{32.8}$$

with, see (32.1),

$$\chi_q = (\partial n_q / \partial \varphi_q)_0 = -\partial n_q^0 / \partial \varepsilon_q = \beta n_q^0 (1 \pm n_q^0) \ . \tag{32.9}$$

So one may also write (32.8) as (32.4) with

$$\gamma_{pq} = -(\partial r_p / \partial n_q)_0 \chi_q \ . \tag{32.10}$$

Coming back to (32.7), this leads to a linear equation for (32.5):

$$\dot{n}_p = \sum r_{pq} \delta n_q \tag{32.11}$$

Such an equation occurs for instance for the scattering of particles (or quasi-particles) by fixed centres (electrons by impurities) or by the thermal excitations of a "heat bath" (phonons, magnons, etc.), which is considered as inert and always in equilibrium. Linear rate equations such as (32.11) are often called *master equations*. The r_{pq} are *transition rates* (transition probabilities per unit time).

The form of the r_{pq} is further restricted by symmetries and conservation laws. The most important conservation laws are, of course,

$$\sum_p \dot{n}_p = -\sum_{p,q} \gamma_{pq} \varphi_q = \sum_{p,q} r_{pq} \delta n_q = 0 \text{ (particle number)}, \tag{32.12}$$

$$\sum_p p \dot{n}_p = -\sum_{p,q} p \gamma_{pq} \varphi_q = \sum_{p,q} p r_{pq} \delta n_q = 0 \text{ (momentum)}, \tag{32.13}$$

$$\sum_p \varepsilon_p \dot{n}_p = -\sum_{p,q} \varepsilon_p \gamma_{pq} \varphi_q = \sum_{p,q} \varepsilon_p r_{pq} \delta n_q = 0 \text{ (energy)} \ . \tag{32.14}$$

Since the r.h.s. have to vanish for arbitrary δn_q this requires

$$\sum_p \gamma_{pq} = \sum_p r_{pq} = 0 , \qquad (32.12a)$$

$$\sum_p p\gamma_{pq} = \sum_p p r_{pq} = 0 , \qquad (32.13a)$$

$$\sum_p \varepsilon \gamma_{pq} = \sum_p \varepsilon r_{pq} = 0 . \qquad (32.14a)$$

One should keep in mind that none of these conservation laws really has to hold in general. For bosons (e.g. phonons or photons) the particle number is not conserved. For energy exchange with a heat bath the energy is nonconserved, and for scattering by fixed impurities the momentum changes. So which of the conservation laws (32.12–14) are valid depends on the circumstances.

Let us assume that particle number is conserved. Then, because of (32.12a), we can express the diagonal elements γ_{pp} or r_{pp} in terms of the off-diagonal elements

$$\gamma_{pp} = -\sum_{q \neq p} \gamma_{qp} \quad \text{and} \quad r_{pp} = -\sum_{q \neq p} r_{qp} . \qquad (32.15)$$

Inserting this into (32.11) yields

$$\boxed{\dot{n}_p = -\sum_q (\gamma_{pq}\varphi_q - \gamma_{qp}\varphi_p) = \sum_q (r_{pq}n_q - r_{qp}n_p) ,} \qquad (32.16)$$

where the δn has been replaced by n and the restriction ($\neq p$) has been omitted, since the additional terms occurring this way cancel out anyway.

The first term on the r.h.s. of these equations may be interpreted as describing "gain" processes by transitions from other states into p; the second term describes "loss" processes from p to all other states. The factors n on the r.h.s. take care of the fact that all these processes can happen only if the *initial* state is occupied. In dense gases and liquids the occupation numbers of the *final* states occur too. For instance, in Fermi gases and liquids processes can occur only if the final state is *unoccupied*. In general one has in dense systems instead of (32.16) the equations

$$\dot{n}_p = \sum_q [r_{pq} n_q (1 \pm n_p) - r_{qp} n_p (1 \pm n_q)] ; \quad \begin{cases} + : & \text{bosons}, \\ - : & \text{fermions} \end{cases} . \qquad (32.17)$$

It is interesting to write down the equilibrium condition in this case explicitly:

$$\sum_q \left[r_{pq} n_q^0 (1 \pm n_p^0) - r_{qp} n_p^0 (1 \pm n_q^0) \right] = 0 . \qquad (32.18)$$

If this relation holds for each term of the sum separately one speaks of *detailed balance*. Indeed this relation is closely related to the original detailed balance relation (8.2), see Chap. 9. Using (32.9,10) and (32.1) it can be simplified to

$$\gamma_{pq} = \gamma_{qp} \quad \text{and} \quad r_{pq}e^{-\beta\varepsilon_q} = r_{qp}e^{-\beta\varepsilon_p} \quad . \tag{32.19}$$

Detailed balance is valid under three conditions: 1. (Trivially:) If the sum in (32.17) contains only one term. (Einstein used this in his derivation of Planck's formula for a two-level system in a radiation field.) 2. In first-order perturbation theory for the transition probabilities (Problem 32.2). 3. If space and time reversal invariance hold (as we shall see immediately). Since this is true almost always, one can say that detailed balance is valid under very general circumstances. Note, however, that the somewhat weaker equilibrium condition (32.18) is valid always, even in solids without inversion symmetry and for strong interactions.

For the sake of completeness we mention that, if detailed balance holds, using (32.1) one can cast the rate equation (32.17) into the more symmetric form

$$\dot{n}_p = \sum r_{pq} e^{\beta(\varepsilon_p - \mu)} n_p n_q (e^{-\beta\varphi_p} - e^{-\beta\varphi_q}) \quad . \tag{32.20}$$

Now let us come to the inversion symmetries and the proof of detailed balance. It is simpler to start with the γ_{pq} since they are the usual Onsager coefficients ("mobilities") and thus have the symmetry properties under time reversal invariance

$$\gamma_{p,q} = \gamma_{-q,-p} \quad . \tag{32.21}$$

If in addition space inversion symmetry holds, one has also

$$\gamma_{p,q} = \gamma_{-p,-q} \quad \text{and} \quad r_{p,q} = r_{-p,-q} \quad . \tag{32.22}$$

Combining both symmetries, one indeed finds

$$\gamma_{pq} = \gamma_{qp} \quad , \tag{32.23}$$

which is the first part of (32.19). The second part follows from (32.17,20).

Let us now discuss several ideas of kinetic gas theory. First the *lifetime of the state p*, τ_p. It occurs for a situation when all δn_q are zero except a single one for $q = p$. Then (32.11) becomes

$$\delta\dot{n}_p = r_{pp}\delta n_p = -\delta n_p/\tau_p \tag{32.24}$$

describing an exponential decay with a lifetime τ_p given by

$$\frac{1}{\tau_p} = -r_{pp} = \sum_{q \neq p} r_{qp} \quad , \tag{32.25}$$

the second equality sign being valid if one has particle number conservation. In general one has in the linear regime

$$\frac{1}{\tau_p} = \frac{\gamma_{pp}}{\chi_p} \quad . \tag{32.26}$$

Next let us remind the reader of the connections between transition rates and scattering cross sections. First we consider first-order perturbation theory for both, and assume an uncorrelated random distribution of scatterers with potentials $w(\bm{r} - \bm{r}_n)$. Then Fermi's "golden rule" yields

$$r_{pq} = 2\pi n_i |w(\bm{p} - \bm{q})|^2 \delta(\varepsilon_p - \varepsilon_q)/(\hbar V) \quad . \tag{32.27}$$

Here n_i is the density of the random scatterers at the positions $\bm{r}_n$, $w(\bm{k})$ is the Fourier transform of the potentials $w(\bm{r})$ (in contrast to the rest of this section, in the context of this formula we indicate three-dimensional vector characters explicitly), and V is the volume of the system.

Now we insert this into (32.25) and use the relation

$$\sigma(\bm{p} - \bm{q}) = m^2 |w(\bm{p} - \bm{q})|^2/(2\pi\hbar^2)^2 \tag{32.28}$$

between scattering cross section and potential valid in the Born approximation. We also convert the p-sums into integrals. Then (32.25) takes the form

$$\frac{1}{\tau_p} = \frac{n_i \sigma_t(p)|p|}{m} \tag{32.29}$$

with the total cross section $\sigma_t(p) = \int \sigma(\bm{p} - \bm{q}) d\Omega_q$.

Equation (32.28) suggests a simple generalization of the perturbation theoretical result (32.27): use (32.29) but replace the Born approximation cross section by the exact one. This corresponds to the neglect of *multiple scattering* (scattering of the scattered wave of *one* centre by *another one*). This approximation is often also called the "impulse" approximation.

In addition to the lifetime τ_p, one may also introduce the *mean free path* $\ell_p = |p|\tau_p/m$. Then (32.29) takes the simple form

$$1/\ell_p = n_i \sigma_t(p) \quad . \tag{32.30}$$

The solution of (32.11) in general requires numerical work. One can, however, make some general statements concerning the form of the result. Incidentally the method used here can be applied to practically all linear kinetic equations. In our case we have the linearized rate equation

$$\chi_p \dot{\varphi}_p = -\sum \gamma_{pq} \varphi_q \quad . \tag{32.31}$$

The idea is to expand φ_p in terms of the eigenvectors $\varphi_{p\mu}$ of the eigenvalue equation

$$\sum_q \gamma_{pq} \varphi_{q\mu} = \gamma_\mu \chi_p \varphi_{p\mu} \quad . \tag{32.32}$$

The γ_μ are then the eigenvalues of the (symmetric) matrix $\chi^{-1/2} \gamma \chi^{-1/2}$. The eigenvectors $\varphi_{p\mu}$, because of (32.32), can be normalized such that

$$\sum \varphi^*_{p\mu} \chi_p \varphi_{p\nu} = \delta_{\mu\nu} \quad . \tag{32.33}$$

The expansion of an arbitrary vector $\varphi_p(t)$ in terms of the eigenvectors then can be written as

$$\varphi_p(t) = \sum_\mu c_\mu(t) \varphi_{p\mu} \quad . \tag{32.34}$$

The expansion coefficients c_μ can be obtained by making use of the orthogonality relations (32.33), multiplying (32.34) by $\varphi^*_{p\mu} \chi_p$ and summing over p. One finds using (32.33), for instance at $t = 0$,

$$c_\mu(0) = \sum \varphi^*_{p\mu} \chi_p \varphi_p(0) \quad . \tag{32.35}$$

Insertion into (32.34) shows that the completeness relation reads

$$\sum_\mu \varphi_{p\mu} \chi_q \varphi^*_{q\mu} = \delta_{pq} \quad . \tag{32.36}$$

Now, because of (32.32), the eigenvectors have an exponential time dependence

$$c_\mu(t) = c_\mu(0) e^{-\gamma_\mu t} \quad . \tag{32.37}$$

Collecting everything together, $\varphi_p(t)$ has the form

$$\varphi_p(t) = \sum \varphi_{p\mu} \chi_q \varphi^*_{q\mu} e^{-\gamma_\mu t} \varphi_p(0) \quad . \tag{32.38}$$

The solution is a sum of exponentials.

The method is familiar from the solution of the time-dependent Schrödinger equation by expansion in terms of its eigenfunctions with the exception of the occurrence of χ. The eigenvalue problem in kinetic theory is for the matrix (mobility matrix)/(susceptibility matrix). Both matrices are symmetric and the second one is positive. This then defines a Hermitian eigenvalue problem.

PROBLEMS

32.1 Discuss and plot the susceptibilities (32.9) $\chi_q = \chi(\varepsilon_q, T)$ as a function of ε_p with T as the parameter. Consider in particular the Fermi case for low T.

32.2 Prove the validity of detailed balance for first-order perturbation theory. Use Fermi's golden rule for the transition rates of interaction with a heat bath. Average over the initial states of the bath with weights $\propto e^{-\beta E}$.

33. Kinetic Transport Equations

Transport equations in the widest sense describe the transport of physical properties (mass, momentum, energy, heat, magnetism, charge, etc.) in space. Diffusion theory and hydrodynamics use transport equations in this sense. In a more restricted sense they describe transport in *phase space*, i.e. (p, r)-space. We are going to use the notation in this more restricted sense. In the heading of this chapter we have indicated this by the term kinetic.

Kinetic transport equations in some sense form a microscopic basis for the more macroscopic equations of, say, hydrodynamics. In fact it will turn out that hydrodynamic equations are the low-frequency, long-wavelength limit of transport equations in phase space.

Nevertheless, we have taken transport equations into account in the second, phenomenological part of this volume, since we are going to treat the collision term in our transport equations more or less as a phenomenological parameter. The actual microscopic theory in Part III then has to deal with the calculation of this parameter.

The prototype of transport equations is the Boltzmann equation [33.1]. It has proved to be very flexible and far reaching. It was originally invented for the treatment of dilute gases, but with only slight modifications turned out to be applicable also to plasmas [33.2], electrons in metals [33.3], phonons in solids [33.4] and other elementary excitations in so-called quantum liquids [33.5,6] (liquid ^{3}He and ^{4}He, and nuclear matter).

Although in principle the Boltzmann equation can be derived from Onsager's prescriptions or the expressions (13.14–16), see Chaps. 52, 53, we have most of the ingredients already in foregoing chapters. We therefore adopt a more heuristic approach. We first define the key quantity of the theory, the *phase space density* $n(p, r, t)$. Multiplied by the volume element $d^3p\, d^3r$ it is the number of particles in this volume element. Hence various other densities in coordinate space can be derived from it, for instance

$$n(r, t) = \int n(p, r, t) d^3p \quad \text{(particle density)} , \quad (33.1)$$

$$g(r, t) = \int p\, n(p, r, t) d^3p \quad \text{(momentum density)} , \quad (33.2)$$

$$\varepsilon_k(r, t) = \int \varepsilon_k(p) n(p, r, t) d^3p \quad \text{(kinetic energy density)} . \quad (33.3)$$

We are going to adopt a pragmatic point of view, taking the justification for using a description in terms of a phase space density with one coordinate and one momentum variable for a many-particle system mainly from its success. But let us briefly consider its validity. The quantum object whose average is the phase space density can be written as $\psi^*(r)\psi(p)\exp(ip\cdot r/\hbar)$ [Ref. 33.7, Sect. 23], where $\psi^*(r)$ is the field operator creating a particle at r, and $\psi(p)$ the operator annihilating a particle with momentum p. Now, while (33.1–3) are exact, this would be no longer be true for the p integrals of products of functions of p and r. Then the noncommutativity of p and r components becomes important. Hence one uses the phase space description usually only in the semiclassical approximation. It turns out that only then can one guarantee that the phase space density is positive, as it should be in a classical description [33.8].

There exists an almost infinite amount of work on the derivation of the Boltzmann equation. The main idea behind the validity of this equation is that in a dilute system the phase space density of the many-particle system, see Chap. 2, Eq. (2.31), apart for some short time intervals in some small regions of phase space when particles undergo collisions, can be factorized into a product of single-particle phase space densities $\prod n(p_i, r_i, t)$. We shall not go into the details of such derivations [33.9,10] but, as already stated, take a more heuristic point of view.

The time derivative of n will contain a reversible and an irreversible part. For the reversible one we just take the single particle form of the Liouville equation (2.29). The irreversible one has to describe the effect of the collisions. Now in a dilute system a collision leads to a change in momentum but essentially occurs at a fixed position. Thus it is plausible that one can just take for the collision rate an expression as derived in the previous chapter, but with the position r as a fixed parameter. This would lead to

$$\dot{n}(p, r, t) = -\left(\frac{\partial \varepsilon}{\partial p} \cdot \frac{\partial n}{\partial r} - \frac{\partial \varepsilon}{\partial r} \cdot \frac{\partial n}{\partial p}\right) + r_p(\cdots, n(q, r, t), \cdots) \quad . \tag{33.4}$$

Boltzmann, having the application to dilute gases in mind, specialized the collision rate r_p in the appropriate way. We do not do this at present since we first want to treat simpler examples.

For charged particles one has to take into account electromagnetic fields. If p were the canonical momentum, then one would just have to subtract the term $-eA/c$ from p and add the electrostatic potential to ε, and (33.4) would remain as it is. In most applications, however, one prefers to keep $p = mv$. Then, if the canonical momentum is called p_c, one has

$$\varepsilon = \varepsilon(p, r, t) = \varepsilon_k(p) + e\varphi(r, t) \tag{33.5}$$

and

$$p = p_c - eA/c \quad . \tag{33.6}$$

In going from canonical (p_c) to mechanical (p) momenta one has to take into account that the $\partial/\partial r$ and $\partial/\partial t$ occurring in the reversible (or "drift") term of the Boltzmann equation keep one or the other of the momenta fixed. So one has with

$$n(\boldsymbol{p}, \boldsymbol{r}, t) = n(\boldsymbol{p}c - e\boldsymbol{A}(\boldsymbol{r},t)/c, \boldsymbol{r}, t) = n_c(\boldsymbol{p}_c, \boldsymbol{r}, t) \tag{33.7}$$

$$dn_c = dn - e(\partial n/\partial \boldsymbol{p}) \cdot d\boldsymbol{A}/c \tag{33.8}$$

and a similar equation for ε. Applying this to the reversible part of the Boltzmann equation one finds after several intermediate steps (using the connections between vector potential and electromagnetic fields)

$$\left(\dot{n} + \frac{\partial \varepsilon}{\partial \boldsymbol{p}} \cdot \frac{\partial n}{\partial \boldsymbol{r}} - \frac{\partial \varepsilon}{\partial \boldsymbol{r}} \cdot \frac{\partial n}{\partial \boldsymbol{p}}\right)_c = \dot{n} + \frac{\partial \varepsilon}{\partial \boldsymbol{p}} \cdot \frac{\partial n}{\partial \boldsymbol{r}} + \left(-\frac{\partial \varepsilon}{\partial \boldsymbol{r}} + \boldsymbol{F}_t\right) \cdot \frac{\partial n}{\partial \boldsymbol{p}}, \tag{33.9}$$

where $\boldsymbol{F}_t$ is the transverse part of the Lorentz force

$$\boldsymbol{F}_t = e(\boldsymbol{E}_t + \boldsymbol{v} \times \boldsymbol{B}/c) . \tag{33.10}$$

Note that the longitudinal part is already contained in the $\partial \varepsilon/\partial \boldsymbol{r}$

$$-\partial \varepsilon(\boldsymbol{p}, \boldsymbol{r}, t)/\partial \boldsymbol{r} = -e\mathrm{grad}\varphi(\boldsymbol{r}, t) = \boldsymbol{F}_l . \tag{33.11}$$

The r.h.s. of (33.9) can be derived more easily using the alternative form, see (2.33), in which the factor of $\partial n/\partial \boldsymbol{p}$ is just $d\boldsymbol{p}/dt = \boldsymbol{F}$, if $\boldsymbol{p}$ is the mechanical momentum $m\boldsymbol{v}$.

Looking at (33.5,11) one may take φ to be some external potential. One can, however, also take into account the internal potential of the charged particles by

$$\varphi(\boldsymbol{r}, t) = \varphi^e(\boldsymbol{r}, t) + \int \frac{e^2 n(\boldsymbol{r}', t)}{|\boldsymbol{r} - \boldsymbol{r}'|} d^3 r' . \tag{33.12}$$

If this is inserted into (33.5 and 4) one obtains the *Vlasov equation* [33.3], which is an important equation in plasma physics. Using (33.1) in (33.12) one sees that in the Vlasov equation the *drift* term is obviously *nonlinear* and the single-particle energy even without external field depends not only on $\boldsymbol{p}$ but also on $\boldsymbol{r}$.

The internal field in (33.12) is, of course, nothing but the Hartree field of the charged particles. From (33.12) it is only a few steps to *Landau's* theory of *Fermi liquids* [33.6]. First of all, also short range forces have a Hartree potential. If the range of the force is short compared to the macroscopic variations which one wants to study one can replace the force by a delta function $v(\boldsymbol{r}-\boldsymbol{r}') = f\delta(\boldsymbol{r}-\boldsymbol{r}')$, then the Hartree field would be given by [inserting v for the Coulomb potential in (33.12)] $fn(\boldsymbol{r}, t)$. Furthermore, there may be exchange contributions to the self-consistent field. They will contribute a $\boldsymbol{p}$-dependent effective force and hence also to the $\boldsymbol{p}$ dependence of ε. *Landau* therefore divides up the single particle energy as

$$\varepsilon(\boldsymbol{p},\boldsymbol{r},t) = \varepsilon_0(\boldsymbol{p}) + \int f(\boldsymbol{p},\boldsymbol{p}')\delta n(\boldsymbol{p}',\boldsymbol{r},t)d^3p' \quad . \tag{33.13}$$

Here $\varepsilon_0(p)$ is the sum of kinetic plus exchange energy in the ground state and the additional term is the change of the self-consistent Hartree–Fock field in a small deviation δn from the ground state. Now strictly speaking *Landau* does not deal just with Hartree–Fock equations. He wanted to apply his theory to liquid He^3, which is not a dilute gas. His key point was, however, that all the complicated many-body effects occurring in denser Fermi systems (at temperatures low compared to the Fermi temperature) tend to renormalize the parameters $\varepsilon_0(p)$ and $f(\boldsymbol{p},\boldsymbol{p}')$, but not the structure (33.13) of the equations. For electrons in metals the Hartree–Fock equation is already an acceptable approximation. Equation (33.13) may then be used for the exchange contribution, one only has to add the Hartree term from (33.12).

Coming now to the collision term r_p we first study two approximations or models (the relaxation time approximation and the Fokker–Planck approximation), and only in Chap. 38 the more complicated case of binary collisions in a gas. As a preparation for the next chapter we introduce the relaxation time approximation.

The simplest ansatz for a linearized collision term would be (32.11) with constant r_{pq}

$$r_p = -\delta n(\boldsymbol{p},\boldsymbol{r},t)/\tau \quad . \tag{33.14}$$

Such an ansatz would not obey any of the conservation laws (32.11–13). For instance

$$\int r_p d^3p = -\delta n(\boldsymbol{r},t)/\tau \tag{33.15}$$

is an expression which in general does not vanish, so particle number conservation will in general be violated. One can, however, restore it by a simple subtraction term in (33.15) by putting

$$r_p = -[\delta n(\boldsymbol{p},\boldsymbol{r},t) - \delta n_l(\boldsymbol{p},\boldsymbol{r},t)]/\tau \tag{33.16}$$

with

$$\delta n_l = \left(\frac{\partial n_0(p)}{\partial \mu}\right)_T \left(\frac{\partial \mu}{\partial n}\right)_T \delta n(\boldsymbol{r},t) \quad . \tag{33.17}$$

First of all, when (33.17) is inserted into (33.16) it is obvious that now the integral (33.15) vanishes. [Note that there is a factor $(2\pi\hbar)^3$ now between n_p^0 (32.1) and $n_0(p)$ since we now use p integration instead of summation, $\int n_0(p)d^3p = n$.] Since the actual variable on which $n_0(p)$ depends is $\beta[\varepsilon(p)-\mu]$, one can also say that the difference on the r.h.s. of (33.16) is the deviation of $n(\boldsymbol{p},\boldsymbol{r},t)$ not from the total equilibrium distribution $n_0(p)$ but from the so-called *local equilibrium* distribution in which the μ in n_0 is replaced by $\mu(\boldsymbol{r},t) = \mu + (\partial\mu/\partial n)_T \delta n(\boldsymbol{r},t)$.

With similar tricks one can satisfy further conservation laws (Problem 33.3). For instance, with a subtraction term $(\partial n_0/\partial\beta)(\partial\beta/\partial\varepsilon)\delta\varepsilon(\boldsymbol{r},t)$, energy conservation is restored. In principle this technique can be systematized in terms of a moment expansion of γ_{pq} of Chap. 32. We do not pause to do so but for the time being just use (33.16) in the next chapter as a model for further studies.

PROBLEMS

33.1 Using (33.8) with $d = \partial_{r'}$ and ∂_t check the validity of (33.9).

33.2 Prove that the drift term in the Boltzmann equation (33.4) is reversible, i.e. that its contribution to the entropy change

$$T\dot{S} = \int \varphi_p(\boldsymbol{r},t)\dot{n}_d(\boldsymbol{p},\boldsymbol{r},t)d^3p\,d^3r = \int \varphi_p\{n,\varepsilon\}d^3p\,d^3r$$

vanishes, where the braces indicate Poisson brackets.

Hint: Use $\int a\{b,c\}d^3p\,d^3r = \int c\{a,b\}d^3p\,d^3r$ (proof?), and $\{-\varepsilon, n\} = 0$ [proof ? – see (32.1)].

33.3 How can momentum conservation be built into a relaxation ansatz? Which term has to be added in $n_0(\boldsymbol{p})$ in order to describe local equilibrium with a nonzero momentum density?

34. The Dynamic Conductivity in the Relaxation Time Model

The simple relaxation time ansatz for the collision term allows a complete solution of the linearized Boltzmann equation for arbitrary frequencies and wave numbers. This is what we are going to consider in this chapter. We discuss the response of the density and current density of electrons in isotropic condensed matter to an electric field E. It is assumed to be the total field and we neglect self-consistent exchange fields. The quantity $\partial \varepsilon / \partial p$ occurring in the drift term (which is only going to be used at the Fermi sphere) will be written as p/m or v, where m is an effective mass, and v the Fermi velocity $|v| = v_f$. Then Boltzmann's equation linearized in the field E in Fourier space takes the form

$$\left(\omega + \frac{i}{\tau} - \frac{p \cdot k}{m} \right) \delta n(p, k, \omega) - \frac{i}{\tau} \delta n_l(p, k, \omega) = -ieE \cdot \frac{\partial n_0}{\partial p} \quad . \tag{34.1}$$

34.1 Longitudinal Excitations

For a longitudinal electric field one can write

$$E = E^l = -ik\varphi(k, \omega) \quad , \tag{34.2}$$

and hence for the r.h.s. of (34.1)

$$-ieE \cdot \frac{\partial n_0}{\partial p} = e\varphi \frac{p \cdot k}{m} \frac{\partial n_0}{\partial \mu} \quad . \tag{34.3}$$

If one inserts this into (34.1) and solves for δn one finds, taking into account (33.18),

$$\delta n(p, k, \omega) = \frac{e\varphi p \cdot k/m + (i/\tau)(\partial \mu / \partial n) \delta n(k, \omega)}{\omega + i/\tau - p \cdot k/m} \left(\frac{\partial n_0}{\partial \mu} \right) \quad . \tag{34.4}$$

In order to determine $\delta n(k, \omega)$ one has to integrate (34.4) over p. We restrict ourselves to temperatures T small compared to the Fermi temperature. Then $(\partial n_0 / \partial \mu)$ is proportional to $\delta(|p - p_f|)$, so the integral is actually a surface integral over the Fermi surface.

The expressions simplify considerably if one introduces the so-called Lindhard function [34.1] L (or more precisely its semiclassical limit) by

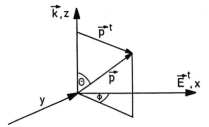

Fig. 34.1. Coordinates and polar angles for the p integral in (34.12)

$$L\left(\frac{\omega + i/\tau}{v_f k}\right) = \int \frac{-v \cdot k}{(\omega + i/\tau - v \cdot k)} \frac{d\Omega}{4\pi} \quad . \tag{34.5}$$

The integral can be done by introducing k as the polar axis, then $v \cdot k = v_f k \cos\Theta$. The result is

$$L(x) = \frac{1}{2} \int_{-1}^{+1} \frac{\cos\Theta \, d\cos\Theta}{\cos\Theta - x} = 1 + \frac{x}{2} \ln\frac{x-1}{x+1} = \begin{cases} -1/3x^2; & |x| \gg 1 \\ 1 + i\pi x/2; & |x| \ll 1 \end{cases} . \tag{34.6}$$

With this function the p integral of (34.4) can then be written as

$$\delta n(k, \omega) = -(\partial n/\partial \mu) L \, e \, \varphi(k, \omega) + (1 - L)\delta n(k, \omega)/(1 - i\omega\tau) \tag{34.7}$$

where L stands for (34.5). One now solves for δn and expresses the final result in the form $\delta n = -\chi \, e \, \varphi$. Then the susceptibility χ is given by

$$\chi(k, \omega) = \frac{\partial n}{\partial \mu} \frac{(1 - i\omega\tau)L}{L - i\omega\tau} \quad . \tag{34.8}$$

We give the result explicitly in the limit of large and small arguments of the Lindhard function (34.6) for the longitudinal conductivity (Chap. 31)

$$\sigma^l = -ie^2 \omega \chi / k^2 \quad . \tag{34.9}$$

Using $\partial n/\partial \mu = 3n/m v_f^2$ one finds after some intermediate steps

$$\sigma^l(k, \omega) = \begin{cases} \frac{e^2 n}{m} \frac{\tau}{[1 - i\omega\tau + iv_f^2 k^2 \tau/3\omega(1 - i\omega\tau)]} ; & kv_f\tau \ll |1 - i\omega\tau| \\ -i\frac{3e^2\omega}{m v_f^2 k^2}(1 - i\pi\omega/2v_f k) ; & kv_f\tau \gg |1 - i\omega\tau| \end{cases} . \tag{34.10}$$

The discussion of these results will be postponed until Sect. 34.3.

34.2 Transverse Excitations

In the transverse case there is no density change. Hence one has $\delta n_l = 0$. The solution of the kinetic equation then can be written down immediately:

$$\delta n(p, k, \omega) = \frac{ie \, p \cdot E^t/m}{(\omega + i/\tau - p \cdot k/m)} \frac{\partial n_0}{\partial \mu} \quad . \tag{34.11}$$

The transverse conductivity can be obtained from this again by integration over p:

$$eJ^t = i\left(\frac{e}{m}\right)^2 \int \frac{p^t(p \cdot E^t)}{(\omega + i/\tau - p \cdot k/m)} \frac{\partial n_0}{\partial \mu} d^3p \quad . \tag{34.12}$$

To carry out the p integral we use a coordinate system as in Fig. 34.1. In these coordinates the conductivity takes the form

$$\sigma^t(k,\omega) = ie^2 \frac{\partial n}{\partial \mu} \int \frac{v_x^2}{(\omega + i/\tau - v \cdot k)} \frac{d\Omega}{4\pi} \quad , \tag{34.13}$$

and finally using $v_x^2 = v_f^2 \sin\Theta \cos^2\alpha$ and $v \cdot k = v_f k \cos\Theta$ one finds

$$\sigma^t(k,\omega) = \frac{3ine^2}{4mv_f k}\left(2x + (x^2 - 1)\ln\frac{x-1}{x+1}\right) \quad . \tag{34.14}$$

Here we have introduced the same variable $x = (\omega + i/\tau)/v_f k$ as in the Lindhard function. We write down again the two limiting cases of small and large x

$$\sigma^t(k,\omega) = \begin{cases} \frac{e^2 n}{m}\frac{\tau}{(1-i\omega\tau)} \; ; & kv_f\tau \ll |1 - i\omega\tau| \quad , \\ \frac{3\pi n e^2}{4mv_f k} \; ; & kv_f\tau \gg |1 - i\omega\tau| \quad . \end{cases} \tag{34.15}$$

34.3 Discussion of $\sigma^l(k,\omega)$ and $\sigma^t(k,\omega)$

We start our discussion with the small x limit and take first the extreme limit $k = 0$. Then one obtains for both longitudinal and transverse conductivity the result of Drude theory

$$\sigma(0,\omega) = \frac{e^2 n}{m}\frac{\tau}{(1 - i\omega\tau)} \quad . \tag{34.16}$$

It has the well-known zero-frequency limit $e^2 n\tau/m$ and goes over to a i/ω behaviour at high frequencies. The ω dependence is nothing but the effect of the inertia of the electrons and follows directly from Newton's equation

$$m\left(\frac{dv}{dt} + \frac{v}{\tau}\right) = eE \quad , \tag{34.17}$$

which, after Fourier transformation, using $J = nv$ is identical with (34.16). The dielectric function in the high-frequency regime is

$$\varepsilon(0,\omega) = 1 + i\frac{4\pi\sigma}{\omega} \to 1 - \frac{4\pi e^2 n}{m\omega^2} \; ; \quad \omega\tau \gg 1 \; , \tag{34.18}$$

again valid for both longitudinal and transverse excitations. This leads, see Chap. 31, to the eigenfrequencies

$$\omega_l^2 = \frac{4\pi e^2 n}{m} = \omega_p^2 \quad \text{and} \quad \omega_t^2 = \omega_p^2 + c^2 k^2 \tag{34.19}$$

of the so-called *plasma vibrations*; ω_p is called the *plasma frequency*.

We now take k nonzero but both $\omega\tau \ll 1$ and $kv_f\tau \ll 1$. This regime is usually called the *hydrodynamic regime*. One is still in the small x limit. Then the transverse conductivity still has the same form (34.16) but in the longitudinal case a k-dependent term occurs in the denominator. In this case it is useful to introduce

$$D = \frac{mv_f^2}{3}\frac{\tau}{m} = n\left(\frac{\partial \mu}{\partial n}\right)_T B = \left(\frac{\partial P}{\partial n}\right)_T B, \tag{34.20}$$

which is nothing but Einstein's relation between the diffusion constant D and mobility $B = \tau/m$ extended to Fermi gases at low T. It is also useful to go back to the original susceptibility χ in this case and finally write

$$\chi(k,\omega) = \frac{\partial n}{\partial \mu}\frac{Dk^2}{(Dk^2 - i\omega)}; \quad \omega\tau \ll 1 \text{ and } kv_f\tau \ll 1. \tag{34.21}$$

So the hydrodynamic regime is characterized by *diffusion behaviour*.

What is the difference between the longitudinal and transverse case, responsible for the presence and absence, respectively, of the k^2 term in the denominator? A glance at (33.20) shows that it is the pressure [or better nonzero compressibility $(\partial P/\partial n)$] in the longitudinal case, occurring as the prefactor of the k^2 term. Since the electron gas has no transverse stiffness there is no k^2 term in the transverse case.

The dielectric function in the static limit $\chi = \partial n/\partial \mu$ is of interest. Going back to (34.9) and $\varepsilon = 1 + 4\pi i\sigma/\omega$

$$\varepsilon^l(k,0) = 1 + \frac{4\pi e^2}{k^2}\left(\frac{\partial n}{\partial \mu}\right) = 1 + (k_s/k)^2 \tag{34.22}$$

with the *screening wave number* k_s given by

$$k_s^2 = 4\pi e^2 \frac{\partial n}{\partial \mu} = \frac{3\omega_p^2}{v_f^2}, \tag{34.23}$$

which for metals is of the order of the Fermi wave number. The physical meaning of this *screening* wave number can be seen most directly by introducing an external positive point charge $|e|$ into the system. Then the potential $\varphi^e = 4\pi e^2/k^2$ of this charge will be screened by the Hartree field of the electrons attracted by the charge. The resulting total field has the potential

$$\varphi(k) = \frac{\varphi^e(k)}{\varepsilon^l(k,0)} = \frac{4\pi e^2}{(k^2 + k_s^2)}, \tag{34.24}$$

which has no $1/k^2$ singularity any more, and hence no longer has an infinite range. Fourier back transformation into coordinate space would yield

$$\varphi^e(r) = \frac{e}{r} \; ; \quad \varphi(r) = \frac{e}{r} \exp(-k_s r) \quad . \tag{34.25}$$

So k_s is just the decay length of the screening charge. Note, however, that the Fourier integral involves large k values, for which the approximation (34.24) is no longer valid. It would be a better approximation to cut off the integral at $k = 2k_f$, because the electron system cannot respond to k values larger than the diameter of the Fermi sphere. This would lead to oscillations of the potential in coordinate space with the cutoff wave number. Indeed the proper quantum treatment of screening leads to such *Friedel oscillations*. Furthermore, at small r (34.25) is also not a very good approximation, because the linear approximation breaks down. Nevertheless, the simple *Thomas Fermi screening* approximation is a reasonable description for many semiquantitative calculations.

Finally we come to the limit of large x: $kv_f\tau \gg |1 - i\omega\tau|$. In this case there are quantitatively different results for longitudinal and transverse excitations. Qualitatively they have in common that there is a real part of σ (which means energy absorption from the external field) without collisions (τ does not occur any more). This is called "collision-free" or *Landau damping*. Classically the zeros of the denominator in (34.4,11) responsible for energy absorption correspond to a resonance of the phase velocity ω/k and the velocity v of the particles: they can "surf ride" on the wave. Quantum mechanically the condition $\omega = \boldsymbol{v} \cdot \boldsymbol{k}$ is nothing but a consequence of energy and momentum conservation for the absorption process $\hbar\omega = \varepsilon(\boldsymbol{p} + \hbar\boldsymbol{k}) - \varepsilon(\boldsymbol{p})$ in the limit of small k.

These processes become experimentally relevant for *ultrasonic attenuation* in metals and, for the transverse part, in the *anomalous skin effect*. Sound waves in metals with a wavelength large compared to the electron mean free path come into equilibrium with the electrons essentially everywhere at all times. The attenuation of the wave then is caused by the viscosity of the electron gas (just as in hydrodynamics). This viscosity is beyond our present scope, since it is due to electron–electron scattering, which would be momentum conserving, whereas our relaxation ansatz conserves only particle number. But, in the other limit, at high frequencies, where the sound wavelength is small compared to the mean free path, the wave does not come into equilibrium with the electrons and the attenuation mechanism discussed above comes into play. It turns out that the ultrasonic absorption coefficient is directly proportional to the imaginary part of the susceptibility multiplied by ω. Such situations occur in metals for frequencies above $\approx 10^6$ Hz at low T ($T < 20$ K).

The skin effect is in some sense similar: at low frequencies one has the normal skin effect with a penetration depth $d = c/(2\pi\sigma\omega)^{1/2}$ determined by Ohmic conductivity. With increasing frequency this depth becomes smaller and smaller, and finally smaller than the mean free path. Above this frequency one comes again into the region of collisionless damping. This then is called the anomalous skin effect. It can be observed at helium temperatures and for frequencies above 10^{10} Hz.

34.4 Quantum Corrections

As already mentioned near the beginning of Chap. 33, the phase space density may be considered as the average values of $\psi^*(r)\psi(p)\exp(i p \cdot r/\hbar)$, or after Fourier transformation: $n(p, k) = \langle \psi^*(p - \hbar k)\psi(p)\rangle$. For anyone not familiar with field operators, the latter means nothing but: (let us call $p - k = q$ for a while) take the statistical operator in the momentum representation, specify its variables as $\langle p, p_2, p_3 \cdots |\varrho| q, p_2, p_3 \cdots \rangle$ and integrate over all p_2, p_3, etc. The result may be called $\langle p|\varrho|q\rangle$ and one has

$$n(p, k, t) = \langle p|\varrho(t)|p - \hbar k\rangle \quad . \tag{34.26}$$

The quantum mechanical generalization of the semiclassical equation (33.5) then would be

$$\dot{n} = -\frac{i}{\hbar}[h, n] + r \quad , \tag{34.27}$$

where the Poisson bracket has turned into a commutator.

We linearize this equation according to $h = \varepsilon + \delta h$ and $n = n_0 + \delta n$, where ε and n_0 are diagonal in momentum space. After Fourier transformation in t one has

$$\left\{\omega - \frac{\varepsilon(p) - \varepsilon(q)}{\hbar}\right\}\langle p|\delta n|q\rangle = \frac{[n_0(p) - n_0(q)]\langle p|\delta h|q\rangle}{\hbar} \quad . \tag{34.28}$$

If one here puts $q = p - \hbar k$ and considers the limit of small k one comes back to the classical limit, obviously. Quantum corrections can be obtained by expanding one step further in k.

As a simple example we calculate the quantum corrections to the transverse static susceptibility (response of current ep^t/m to $a^t = A/c$ with $\delta h = -ep^t \cdot a^t/m$)

$$\chi^t(k, 0) = -\left(\frac{e}{m}\right)^2 \int (p^t)^2 \frac{n_0(p) - n_0(p - \hbar k)}{\varepsilon(p) - \varepsilon(p) - \hbar k} d^3p \quad . \tag{34.29}$$

Power expansion in $\hbar k$ up to second order yields

$$\chi^t(k, 0) = -\left(\frac{e}{m}\right)^2 \int (pt)^2 \left(n_0' - \frac{\hbar^2 k^2}{4m} n_0'' + \frac{\hbar^2 (p \cdot k)^2}{6m} n_0'''\right) d^3p \quad . \tag{34.30}$$

Here the primes on n_0 mean derivatives with respect to μ. The p integral can be carried out using the variables introduced in Fig. 34.1. One finds

$$\chi^t(k, 0) = \frac{e^2 n}{m}[1 - (\hbar k/2p_f)^2 + \cdots] \quad . \tag{34.31}$$

The interpretation of this result is very interesting. We saw in our discussion above that a classical gas has no transverse stiffness. We shall now see that quantum mechanically there is a small stiffness $\propto k^2$. For this we have to compare (34.31) to the transverse conductivity. The general connection is, see (31.24)

$$\chi(k,\omega) = \frac{e^2 n}{m} + i\omega\sigma(k,\omega) \quad . \tag{34.32}$$

Comparing the last two formulae, one sees that the quantity $i\omega\sigma$ in the static limit picks up a k-dependent additional term. Or else in the dielectric function $\varepsilon = 1 + 4\pi i\,\sigma/\omega$ one has an additional term $-4\pi e^2 n(\hbar k)^2/4p_{\mathrm{f}}^2 \omega^2 m$. Hence one may write

$$\varepsilon^t(k,\omega) = \varepsilon^t(k,\omega)_{\mathrm{class}} + 4\pi c^2 k^2 \chi_{\mathrm{m}}/\omega^2 \tag{34.33}$$

with

$$\chi_{\mathrm{m}} = -\mu_{\mathrm{B}}^2 mn/p_{\mathrm{f}}^2 \quad . \tag{34.34}$$

Looking at (31.41) we see that the electrons have a negative (diamagnetic) magnetic susceptibility (34.34), which is nothing but *Landau diamagnetism* [34.2]. This is how one calculates a diamagnetic susceptibility from a dielectric function. Concluding, one may say the transverse electronic conductivity in quantum mechanics has a small transverse stiffness contribution due to Landau diamagnetism.

Something similar happens in the longitudinal case at high frequencies ($\omega\tau \gg 1$). In this case a glance at (34.10) shows that the compressibility term $\propto k^2$ in the denominator is suppressed as compared to the Drude term $i\omega\tau$: at frequencies large compared to collision frequencies no equilibrium can be established, and the pressure term vanishes. Since then the longitudinal dielectric

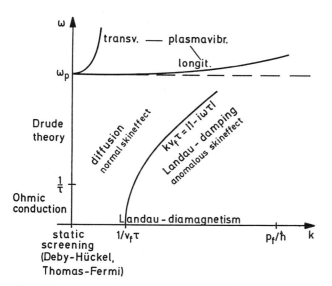

Fig. 34.2. Various effects and regimes in the electromagnetic response of metals that can be treated by the relaxation model, plotted in the (ω, k)-plane

function becomes independent of k, the longitudinal plasma frequency will have no dispersion. Of course one might blindly apply the hydrodynamic limit for the plasma vibrations, then there would be dispersion. The relaxation ansatz and its consequences (34.10) show that hydrodynamics breaks down in the plasma frequency region.

If quantum mechanics is taken into account, the situation changes again. It turns out that the dielectric function again receives an additional term, which is k dependent and restores some longitudinal stiffness,

$$\varepsilon^l(k,\omega) = 1 - \frac{\omega_p^2}{\omega^2}\left[1 + \frac{3}{5}\left(\frac{kp_f}{m\omega}\right)^2 + \cdots\right] . \qquad (34.35)$$

For small k this leads to a dispersion

$$\omega_l^2(k) = \omega_p^2 + 3(v_f k)^2/5 + \cdots . \qquad (34.36)$$

Figure 34.2 is a survey of all the effects we have discussed and their locations in the (ω, k)-plane.

PROBLEMS

34.1 Using (34.31,32) determine $\chi^t(k,\omega)$ for small ω. Discuss the transition to a superconductor ($\tau \to \infty$) in terms of χ^t. Note that the static limit and the limit to superconductivity must not be interchanged.

34.2 Calculate the longitudinal susceptibility in quantum mechanics. Try to verify (34.35).

35. Zero Sound

In the previous chapter we saw that in the long-wavelength limit $k\ell \ll 1$ ($\ell = v_f\tau$) one has to distinguish the low-frequency ($\omega\tau \ll 1$) and high-frequency ($\omega\tau \gg 1$) limits. In the first case one has local equilibrium with a hydrodynamic behaviour (diffusion in the case of electrons). In the second case one can neglect collisions. Nevertheless one has collective excitations, the so-called plasma vibrations. They occur as a consequence of the self-consistent Hartree field.

A physically very similar type of excitation is "zero sound", predicted by *Landau* [35.1]. It can be observed in liquid ^{3}He at low temperatures. In helium the collision term is not only particle number conserving but also momentum and energy conserving. As a consequence (in particular due to the momentum conservation), one finds sound waves in the hydrodynamic regime instead of diffusion.

In the high-frequency regime $\omega\tau \gg 1$ one can again neglect the collisions to first order. Nevertheless, there is interaction between the particles via the self-consistent field. The main difference as compared to the Coulomb case $e^2/|\mathbf{r} - \mathbf{r}'|$ is that the interaction $v(|\mathbf{r} - \mathbf{r}'|)$ is short ranged. We have considered this already in (33.13). The simplest way to take a self-consistent "molecular" field into account is by a corresponding shift of the external field (or the chemical potential)

$$\delta n(\mathbf{k},\omega) = \chi(k,\omega)[\delta\mu(\mathbf{k},\omega) - v(k)\delta n(\mathbf{k},\omega)] \quad . \tag{35.1}$$

Here $\chi(k,\omega)$ is the density response function of the bare system and $v(k)$ is the Fourier transform of $v(r)$. Note that the shift of the chemical potential has the opposite sign to the energy shift (33.13). Equation (35.1) can be solved for the density response $\chi^e(k,\omega) = \delta n/\delta\mu$ to the "external field" $\delta\mu(\mathbf{k},\omega)$

$$\chi^e(k,\omega) = \frac{\chi(k,\omega)}{1 + v(k)\chi(k,\omega)} \quad . \tag{35.2}$$

This equation is a simple generalization of the one in Problem 31.2 in which $v(k) = 4\pi e^2/k^2$ is the Coulomb potential.

The eigenfrequencies are then given by the zeros of the denominator of (35.2):

$$1 + v(k)\chi(k,\omega_l(k)) = 0 \quad . \tag{35.3}$$

In the high-frequency (collisionless) limit, see (34.8), $\chi = (\partial n/\partial\mu)L(x)$, where the argument of the Lindhard function is simply $x = \omega/v_f k$. For small k one can approximate $v(k)$ by $v(0)$ and write

$$f_0 = v(0)(\partial n/\partial \mu) \tag{35.4}$$

as in (33.13). Equation (35.3) then takes the form

$$1 + f_0 L(x) = 0 \ . \tag{35.5}$$

For sufficiently large f_0 the solution occurs in the large x limit. For a correct comparison of the solution of (35.5) with ordinary ("first") sound one has to go one step further, however, as in the expansion (34.6)

$$L(x) = -\frac{1}{3x^2} - \frac{1}{5x^4} + \cdots \ ; \quad |x| \gg 1 \ . \tag{35.6}$$

Insertion into (35.5) yields

$$\omega_l(k) = (f_0/3 + 9/15)v_f k = c_0 k \ , \tag{35.7}$$

hence a linear relation as in the case of first sound. However, the velocities c_0 and c_1 of the two sound waves are different. The static compressibility $mc_1^2 = \partial P/\partial n$ determines c_1. Actually the derivative should be the adiabatic one, but in a Fermi system at low T the difference from the isothermal one is negligible. Thus writing (35.2) for $\omega = 0$ and small k one has

$$\frac{1}{n}\left(\frac{\partial n}{\partial \mu}\right)^e = \frac{\partial n}{\partial P} = \frac{1}{(1+f_0)}\frac{1}{n}\frac{\partial n}{\partial \mu} = \frac{1}{mc_1^2} \ . \tag{35.8}$$

Comparing with (35.7) one finds using $(\partial n/\partial \mu) = 3nm/p_f^2$

$$c_0 = c_1 \left(1 + \frac{4}{10(1+f_0)} + \cdots\right) \ . \tag{35.9}$$

By measuring the specific heat and compressibility of liquid He one finds a value of

$$f_0 \approx 11 \ . \tag{35.10}$$

First of all this means that the expansions above are justified and secondly that the zero sound velocity should be about 3% higher than the first sound velocity.

Figure 35.1 shows experimental results for sound velocity and attenuation for liquid ^{3}He in the region between 1 and 200 mK for two different frequencies. With increasing T the collision time decreases $\propto T^{-2}$. Below about 10 mK first sound goes into zero sound. Besides the few per cent increase in the velocity there is a marked change in the attenuation.

We have so far not discussed the attenuation, but will come back to it in Chaps. 52 and 53. In the first sound regime one has, of course, the usual hydrodynamic attenuation (Chap. 24). Since viscosity and the heat diffusion

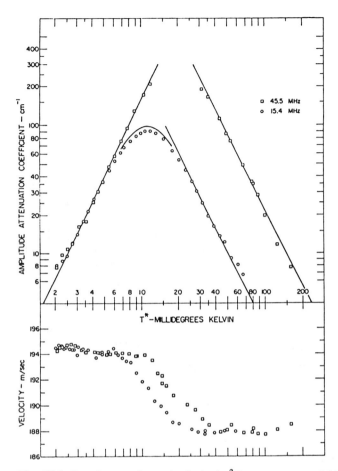

Fig. 35.1. Sound attenuation and velocity in ^{3}He at a pressure of 0.32 atm for frequencies of 15.4 and 45.5 MHz. The low-T straight line shows a T^2 behaviour; the high-T lines show an ω^2/T^2 behaviour [35.2]

coefficient are proportional to τ, the attenuation first increases with decreasing T. It reaches a maximum near the transition region to zero sound and then decreases with decreasing T as it becomes proportional to $1/\tau$.

In our derivation of (35.9) we have used a rather simplified version of Landau's theory, in particular we have completely neglected the dependence of the Landau parameters $f_{pp'}$ on p and p'. It turns out, fortunately, that (35.9) is a rather good approximation [35.2].

PROBLEMS

35.1 Using the relation between ε^l and χ for a system with charged particles, show that the conditions $\varepsilon_l = 0$ and $\chi^e = \infty$ agree.

36. The Fokker–Planck Approximation

Often one calls kinetic differential equations that are first order in time and second order in some other variable Fokker–Planck equations [36.1,2]. We are going to consider the Fokker–Planck equation in a more restricted sense as an approximation for the collision term.

The relaxation ansatz of Chap. 34 corresponds to spherically symmetric scattering. One can expect it to be a good approximation for the scattering of light particles by heavy neutral centres, for instance for the scattering of electrons by neutral atoms or defects in solids. In some sense the opposite limit occurs for scattering with only small momentum changes, for instance for scattering of heavy particles by light ones or scattering of electrons by charged centres.

For the description of such processes an approximation of the collision term is useful which was first given by *Fokker* and *Planck* [36.1]. For its derivation we start directly from (32.16) for a dilute system with a collision term that conserves only the particle number

$$\dot{n}_p = \sum_q (r_{pq} n_q - r_{qp} n_p) \quad . \tag{36.1}$$

If there are only small momentum changes during the scattering it means that r_{pq} is close to a delta function in $p-q$. Then one can try to expand n_q in powers of $q-p$ at p. Restricting ourselves to one dimension we expect an expansion of (36.1) of the form

$$\dot{n}_p = \gamma_0(p) n_p + \gamma_1(p)\frac{\partial n_p}{\partial p} + \gamma_2(p)\frac{\partial^2 n_p}{\partial p^2} + \cdots \quad . \tag{36.2}$$

The $\gamma_i(p)$ are certain moments of the original r_{pq}. We do not need the explicit form of these moments. We note only two general properties:

$$\gamma_i(p) = (-1)^{-i}\gamma_i(-p) \tag{36.3}$$

following from inversion symmetry, and

$$0 = \left(\gamma_0(p) n_p + \gamma_1(p)\frac{\partial n_p}{\partial p} + \gamma_2(p)\frac{\partial^2 n_p}{\partial p^2}\right)_0 \quad , \tag{36.4}$$

where the label "0" refers to "equilibrium". This equation is nothing but the equilibrium condition.

The Fokker–Planck approximation corresponds to (36.2) and (a) $\gamma_i = 0$ for $i > 2$, (b) a power expansion of γ_0, γ_1, γ_2 in p up to the lowest power compatible with (36.3,4). This means $\gamma_0 = \gamma = \text{const}$, $\gamma_1(p) = \gamma_1 p$, and $\gamma_2 = \text{const}$. If we restrict ourselves to high temperatures, where $(n_p)_0$ is the Maxwell distribution, one finds

$$\gamma_1(p) = \gamma p, \quad \gamma_2(p) = mkT\gamma \quad . \tag{36.5}$$

It is difficult to obtain a consistent theory taking higher than second derivatives into account (see the end of this chapter). In the Fokker–Planck approximation we now have

$$\boxed{\dot{n}_p = \gamma \frac{\partial}{\partial p}\left[\left(p + mkT\frac{\partial}{\partial p}\right)n_p\right]} \quad . \tag{36.6}$$

Before we add a drift term to this equation, in order to get an approximation for a transport equation, we consider (36.6) alone. It can be reduced to a simple diffusion equation by introducing the new variables

$$q = pe^{\gamma t}, \quad \Theta = (e^{2\gamma t} - 1)/(2\gamma), \quad \nu(q, \Theta) = n_p(t)e^{-\gamma t} \quad . \tag{36.7}$$

Then (36.6) is transformed into

$$\frac{\partial \nu}{\partial \Theta} = mkT\gamma \frac{\partial^2 \nu}{\partial q^2} \quad . \tag{36.8}$$

In order to construct the general solution one can start out from the so-called "source" solution

$$\nu = (4\pi mkT\gamma\Theta)^{-1/2} \exp[-(q - q_0)^2/(4mkT\gamma\Theta)] \quad . \tag{36.9}$$

After inversion of the transformation (36.7), this goes into

$$n(p, t; p_0) = \frac{n}{[2\pi mkT(1 - e^{-2\gamma t})]} \exp\left(-\frac{(p - p_0 e^{-\gamma t})^2}{2mkT(1 - e^{-\gamma t})}\right) \quad . \tag{36.10}$$

The normalization is now such that $\int n(p, t; p_0) dp = n$. Equation (36.10) describes the situation that a distribution which at $t = 0$ is sharply peaked (like a delta function) at $p = p_0$ relaxes exponentially to a centre at $p = 0$, and during that time broadens into a Maxwell distribution. The general solution to an arbitrary initial distribution $n_0(p)$ can be obtained by a convolution with this distribution.

Let us now add a drift term to (36.6) and assume the particles move in a potential $W(x)$. Then we obtain a transport equation

$$\dot{n}(p, x, t) + \frac{p}{m}\frac{\partial n}{\partial x} - \frac{\partial W(x)}{\partial x}\frac{\partial n}{\partial p} = \gamma \left[\left(p + mkT\frac{\partial}{\partial p}\right)n\right] \quad . \tag{36.11}$$

The potential $W(x)$ complicates the solutions considerably. General solutions are known [36.3] for the case of a harmonic oscillator $W(x) = mx^2\omega^2/2$. In general one has to make use of approximations [36.4].

Let us consider an approximation scheme that is often used in the context of transport equations: the "moment method". It consists in multiplying (36.11) with increasing powers p^i of p and integrating. The first four moments $M_i(x,t) = \int p^i n(p,x,t)dp$ ($i = 0, 1, 2, 3$) have a direct physical interpretation namely

$n(x,t) = \int n(p,x,t)dp$	particle density,	(36.12)
$g(x,t) = \int p n(p,x,t)dp$	momentum density,	(36.13)
$\Pi(x,t) = \int p^2 n(p,x,t)dp/m$	momentum flux density,	(36.14)
$\varepsilon(x,t) = \Pi/2 + W(x)n(x,t)$	energy density,	(36.15)
$j_\varepsilon(x,t) = \int \frac{p^3}{2m^2} n(p,x,t)dp + \frac{1}{m}W g(x,t)$	energy current density.	(36.16)

Multiplication of the kinetic equation (36.11) by the corresponding powers of p and integration over p leads to the equations

$$m\dot{n} + \frac{\partial g}{\partial x} = 0 \;, \tag{36.17}$$

$$\dot{g} + \frac{\partial \Pi}{\partial x} = -n\frac{\partial W}{\partial x} - \gamma g \;, \tag{36.18}$$

$$\dot{\varepsilon} + \frac{\partial j_\varepsilon}{\partial x} = \gamma(nkT - \Pi) \;. \tag{36.19}$$

They are, obviously, nothing but the balance equations for particle number, momentum and energy. This system of equations is exact, but not closed. There are three equations for four variables. Of course one could consider more moments of (36.11) but one would always have one more moment than equations.

The three equations (36.17–19) become approximately closed in the limit of large $\gamma (\gg \omega)$. Since the l.h.s. of (36.19) is independent of γ, the r.h.s. has to vanish. This leads to $\Pi = nkT$. In (36.18) the $\partial g/\partial t$ can be neglected as compared to γg. Equation (36.18) with $\Pi = nkT$ then becomes Fick's law for the diffusion of a particle in a potential $W(x)$. We see that the moment method leads to a derivation of hydrodynamic equations from kinetic transport theory including a determination of transport coefficients (in this case of the diffusion coefficient) from the collision term.

The derivation of the Fokker–Planck equation by an approximation of the collision term in the Boltzmann equation is the first step of the *Kramers–Moyal expansion* [36.5,6]. It seems obvious that one can improve the Fokker–Planck equation by taking higher derivatives of the collision term into account. However, it turns out [36.7] that one runs into certain difficulties with the requirement of positivity of $n(p,r,t)$. On the other hand, this need not be of practical importance in general [36.8]. If one takes into account higher than second order derivatives, the distribution function becomes weakly negative in certain regions, but is nevertheless improved in most cases. Therefore it seems "natural" to stop the expansion of the collision term after the second derivatives from a principle view point, but practically it may well be reasonable to include higher terms.

PROBLEMS

36.1 Consider a generalized one-dimensional Fokker–Planck equation in momentum space:

$$\dot{n}(p,t) = \sum_{n=0}^{N} \gamma_n(p) \frac{\partial^n n(p,t)}{\partial p^n} \quad ; \quad N > 2.$$

Demonstrate that for arbitrary $\gamma_n(p)$ it is always possible to find solutions that are normalizable with $n(p,0) \geq 0$, for which $n(0,0) = 0$ and $\partial n/\partial t(0,0) < 0$. Hint: try an ansatz $(p + cp^{N-1})^2/(1 + b^2 p^{2N})$.

37. Brownian Motion and Diffusion*

Let us pause in this chapter for a little intermezzo with a discussion of the generalized diffusion equation (36.18). We express the momentum density g in terms of the particle current density j as $g(x,t) = mj(x,t)$. Then (36.17–19) in the limit of large γ take the form

$$\partial n/\partial t + \partial j/\partial x = 0 \quad , \tag{37.1}$$

$$m\partial j/\partial t + \partial \Pi/\partial x = -n\partial W/\partial x - m\gamma j \quad , \tag{37.2}$$

$$\Pi = P = nkT \quad , \tag{37.3}$$

where T = const is the temperature of the "bath". For constant T the change of the chemical potential $d\mu_0 = -sdT + dP/n$ obeys $nd\mu_0 = dP$. We have introduced a label "0" at μ in order to indicate that it does not contain the potential $W(x)$, whereas

$$\mu = \mu_0(P, T) + W(x) \tag{37.4}$$

is the total chemical potential. Using this definition (37.2) can be rewritten as

$$\boxed{m\partial j/\partial t = -n\partial \mu/\partial x - m\gamma j \quad ,} \tag{37.5}$$

an equation which plays a role in many areas of physics.

Let us begin with the *static case* $j = 0$, then (37.5) becomes the well-known starting equation for the derivation of the *barometric pressure equation*

$$\mu = \mu_0(P, T) + W(x) = \text{const} \quad . \tag{37.6}$$

In the *stationary case* one neglects only the acceleration term $m\partial j/\partial t$. Then one has either *Fick's law* or (in the electric case) *Ohm's law*

$$j = -[n/(m\gamma)]\partial \mu/\partial x \quad . \tag{37.7}$$

If the two terms contained in μ are treated separately one usually writes the r.h.s. as

$$j = nv = -D\partial n/\partial x - nB\partial W/\partial x \quad , \tag{37.8}$$

where

$$D = n(\partial\mu/\partial n)/(m\gamma) \quad \text{and} \quad B = 1/(m\gamma) \quad \text{with} \quad D = BkT \tag{37.9}$$

are the diffusion constant and mobility, respectively, obeying the *Einstein relation* [37.1]. In fact, it was the balance of the two terms on the r.h.s. in barometric equilibrium in the gravitational field which, historically, led *Einstein* to his relation in 1905.

If one differentiates (37.8) with respect to x and uses the continuity equation (37.1) one obtains the *Smoluchowski* equation [37.2] for the diffusion in potential fields

$$\dot{n} = D\frac{\partial}{\partial x}\left(\frac{n}{kT}\frac{\partial W}{\partial x} + \frac{\partial n}{\partial x}\right) \quad . \tag{37.10}$$

It is another example of a kinetic equation of the *Fokker–Planck* type (first order in t, second order in x). A special case is, of course, the normal diffusion equation of Chap. 20 for $W = 0$.

The full equation (37.5) including the inertial term has found an application in *Drude's theory* [37.3] of electronic conduction. One only has to remember

$$-\partial\mu/\partial x = eE, \quad ej : \text{electric current density}, \quad \tau = 1/\gamma \quad , \tag{37.11}$$

introducing the electric field E and collision time τ, then (37.5) takes the well-known form of the equation of motion of the Drude theory (34.17):

$$\partial ej/\partial t = \frac{e^2 n}{m}E - ej/\tau \quad . \tag{37.12}$$

After this random walk through various fields of physics let us come back to the relations between Brownian motion and diffusion as promised in the heading of this chapter. We introduce the probability density $\varrho(x,t)$, which is proportional to $n(x,t)$ but normalized to 1 instead of N. Again combining (37.2) with the continuity equation (37.1) and leaving out the potential W for simplicity we obtain

$$\ddot{\varrho} + \gamma\dot{\varrho} = \frac{kT}{m}\varrho'' \quad . \tag{37.13}$$

The probability density ϱ can be used to calculate various averages, in particular those of x and x^2,

$$X(t) = \overline{x(t)} = \int x\varrho(x,t)dx \quad \text{and} \quad \overline{x^2(t)} = \int x^2\varrho(x,t)dx \quad . \tag{37.14}$$

Rather than first integrating (37.13) and then calculating (37.14) it is easier to first average (37.13) and then integrate. So we multiply (37.13) by x and integrate. The r.h.s. of (37.13) then (after partial integration) yields zero and there remains

$$\ddot{X}(t) + \gamma\dot{X}(t) = (kT/m)\int x\varrho'' dx = 0 \quad . \tag{37.15}$$

This, because $mdX/dt = P$, is nothing but $dP/dt + \gamma P = 0$.

Similarly one can now multiply (37.13) by x^2 and integrate. Taking into account that after partial integration one has

$$\int x^2 \varrho'' dx = 2 \int \varrho dx = 2 \quad , \tag{37.16}$$

one finds a differential equation for the average of x^2

$$\frac{d^2 \overline{x^2}}{dt^2} + \gamma \frac{d\overline{x^2}}{dt} = 2kT/m \tag{37.17}$$

with the solution

$$\overline{x^2} = \sigma_0 + \sigma_1 e^{-\gamma t} + 2kTt/(m\gamma) \quad . \tag{37.18}$$

To fix the constants $\sigma_{0,1}$ of the integration we consider a situation where at $t = 0$ ϱ is concentrated at $x = 0$ and the particles start with random velocities δu (according to a Maxwell distribution) from this point. Then for $t \ll \tau$, i.e. before collisions start to play a role, one expects

$$\overline{x^2} = \langle (\delta u)^2 \rangle t^2 \quad ; \quad t \ll \tau = 1/\gamma \quad . \tag{37.19}$$

At any rate one has

$$\overline{x^2} = 0, \quad \frac{d\overline{x^2}}{dt} = 0 \quad \text{for} \quad t = 0 \quad . \tag{37.20}$$

The solution (37.18) then may be written as

$$\overline{x^2} = 2\frac{kT}{m\gamma} \left(t - \frac{1}{\gamma}(1 - e^{-\gamma t}) \right) \quad ; \quad t \geq 0 \quad . \tag{37.21}$$

The same equation was derived before in (4.14), considering the motion of a single particle in equilibrium. This is one of the many quantitative relations between diffusion and Brownian motion. A more profound one will be considered in the next chapter.

The two limiting cases for large and small t have been considered already in (4.15). It is instructive to derive them once again directly from (37.13) neglecting either the first or the second term on the l.h.s. of this equation. As one can see, the second-order (acceleration) term is needed for the correct behaviour of (37.21) at small t; Fick's law would not be sufficient for that. Equation (37.13) describes the time evolution of a *nonequilibrium* situation, namely a point singularity in the density. In the initial stages it describes an isothermal shock wave, and afterwards the diffusion of a point source. On the other hand the interpretation of (37.21) from the point of view of the Brownian motion of a single particle concerns, in principle, an *equilibrium* situation. The fluctuation of x is interpreted as a time average along the path $x(t)$ of a single particle according to

$$\overline{x^2} = \int x^2 \varrho(x,t) dx = \langle [x(t) - x(0)]^2 \rangle \quad . \tag{37.22}$$

This is, of course, nothing but a special case of the ergodic theorem (time integral equal to ensemble average). For practical purposes the time integral is replaced by a sum

$$\langle [x(t) - x(0)]^2 \rangle = \sum [x(t + i\Delta t) - x(i\Delta t)]^2 / I \qquad (37.23)$$

where the sum runs from $i = 1$ to $i = I$ and Δt is a convenient time interval for the observations of the position of the particle.

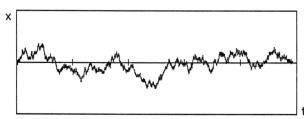

Fig. 37.1. Position $x(t)$ of a Brownian particle calculated from a simple set of random numbers (compare Problem 37.1)

Finally we consider a simplified model for the equality (37.22). Figure 37.1 exhibits a curve $x(t)$ of a random walk (Problem 37.1). The collision times are all chosen equal to the relaxation time τ. The figure covers a length of 7000 collision times. From this curve the quantity $\Delta x(t)^2 = \langle [x(t) - x(0)]^2 \rangle$ was computed for $I = 1000, 330, 110$. The result is plotted in Fig. 37.2.

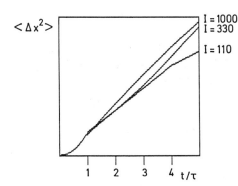

Fig. 37.2. Mean square displacement for the simplified Brownian motion model described in the text as a function of time for three different lengths I of averaging interval

PROBLEMS

37.1 Figure 37.1 was produced by generating random numbers between, say, -5 and +5 and interpreting them as velocities between two collisions. Use a computer to generate your own random numbers and plots like Figs. 37.1 and 2.

37.2 If one writes p on the r.h.s. of (36.14) as $p = mv + \delta p$ and assumes the δp are Maxwell distributed while v is a constant drift velocity, one obtains an improved momentum current density $\Pi = mv^2 + P$. What happens if you apply the same procedure to the energy current density (36.16)?

38. Fokker–Planck and Langevin Equations

There are some interesting relations between the Fokker–Planck equation and the Langevin equation, which we want to discuss in this chapter. We remind the reader, see Chap. 14, that from Langevin's equation

$$\dot{p}(t) + \gamma p(t) = F(t) \tag{38.1}$$

and the condition

$$\overline{F(t)} = 0 \tag{38.2}$$

for the average value of the stochastic force in an arbitrary nonequilibrium situation, one can easily read off an equation for the time evolution of the average momentum $\bar{p}(t)$

$$\dot{\bar{p}}(t) + \gamma \bar{p}(t) = 0 \quad . \tag{38.3}$$

Furthermore, if one requires

$$\langle F(t)F(0) \rangle = \gamma m k T \delta(t) \tag{38.4}$$

for the equilibrium correlation function of the stochastic force one can similarly derive an equation (38.3) also for the equilibrium correlation function $\langle p(t)p(0) \rangle$ of the momenta with the initial condition $\langle p(0)^2 \rangle = mkT$.

Now exactly the same results can be derived from the Fokker–Planck equation

$$\frac{\partial n(p,t)}{\partial t} = \frac{\partial}{\partial p}\left[\left(\gamma p + \gamma m k T \frac{\partial}{\partial p}\right) n(p,t)\right] \tag{38.5}$$

after multiplcation by p and integration over p

$$\bar{p}(t) = \int p n(p,t) dp \quad . \tag{38.6}$$

The friction term in (38.3) results from the first term in the bracket on the r.h.s. of (38.5) by partial integration, while the second term vanishes.

The correlation function of the momenta can be obtained from the source solution (36.10) of the Fokker–Planck equation as

$$\langle p(t)p(0) \rangle = \int dp \int dp_0 \, w(p, p_0, t) n_0(p_0) \quad . \tag{38.7}$$

Here $n_0(p_0)$ is a Maxwell distribution of the momenta at $t = 0$ and the source solution $w(p, p_0, t) = n(p, p_0, t)/n$ is the probability distribution of momenta p at time t which had the value p_0 at time $t = 0$. Since w is a (special) solution of the Fokker–Planck equation, so is every convolution of it with a function of p_0. Hence the correlation function (38.7) obeys the same equation as (38.6).

From the source solution (36.10) one can also derive higher correlation functions of the momenta. From the Langevin equation this would only be possible if higher correlation functions of the fluctuating forces were known.

The remaining contents of this chapter will now run as follows: first we derive the distribution function of the fluctuating forces compatible with the Fokker–Planck equation. It will turn out that the stochastic forces have a Gaussian distribution. We will then prove that under the conditions of such a Gaussian distribution one can also derive the Fokker–Planck equation from the Langevin equation, so in this sense there is a one-to-one correspondence between the two descriptions.

We first use the source solution (36.10) to determine the distribution function $n(p, t_1)$ from the one at an earlier time $n(p, t_0)$ by

$$n(p_1, t_1) = \int w(p_1, p_0, t_1 - t_0) n(p_0, t_0) dp_0 \quad . \tag{38.8}$$

We now apply this equation n times and call $p_n = p$, $t_n = t$. Then one obtains the so-called "compatibility condition"

$$w(p, p_0, t - t_0) = \int \cdots \int w(p, p_{n-1}, t - t_{n-1}) \cdots w(p_1, p_0, t_1 - t_0) dp_{n-1} \cdots dp_1 \quad . \tag{38.9}$$

The integrand on the r.h.s. has the meaning of the probability of finding the momenta $p_1, \cdots, p_n$ at times $t_1, \cdots, t_n$, if at time t_0 the momentum was p_0. Or, one may say, if one picks a special sequence $p_i = p(t_i)$, then the integrand of (38.9) gives the probability of finding a path $p = p(t_i)$, starting with $p(t_0) = p$.

For reasons which will soon become evident we consider the limit $n \to \infty$ for fixed p and p_0, such that the separation of adjacent p_i's approaches zero. A simple choice would be, see Fig. 38.1,

$$t_i = t_0 + i \Delta t, \quad \text{with} \quad \Delta t = (t - t_0)/n \quad . \tag{38.10}$$

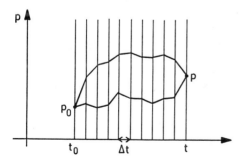

Fig. 38.1. Two "paths" in the p, t diagram going from p_0 to p

The great simplification occurring in the limit $n \to \infty$ (i.e. $\Delta t \to 0$) is that the $\exp(-\gamma \Delta t)$ in the source solutions can be replaced by $1 - \gamma \Delta t$:

$$w(p_i, p_{i-1}, \Delta t) = \frac{\exp\{[-(\Delta p_i/\Delta t + \gamma p_{i-1})^2/(4mkT\gamma)]\Delta t\}}{(4\pi mkT\gamma \delta T)^{1/2}} \quad , \quad (38.11)$$

where $\Delta p_i = p_i - p_{i-1}$. In the limit $\Delta t \to 0$ one has $\Delta p_i/\Delta t \to dp(t_i)/dt$.

Using the Langevin equation (38.1) one may also introduce the fluctuating forces in the exponent by

$$\Delta p_i/\Delta t + \gamma p_{i-1} = F(t_i) \quad . \quad (38.12)$$

For the limit $n \to \infty$ of the r.h.s. of (38.9) we have introduced the suggestive notation of a "path integral", known to physicists from Feynman's path integral method of integrating the Schrödinger equation:

$$w(p, p_0, t - t_0) = \int_{p_0}^{p} \left(\exp \int_{t_0}^{t} \{-[\dot{p}(\tau) + \gamma p(\tau)]^2/(4mkT\gamma)\}d\tau \right) Dp(\tau) \quad . \quad (38.13)$$

The exponential function now gives the probability of the occurrence of a path $p(\tau)$. As already mentioned, one may go from these probabilities to the probabilities for the fluctuating forces $F(\tau)$. One only has to determine the functional determinant $|Dp(\tau)/DF(\tau')|$. Now since the relation between the $p(t_i)$ and $F(t_j)$ is linear, because of (38.12), this determinant has to be a constant independent of $p(\tau)$ and $F(\tau)$, say, C, which in principle is determined by the normalization condition. Hence we can write for the probability distribution of the fluctuating forces $W[F(\tau)]$

$$\boxed{W(F(\tau)) = C \exp \int_0^\infty [-F(\tau)^2/(4mkT\gamma)]d\tau} \quad (38.14)$$

where C has to be determined from

$$\int W(F(\tau)) DF(\tau) = 1 \quad . \quad (38.15)$$

The limits of the integral in (38.13), t_0 and t, have been replaced by 0 and ∞ respectively.

For the determination of correlation functions of $F(\tau)$ one can avoid the explicit determination of the normalization constant C if one first introduces the generating functional

$$\left\langle \exp\left[i \int_0^\infty q(t) F(t) dt\right] \right\rangle = \exp\left[-mkT\gamma \int_0^\infty q(t)^2 dt\right] \quad . \quad (38.16)$$

From this one then obtains by differentiation with respect to $q(t)$, $q(t)q(t')$, etc. at $q(t) = 0$ [note: $\delta q(t)/\delta q(t') = \delta(t-t')$]:

$$\langle F(t) \rangle = 0 \quad , \tag{38.17a}$$

$$\langle F(t)F(t') \rangle = mkT\gamma\delta(t-t') \quad , \tag{38.17b}$$

$$\langle F(1)F(2)F(3) \rangle = 0, \quad (\text{calling } t_i = i) \quad , \tag{38.17c}$$

$$\langle F(1)F(2)F(3)F(4) \rangle = \langle F(1)F(2) \rangle \langle F(3)F(4) \rangle$$
$$+ \langle F(1)F(3) \rangle \langle F(2)F(4) \rangle \tag{38.17d}$$
$$+ \langle F(1)F(4) \rangle \langle F(2)F(3) \rangle \quad ,$$

etc. Equations (38.14,16) or the infinite system of equations (38.17) are equivalent descriptions of the fact that the forces at different times are uncorrelated and otherwise have a Gaussian distribution. One calls this *Gaussian white noise*.

Now we turn the argument around and try to prove that for the Langevin equation (38.12) and with Gaussian white noise for the stochastic forces the momentum distribution obeys a Fokker–Planck equation. Rather than going along the line of reasoning above stepwise backwards we use an interesting independent way.

We consider the momentum distribution function $\delta(p - p(t; F(\tau)))$ belonging to a particular force $F(t) = dp(t)/dt + \gamma p(t)$ and take the average

$$w(p,t) = \langle \delta(p - p(t; F(\tau))) \rangle \tag{38.18}$$

of this distribution over all $F(\tau)$ with the Gaussian distribution (38.14). Then one can prove that this function obeys the Fokker–Planck equation. For this purpose one first differentiates (38.18) with respect to t and uses the Langevin equation ($\partial_p = \partial/\partial p$)

$$\dot{w}(p,t) = \langle [\gamma\partial_p p - F(t)\partial_p]\delta(p - p(t)) \rangle \quad . \tag{38.19}$$

Now let us call the differential operator in the bracket L and decompose it into

$$L_0 = \gamma\partial_p p \quad \text{and} \quad L_1 = -F(t)\partial_p \quad . \tag{38.20}$$

Next we try a formal solution of (38.19) similar to the solution of the Schrödinger equation in the interaction picture, L_1 being the interaction term:

$$w(p,p_0,t) = e^{L_0 t} \left\langle \hat{T} \exp \int_0^t \hat{L}_1(\tau) d\tau \right\rangle \delta(p - p_0) \quad , \tag{38.21}$$

where $\hat{T}$ is the time ordering operator familiar from quantum mechanics, and

$$\hat{L}(\tau) = -F(\tau) e^{-L_0 \tau} \partial_p e^{L_0 \tau} = -F(\tau) \hat{\partial}_p(\tau) \quad . \tag{38.22}$$

Now the forces $F(\tau)$ occurring in $W(F(\tau))$ commute at different times, so the averaging with W can be commuted with the $\hat{T}$ symbol. Furthermore, from (38.21,16,22) we see

$$\left\langle \exp\left[-\int_0^t \hat{\partial}_p(\tau) F(\tau) d\tau\right] \right\rangle = \exp\left[mkT\gamma \int_0^t \hat{\partial}_p(\tau)^2 d\tau\right] \quad . \tag{38.23}$$

Hence we may finally write

$$w(p, p_0, t) = e^{L_0 t} \hat{T} \exp\left[mkT\gamma \int_0^t \hat{\partial}_p(\tau)^2 d\tau\right] \delta(p - p_0) \quad . \tag{38.24}$$

But this is just the formal solution of the Fokker–Planck equation

$$\dot{w} = (\gamma \partial_p + mkT\gamma \partial_p^2)w \tag{38.25}$$

in the interaction picture with the initial condition of the source solution.

Thus, the results of this chapter may be cast into the formula: Langevin equation with Gaussian white noise for the stochastic force = Fokker–Planck equation. The friction term of the Langevin equation leads to the first-order differential term in the Fokker–Planck equation, and the stochastic force to the second-order term and hence to the diffusive broadening of an originally sharp distribution. The advantage of the derivation of the Fokker–Planck equation of this section lies in an easy generalization to all possible variables that obey an equation similar to a Langevin equation.

39. Transport Equations in the Hydrodynamic Regime

In this chapter we consider the solution of transport equations in the long wavelength, low frequency limit. Section 39.1 describes the general method, Sect. 39.2 applies it to the diffusion of particles and heat, and Sect. 39.3 to viscosities.

39.1 The Hydrodynamic Approximation

Kinetic transport equations are in general nonlinear integrodifferential equations with very complicated kinds of solutions. For instance, the solutions of nonlinear hydrodynamics (turbulence, etc.!) form only a small subset of all solutions. We do not plan to go into such details.

In this chapter we describe a method for an approximate solution of the linearized transport equation given by *Enskog* [39.1] and later generalized to the nonlinear case [39.2]. The method yields a derivation of hydrodynamic equations from the transport equations, including the conditions of their validity, as well as explicit expressions for the transport coefficients.

Important concepts in the context of the method are the "collision invariants" and "local equilibrium", which was introduced already in Chap. 34. Collision invariants are the conserved quantities (particle number, momentum and energy) which we introduced in (32.12–14). In connection with the symmetry relations $\gamma_{pq} = \gamma_{qp}$, see (32.19), one can write

$$\sum_q \gamma_{pq} = 0 \quad , \tag{39.1}$$

$$\sum_q \gamma_{pq}\varepsilon_q = 0 \quad , \tag{39.2}$$

$$\sum_q \gamma_{pq} q_i = 0 \quad ; \quad i = 1, 2, 3 \quad . \tag{39.3}$$

In the sense of the eigenvalue equations (32.32) one may also say the quantities $\varphi_p = 1$, $\varphi_p = \varepsilon_p$ and $\varphi_p = p_i$ are eigenvectors of the collision operator with eigenvalue 0. Now rather than expanding the solutions $\varphi(p, r, t)$ and hence the deviation of the distribution function $n(p, r, t)$ from the equilibrium distribution $n_0(p)$ to first order

$$n(\boldsymbol{p},\boldsymbol{r},t) - n_0(p) = -(\partial n_0/\partial \varepsilon_p)\varphi(\boldsymbol{p},\cdots \boldsymbol{r},t) = \chi(p)\varphi(\boldsymbol{p},\boldsymbol{r},t) \qquad (39.4)$$

into the complete set of eigenstates of γ, one first considers only a decomposition of φ into a component φ_0 in the collision invariant subspace and φ_1 in the subspace orthogonal to it, with $\langle \varphi_0 | \chi | \varphi_1 \rangle = 0$.

As in Chap. 33, the function $\varphi_0(\boldsymbol{p},\boldsymbol{r},t)$ can be expressed by the distribution function $n_l(\boldsymbol{p},\boldsymbol{r},t)$ of *local equilibrium*. To first order

$$n_l(\boldsymbol{p},\boldsymbol{r},t) - n_0(p) = \chi(p)\varphi_0(\boldsymbol{p},\boldsymbol{r},t) \quad . \qquad (39.5)$$

The function n_l may be fixed by the conditions that for it the average values of the collision invariants $\varphi_\mu(\boldsymbol{p})$ agree with their exact values

$$\int \varphi_\mu(\boldsymbol{p})[n(\boldsymbol{p},\boldsymbol{r},t) - n_l(\boldsymbol{p},\boldsymbol{r},t)]d^3p = \langle \varphi_\mu|\chi|\varphi\rangle - \langle \varphi_\mu|\chi|\varphi_0\rangle = 0 \quad . \qquad (39.6)$$

One needs for this just as many Lagrange parameters in n_l as there are collision invariants. In general these parameters are the chemical potential $\mu(\boldsymbol{r},t)$, the temperature $T(\boldsymbol{r},t)$ and the drift velocities $u_i(\boldsymbol{r},t)$. One then has

$$n_l(\boldsymbol{p},\boldsymbol{r},t) = n_0((\varepsilon_p - \mu - \boldsymbol{p}\cdot\boldsymbol{u})/T) \quad . \qquad (39.7)$$

If now the l.h.s. of (39.5) is expanded to first order around equlibrium and compared with the r.h.s. one obtains

$$\varphi_0 = (\varepsilon_p - \mu)\delta T/T + \delta\mu + \boldsymbol{p}\cdot\boldsymbol{u} \quad . \qquad (39.8)$$

In principle, for Fermi liquids the linearization also produces terms from the self-consistent fields. It turns out, fortunately, that they are without consequences for the determination of transport coefficients [39.3]. We have thus neglected them.

Now let us denote the linearized transport equation by $L\varphi = 0$, with a linear operator L which we are going to write down below. Then the decomposition of this equation into the two subspaces 0, 1 is

$$L_{00}\varphi_0 + L_{01}\varphi_1 = 0 \quad , \quad L_{10}\varphi_0 + L_{11}\varphi_1 = 0 \quad . \qquad (39.9)$$

One can now determine φ_1 from the second equation

$$\varphi_1 = -\frac{1}{L_{11}}L_{10}\varphi_0 \qquad (39.10)$$

and insert it into the first one. Then one obtains an equation for φ_0 alone

$$\left(L_{00} - L_{01}\frac{1}{L_{11}}L_{10}\right)\varphi_0 = 0 \quad . \qquad (39.11)$$

Up to now no approximations have been made, except the linearization. We are now in a position to introduce the so-called *hydrodynamic approximation*. For this purpose we write down the operator L of the linearized transport equation (33.4) with the collision operator (32.4) in Fourier space

$$L = (\omega - \boldsymbol{v} \cdot \boldsymbol{k})\chi + i\gamma \quad . \tag{39.12}$$

Here we have introduced the group velocity $\boldsymbol{v} = \boldsymbol{v}(\boldsymbol{p}) = \partial \varepsilon_p / \partial \boldsymbol{p}$.

The *hydrodynamic regime* can now be defined if the eigenvalue spectrum of the collision operator γ is separated from the lowest eigenvalue zero by a gap, say, τ. Then the hydrodynamic regime can be defined as that region of frequencies and wave numbers that obey

$$\omega \tau \ll 1 \quad \text{and} \quad k\ell \ll 1 \quad . \tag{39.13}$$

Here $\ell = \bar{v}\tau$ is the mean free path corresponding to the average velocity $\bar{v}$. So the hydrodynamic regime is just the low-frequency, long-wavelength regime. In this regime one can neglect in L_{11} the ω- and k-dependent terms as compared to γ, hence

$$L_{11} = i\gamma_{11} \quad \text{(hydrodynamic regime)} \quad . \tag{39.14}$$

Also L_{10} and L_{01} simplify considerably since

$$\gamma_{01} = \gamma_{10} = \omega_{01} = \omega_{10} = 0 \quad , \tag{39.15}$$

and since $\gamma_{00} = 0$ anyway, the *hydrodynamic approximation* of (39.11) finally takes the form

$$\left([(\omega - \boldsymbol{v} \cdot \boldsymbol{k})\chi]_{00} + i(\boldsymbol{v} \cdot \boldsymbol{k}\chi)_{01} \frac{1}{\gamma_{11}} (\boldsymbol{v} \cdot \boldsymbol{k}\chi)_{10} \right) \varphi_0 = 0 \quad . \tag{39.16}$$

This, together with (39.8), is a system of equations for the basic quantities $T(\boldsymbol{k},\omega)$, $\mu(\boldsymbol{k},\omega)$ and $u_i(\boldsymbol{k},\omega)$. For dilute gases (39.16) is, indeed, essentially identical with the hydrodynamic equations.

39.2 Diffusion of Particles and Heat

We consider in this section only scalar collision invariants. Then the vectorial averages $(\boldsymbol{v}\chi)_{00}$ in the first term on the l.h.s. of (39.16) vanish. Let us first consider only a single invariant (particle number or energy). Then (39.16) simplifies to

$$(\omega + i \sum D_{ij} k_i k_j) \varphi_0(\boldsymbol{k},\omega) \quad . \tag{39.17}$$

This obviously represents a diffusion equation with a tensor D_{ij} of diffusion coefficients for either particles or heat

$$D_{ij} = \langle 0 | v_i \chi \frac{1}{\gamma_{11}} \chi v_j | 0 \rangle / \langle 0 | \chi | 0 \rangle \quad . \tag{39.18}$$

Here $|0\rangle$ is the scalar collision invariant. For cubic symmetry and isotropy the off-diagonal elements of D_{ij} and the diagonal elements are independent of i, i.e. $D_{ij} = D\delta_{ij}$ with $D = \sum D_{ii}/3$. If one calls $(1/\gamma_{11})\chi = \tau$ one may then write (39.18) in a shorthand form which is also useful for order of magnitude estimates:

$$D = \sum \langle 0|v_i \chi \tau v_i|0\rangle /3 \approx \overline{v^2}\bar{\tau}/3 \approx \bar{v}\ell/3 \quad , \tag{39.19}$$

where $\overline{v^2} = \langle 0|v^2\chi|0\rangle$ and $\ell \approx \bar{v}\bar{\tau}$ is a mean free path.

Besides the diffusion constant one also uses mobilities or conductivities as transport coefficients. The mobility B is connected to D_n for particle transport as usual by the Einstein condition

$$D_n = n(\partial \mu/\partial n) B \quad . \tag{39.20}$$

The electrical conductivity is essentially given by nB, apart from a factor e^2. The conductivity is often also written in terms of a collision time such that

$$\sigma_e = e^2 nB = e^2 n\tau/m^* \quad . \tag{39.21}$$

The connection between thermal diffusion constant D_T and thermal conductivity κ follows from their definitions $j_q = -\kappa \operatorname{grad} T$ and $\partial q/\partial t = -\operatorname{div} j_q = c_p \partial T/\partial t$ as

$$\kappa = c_p D_T \quad , \tag{39.22}$$

where $c_p = C_p/V$ is the specific heat per unit volume.

Let us finally mention that the form (39.17) can be kept "by force" also beyond the hydrodynamic regime, in the case of a single scalar collision invariant. One just has to keep L_{11} instead of the approximation $i\gamma_{11}$. The diffusion constants D, however, then become functions of k and ω: $D = D(k, \omega)$ whose low-frequency, long-wavelength limit $D(0,0)$ is equal to the hydrodynamic D.

If one is only interested in the transport coefficients, one does not have to go all the way to the hydrodynamic equations (39.16) but can stop at the constitutive equations. In the case of diffusion this is Fick's law. The complete hydrodynamic equations are then obtained by combining the constitutive equation with a conservation law. It turns out that the constitutive law is essentially contained in (39.10). Let us consider this in more detail, but in order not to repeat most of what we said above, let us work in (r, t)-space. The second line of (39.9) from which (39.10) was derived then reads

$$\chi(p)[\partial/\partial t + v(p) \cdot \operatorname{grad}]\varphi_0(p, r, t) = -\int \gamma_{pq}\varphi_1(q, r, t) d^3 q \quad . \tag{39.23}$$

If one multiplies this equation by a collision invariant and integrates over p one obtains a continuity equation (with a zero on the r.h.s.) corresponding to the conservation law of the collision invariant. Such conservation laws are, of course, valid in general beyond the linear regime (Problem 39.1). In the hydrodynamic limit one can solve (39.23) for φ_1. This corresponds to the solution of a linear inhomogeneous integral equation. Suppose we know the solution φ in terms of the l.h.s. of (39.23), i.e. the parameters δT, $\delta\mu$ and $\boldsymbol{u}$, then we can multiply φ_1 by a collision invariant and $\boldsymbol{v}\chi$ to obtain the corresponding current density by integration over p

$$\int \varphi_{0\mu}(p)\boldsymbol{v}\chi\varphi_1(\boldsymbol{p},\boldsymbol{r},t)d^3p = \boldsymbol{j}_\mu(\boldsymbol{r},t) \quad . \tag{39.24}$$

This then is a linear relation between a current density and the parameters occurring on the l.h.s. in φ_1. The corresponding coefficients then can be identified as transport coefficients.

Let us check that in the case of particle diffusion one arrives at the same results as above. For the constitutive equations it is sufficient to consider the stationary case. Then on the l.h.s. of (39.23) the $\partial/\partial t$ can be neglected. The formal solution then is

$$\varphi_1 = -\frac{1}{\gamma_{11}}\chi\boldsymbol{v}\cdot\operatorname{grad}\mu \quad , \tag{39.25}$$

and for the particle current density one finds according to (39.24) a result $j_{ni} = -n\sum B_{ij}\partial\mu/\partial x_j$ with the mobility tensor

$$nB_{ij} = \int v_i(\boldsymbol{p})\chi(p)\frac{1}{\gamma_{11}}\chi(p)v_j(\boldsymbol{p})d^3p \quad . \tag{39.26}$$

For the calculation of the heat conductivity we consider a stationary situation with constant pressure. Then the $\delta\mu$ in (39.8) can be replaced by $\delta\mu = -s\delta T$ and one obtains with the enthalpy $i = e + Pv$ per particle for (39.23) the solution

$$\varphi_1 = -\frac{1}{\gamma_{11}}\chi\boldsymbol{v}\cdot(\varepsilon_p - i)(\operatorname{grad}T)/T \quad . \tag{39.27}$$

Using (20.26) and (39.24) one can then calculate the heat current density. One finds $j_{qi} = -\sum \kappa_{ij}\partial T/\partial x_j$ with the tensor of heat conductivities

$$\kappa_{ij} = \frac{1}{T}\int (\varepsilon_p - i)v_i(\boldsymbol{p})\chi(p)\frac{1}{\gamma_{11}}\chi(p)v_j(\boldsymbol{p})(\varepsilon_p - i)d^3p \quad . \tag{39.28}$$

If one wants to compare the mobilities and conductivities with the diffusivities one has to calculate the "normalization integrals" $\langle 0|\chi|0\rangle$ occurring in (39.18). Compare also (32.33). After introducing the corresponding integrals for the scalar products, one has the relations

$$nB_{ij} = D_{nij}\int \chi(p)d^3p \tag{39.29}$$

and

$$\kappa_{ij} = D_{Tij}\int (\varepsilon_p - i)^2\chi(p)d^3p \quad . \tag{39.30}$$

These are again the typical Einstein-type relations between the Onsager coefficients B and κ and the difussivities D, differing by a factor "susceptibility". In order to have agreement with (39.20,22) the relations

$$\int \chi(p) d^3 p = \partial n/\partial \mu \;, \tag{39.31}$$

$$\int (\varepsilon_p - i)^2 \chi(p) d^3 p = c_p \tag{39.32}$$

have to hold. The first one is evident because $\chi(p) = \partial n_0(p)/\partial \mu$. For the second one, see Problem 39.2.

For the sake of completeness we consider the case of two scalar invariants. This leads to the phenomena of thermodiffusion and thermoelectricity. See Chaps. 20,21. A useful choice of variables is [see again (21.4)] one where grad μ and grad T are the driving forces. Consequently one writes (39.8) in the form

$$\varphi_0 = \alpha_1 \delta\mu + \alpha_2 \delta T \tag{39.33}$$

with

$$\alpha_1 = 1 \quad \text{and} \quad \alpha_2 = (\varepsilon_p - \mu)/T \;. \tag{39.34}$$

Then, after the usual intermediate steps in complete analogy to the case of a single scalar invariant, one finds for the Onsager coefficients of (21.4)

$$\mu_{\kappa\lambda} = \int \alpha_\kappa v \chi \frac{1}{\gamma_{11}} \chi v \alpha_\lambda d^3 p \;. \tag{39.35}$$

For the sake of simplicity we assumed isotropy. Then v stands for any of the components v_i of the velocity.

39.3 The Viscosities

Finally we consider the case where all collision invariants play a role. It is realized for the hydrodynamics of dilute gases and Fermi liquids. Then there are three transport coefficients (Chap. 29): the heat conductivity κ, the longitudinal (ζ) and transverse (η) viscosities. The treatment of the heat conductivity runs as above and need not be repeated here. For the discussion of the viscosities it is sufficient to consider homogeneous (grad P = grad T = grad n = 0), stationary ($\partial u/\partial t = 0$) flows. Then because of (39.8) one has

$$v \cdot \text{grad } \varphi_0 = \sum v_i p_j \partial u_i / \partial x_j \;. \tag{39.36}$$

For isotropic systems momentum p and velocity v have the same direction and the r.h.s. of this eq. can be decomposed into longitudinal and transverse components

$$v \cdot \text{grad } \varphi_0 = \frac{1}{3} v \cdot p \, \text{div } u + \sum \frac{1}{2} \left(v_i p_j - \frac{1}{3} v \cdot p \delta_{ij} \right) \left(\frac{\partial u_j}{\partial x_i} + \frac{\partial u_i}{\partial x_j} - \frac{2}{3} \text{div } u \delta_{ij} \right) . \tag{39.37}$$

For the calculation of $\partial \varphi_0/\partial t$ we consider φ_0 as a function of entropy s and density n. Now for constant density the entropy can change only because of dissipative effects, but this would be an effect of higher order in the small quantities $\omega\tau$ and $k\ell$. Hence φ_0 can be considered as a function of n and the time derivatives can be expressed in terms of div u via the linearized continuity equation $\partial n/\partial t + n \text{div}\, u = 0$ as

$$\dot\varphi_0 = (\partial\varphi_0/\partial n)_s \dot n = -n(\partial\varphi_0/\partial n)_s (\text{div}\, u) \quad . \tag{39.38}$$

Then one can use (39.8) and $d\mu = -sdT + vdP$ to obtain

$$\left(\frac{\partial\varphi_0}{\partial n}\right) = \frac{\varepsilon_p - \mu}{T}\left(\frac{\partial T}{\partial n}\right)_s + \left(\frac{\partial\mu}{\partial n}\right)_s = \frac{(\varepsilon_p - i)}{T}\left(\frac{\partial T}{\partial n}\right)_s + v\left(\frac{\partial P}{\partial n}\right)_s \quad . \tag{39.39}$$

Let us now first consider *longitudinal* flow and start with Fermi liquids at low temperatures. Then φ_0 is needed only in the vicinity of the Fermi energy. Mathematically this is a consequence of the occurrence of the function $\chi(p)$, which at low temperatures in lowest order of the Sommerfeld expansion [Ref. 39.4, Eq. (41.8)] is proportional to a delta function $\chi \propto \delta(\varepsilon_p - \mu) + O((T/\mu)^2)$ at the Fermi energy. Hence in the first of the two equations (39.39) one can neglect the first term on the r.h.s. and take the second one for $T = 0$, where adiabatic and isothermal derivatives are equal

$$(\partial\mu/\partial n)_s = (\partial\mu/\partial n)_0 = p_\text{f} v_\text{f}/3n \quad . \tag{39.40}$$

This term inserted into (39.38) at the Fermi energy is just the opposite of the first term on the r.h.s. of (39.37) (note that v and p are parallel). Hence the spatial and time derivatives on the l.h.s. of (39.23) cancel. This means φ_1, and hence the viscosity, vanishes in the longitudinal case.

For dilute gases it is better to start from the second equation of (39.39). One uses the thermodynamic relations

$$(\partial P/\partial n)_s = (\partial P/\partial n)_\text{T}(c_p/c_v) = kT c_p/c_v \quad , \tag{39.41}$$

$$(\partial T/\partial n)_s = (\partial P/\partial T)_n/(nc_v) = kT/c_v \quad . \tag{39.42}$$

The c's are again the specific heats per unit volume. If one takes into account that $ni = c_p T$ and inserts everything into (39.23) combined with (39.37,38) one finds

$$\partial\varphi_0/\partial t + v \cdot \text{grad}\,\varphi_0 = [pv(p)/3 - nk\varepsilon_p/c_v]\text{div}\, u \quad . \tag{39.43}$$

Hence, for monoatomic gases with $c_v = 3nk/2$ and nonrelativistic $\varepsilon_p = p^2/(2m)$ the r.h.s. of (39.43) vanishes again. There is no longitudinal viscosity. The same is true in the ultrarelativistic case $v = c$, $\varepsilon_p = cp$, $c_v = 3nk$. Otherwise and in the case of molecules where $c_v \ne 3nk/2$ the longitudinal viscosity is nonzero. In the latter case, however, one usually has to take into account the internal degrees of freedom of the molecules explicitly in the kinetic equation. The longitudinal viscosity is also nonzero at higher densities, in particular in liquids. But for them the Boltzmann equation is not sufficient.

Finally we consider the *transverse* viscosity. It comes from the second term on the r.h.s. of (39.37). It is then sufficient to consider only one typical term, for instance a flow where only $\partial u_1/\partial x_2$ is nonzero. Then the solution of (39.23) takes the form

$$\varphi_1 = -\frac{1}{\gamma_{11}} \chi p_1 v_2 \partial u_1/\partial x_2 \quad . \tag{39.44}$$

For the viscous momentum flux tensor Π'_{ij} one finally obtains

$$\Pi'_{12} = \int p_1 v_2 \chi \varphi_1 d^3 p = -\eta \partial u_1/\partial x_2 \tag{39.45}$$

with the transverse viscosity

$$\boxed{\eta = \int p_1 v_2 \chi \frac{1}{\gamma_{11}} \chi p_1 v_2 d^3 p \quad .} \tag{39.46}$$

PROBLEMS

39.1 Derive the hydrodynamic conservation laws (29.6–8) by multiplication of the kinetic equation (33.4.) with the collision invariants 1, p_i and ε_p. What are the expressions for the particle current density $n\boldsymbol{u}$, the momentum flux tensor Π_{ij} and the energy current density $\boldsymbol{j}_\varepsilon$?

39.2 Verify (39.32). Hint: Starting from (20.16) with $\sigma = ns$ one finds $Tnds = \int(\varepsilon_p - i)dn_0 d^3p$. Note, when differentiating n_0 with respect to T for fixed P the chemical potential μ in n_0 has to be differentiated as well!

39.3 Convince yourself that because of Euler's equations of hydrodynamics on the l.h.s. of (39.23,27,37,38), no new terms occur if nonstationary, inhomogeneous flows are considered.

40. The Minimum Entropy Production Variational Principle

In this chapter we consider a very useful variational principle. Section 40.1 describes the principle generally and Sect. 40.2 applies it to the calculation of transport coefficient.

40.1 The Principle of Minimum Entropy Production

The calculation of transport coefficients from (39.26,28,46) requires the inversion of the collision operator γ_{11}, which in practice amounts to the solution of integral equations. The formal solutions of these equations can be written down, see (39.25,27,44), but they are of little use for practical calculations. There is, however, a variational principle that allows the reduction of the calculations to a quadrature, and this has proved to be a powerful tool for the determination of transport coefficients from kinetic equations.

In order to formulate it let us introduce the collision invariants $|0i\rangle$ with the representations $\varphi_{0i}(p)$ and the vectors $|i\rangle$ with the representations

$$\varphi_i(\boldsymbol{p}) = v_x(\boldsymbol{p})\,\varphi_{0i}(\boldsymbol{p}) = \begin{cases} v_x(\boldsymbol{p}) \ , \\ v_x(\boldsymbol{p})(\varepsilon_p - i) \ , \\ v_x(\boldsymbol{p})v_y(\boldsymbol{p}) \ . \end{cases} \tag{40.1}$$

We restrict ourselves to the isotropic case, where it is sufficient to consider only the x component v_x of the currents.

The Onsager mobilities, conductivities and viscosities μ_i and the corresponding diffusivities $\nu_i = \mu_i/\langle i0|\chi|0i\rangle$ can then be written as

$$\mu_i = \nu_i \langle i0|\chi|0i\rangle = \left\langle i \left| \chi \frac{1}{\gamma_{11}} \chi \right| i \right\rangle \ . \tag{40.2}$$

The variational principle now results from a simple application of Schwartz's inequality to this equation, which is valid, since γ_{11} is a positive definite operator. The inequality takes the form

$$\boxed{\mu_i = \left\langle i \left| \chi \frac{1}{\gamma_{11}} \chi \right| i \right\rangle \geq \frac{|\langle i|\chi|\varphi\rangle|^2}{\langle \varphi|\gamma|\varphi\rangle} \ ; \quad \langle \varphi|0i\rangle = 0 \ .} \tag{40.3}$$

Here φ may be any vector orthogonal to the collision invariants $|0i\rangle$. The r.h.s. of this inequality thus yields a lower limit of the transport coefficients. Equation (40.3) may be used for a variational calculation in the usual way: if the variational solution φ contains parameters, they may be varied so as to maximize the r.h.s. of (40.3). This then yields the best lower limit for μ_i. If φ is allowed to vary arbitrarily in the subspace orthogonal to the collision invariants, the Euler equations of the corresponding variational problem can easily be shown to lead to the exact solution $|\varphi\rangle \propto (\gamma_{11})^{-1}\chi|i\rangle$ of the integral equations. See Problem 40.1.

The numerator and denominator on the r.h.s. of (40.3) have a simple physical interpretation. First of all, because $\chi(\mathbf{p})\varphi(\mathbf{p}) = \delta n(\mathbf{p})$ one has

$$\langle i|\chi|\varphi\rangle = \langle i0|v_x|\delta n\rangle = \int \varphi_{0i}(\mathbf{p})v_x(\mathbf{p})\delta n(\mathbf{p})d^3p = j_i \quad , \qquad (40.4)$$

which is the x component of the current density associated with the conserved quantity i. Secondly, the denominator in (40.3), apart from a factor $1/T$, is nothing but the entropy production occurring in connection with the deviation $\delta n = \varphi/\chi$ of the distribution function from equilibrium

$$\dot{\sigma}_\varphi = \langle \varphi|\gamma|\varphi\rangle/T \quad . \qquad (40.5)$$

Using (40.3,4) one can write

$$T\dot{\sigma} \leq \frac{\langle\varphi|\gamma|\varphi\rangle}{|\langle i|\chi|\varphi\rangle|^2}j_i^2 \quad . \qquad (40.6)$$

The correct solution of the kinetic equation corresponds to a minimum of the entropy production for a given current density j_i as described by (40.6).

The simplest application of (40.3) consists of a single ansatz for φ without any variational parameters. An ansatz which has been quite successful is $|\varphi\rangle = |i\rangle$, leading to

$$\mu_i \geq \frac{|\langle i|\chi|i\rangle|^2}{\langle i|\gamma|i\rangle} \quad . \qquad (40.7)$$

We are now going to use the r.h.s. of this inequality as an approximation for the transport coefficients in a number of examples.

40.2 The Classical Boltzmann Gas

The irreversible rate of change of $n(\mathbf{p}, \mathbf{r}, t)$ occurs now by two-particle collisions. Instead of (32.16) we write, following Boltzmann,

$$[\dot{n}(\boldsymbol{p}_1)]_c = \int r(\boldsymbol{p}_1,\boldsymbol{p}_2;\boldsymbol{q}_1,\boldsymbol{q}_2)[n(\boldsymbol{q}_1)n(\boldsymbol{q}_2) - n(\boldsymbol{p}_1)n(\boldsymbol{p}_2)]d^3p_2 d^3q_1 d^3q_2 \ . \tag{40.8}$$

Here we have not indicated explicitly that all distribution functions depend on the same variables $(\boldsymbol{r},t)$ in addition to the momenta. Furthermore, we have made use of the symmetry of r for interchange of $\boldsymbol{p}$ and $\boldsymbol{q}$. This is a consequence of space and time reversal invariance. The explicit expression for r, in analogy to (32.27), is

$$r(\boldsymbol{p}_1,\boldsymbol{p}_2;\boldsymbol{q}_1,\boldsymbol{q}_2) = (2\pi/\hbar)|t(\boldsymbol{p},\boldsymbol{q})|^2 \delta(\boldsymbol{P}-\boldsymbol{Q})\delta(p^2-q^2)m \ . \tag{40.9}$$

Here t is the transition matrix element for the two-particle scattering process (we use a quantum mechanical expression, since for light particles such as H_2 and He the scattering may have to be treated quantum mechanically although everything else is classical). Furthermore, we have introduced the appropriate variables for the two-particle scattering

$$\boldsymbol{P} = \boldsymbol{p}_1 + \boldsymbol{p}_2 \ ; \quad \boldsymbol{p} = (\boldsymbol{p}_1 - \boldsymbol{p}_2)/2 \ , \tag{40.10}$$

the total and relative momentum before the scattering, and similarly for the momenta $\boldsymbol{q}$ after the scattering. The two δ-functions in (40.9) express the conservation of momentum and energy in the scattering event and allow one to carry out four of the six integrations in (q_1,q_2)-space. Then $\boldsymbol{q}_1$ and $\boldsymbol{q}_2$ can be expressed in terms of the initial momenta as

$$\begin{aligned}\boldsymbol{q}_1 &= p\boldsymbol{e} + \boldsymbol{P}/2 \ , \\ \boldsymbol{q}_2 &= -p\boldsymbol{e} + \boldsymbol{P}/2 \ ,\end{aligned} \tag{40.11}$$

where $\boldsymbol{e}$ is a unit vector in $\boldsymbol{q}$-space with the components

$$\boldsymbol{e} = (\sin\vartheta\cos\varphi, \sin\vartheta\sin\varphi, \cos\vartheta) \tag{40.12}$$

and we have chosen the vector $\boldsymbol{p}$ as the polar axis in $\boldsymbol{q}$-space. The two remaining integrations then run over the unit sphere in $\boldsymbol{q}$-space with volume element $d\Omega = \sin\vartheta d\vartheta d\varphi$.

Introducing [in analogy to (32.28)] the differential scattering cross section $\sigma(p,\vartheta)$ by

$$\sigma(p,\vartheta) = (2\pi m^2/\hbar)|t(\boldsymbol{p},p\boldsymbol{e})|^2 \tag{40.13}$$

one can finally write for (40.8)

$$\boxed{[\dot{n}(\boldsymbol{p}_1)]_c = \int (2p/m)\sigma(p,\vartheta)[n(\boldsymbol{q}_1)n(\boldsymbol{q}_2) - n(\boldsymbol{p}_1)n(\boldsymbol{p}_2)]d^3p_2 d\Omega \ .} \tag{40.14}$$

In order to find the collision operator γ one has to linearize (40.14) according to $n(\boldsymbol{p}) = n_0(p) + \beta n_0(p)\varphi(\boldsymbol{p})$ since for the Boltzmann distribution $\chi(p) = \beta n_0(p)$.

Because of energy conservation one has $n_0(q_1)n_0(q_2) = n_0(p_1)n_0(p_2)$. The final result for the operator γ then reads

$$\gamma\varphi(\boldsymbol{p}_1) = \frac{2}{m} \int \beta p\sigma(p,\vartheta) n_0(p_1) n_0(p_2) [\varphi(\boldsymbol{q}_1) + \varphi(\boldsymbol{q}_2) - \varphi(\boldsymbol{p}_1) + \varphi(\boldsymbol{p}_2)] d^3 p_2 d\Omega \quad . \tag{40.15}$$

Now for (40.3) one needs $\langle \varphi|\gamma|\varphi \rangle = \int \varphi(\boldsymbol{p}_1) \gamma\varphi(\boldsymbol{p}_1) d^3 p_1$. Since (40.15) is symmetric with respect to interchange of 1 and 2 as well as $\boldsymbol{p}$ and $\boldsymbol{q}$ (the latter again because of energy conservation) one has the symmetrized result

$$\langle \varphi|\gamma|\varphi \rangle = \frac{\beta}{2m} \int p\sigma(p,\vartheta) n_0(p_1) n_0(p_2)$$
$$\times [\varphi(\boldsymbol{q}_1) + \varphi(\boldsymbol{q}_2) - \varphi(\boldsymbol{p}_1) - \varphi(\boldsymbol{p}_2)]^2 d^3 p_1 d^3 p_2 d\Omega \quad . \tag{40.16}$$

According to Chap. 39 the longitudinal viscosity vanishes. Transverse viscosity and heat conductivity are the only remaining transport coefficients. Let us start with the transverse viscosity. For this one has from (40.1)

$$\varphi(\boldsymbol{p}) = \varphi_i(\boldsymbol{p}) = p_x p_y / m^2 \quad . \tag{40.17}$$

Since the equilibrium distribution functions occur quadratically in the numerator as well as the denominator of (40.7) the normalization of n_0 can be chosen ad libitum. We choose

$$n_0(p) = [\beta/2m\pi]^{3/2} \exp[-\beta p^2/2m] \quad . \tag{40.18}$$

Then the integral of n_0 over p is normalized to unity, independent of the density n of the gas. Consequently the viscosity also is *independent of the density*. The numerator of (40.7) then can be evaluated easily

$$\langle i|\chi|i \rangle = (\beta/m^4) \int p_x^2 p_y^2 n_0 d^3 p = kT/m^2 \tag{40.19}$$

and for the denominator (40.16) leads to

$$\langle i|\gamma|i \rangle m^4 = \pi^{-3} \left(\frac{\beta}{2m}\right)^4 \int d^3 p_1 d^3 p_2 d\Omega p\sigma(p,\vartheta)$$
$$\times \exp\left(-\frac{\beta(p_1^2 + p_2^2)}{2m}\right) (q_{1x}q_{1y} + q_{2x}q_{2y} - p_{1x}p_{1y} - p_{2x}p_{2y})^2 \quad . \tag{40.20}$$

Instead of $\boldsymbol{p}_1$ and $\boldsymbol{p}_2$ we now use total and relative momentum according to (40.10,11). Then the $\boldsymbol{P}$ integration can be carried out, leaving

$$\langle i|\gamma|i \rangle m^4 = \frac{2}{\pi^{3/2}} \left(\frac{\beta}{m}\right)^{5/2} \int d^3 p d\Omega p\sigma(p,\vartheta)$$
$$\times \exp(\beta p^2/m)(q_x q_y - p_x p_y)^2 \quad . \tag{40.21}$$

Because of the symmetry of the integrand one may replace the square in the

204

second line by $2p_xp_y(p_xp_y - q_xq_y)$. Next one can simplify the integral using

$$\int d\Omega \sigma(p,\vartheta)(p_xp_y - q_xq_y) = \frac{3}{2}p_xp_y \int \sigma(p,\vartheta)(1 - \cos^2\vartheta)d\Omega \quad . \qquad (40.22)$$

This equation can be derived, for instance, by considering the l.h.s. of it as the x, y component of the tensor $I_{ij} = \int(p_ip_j - q_iq_j)\sigma d\Omega = \alpha(p_ip_j - p^2\delta_{ij}/3)$. This tensor has to have zero trace (as indicated in the ansatz) because of energy conservation $(p^2 - q^2) = 0$. The constant α can then be determined from $\sum p_i I_{ij} p_j = \int[p^4 - (\mathbf{p}\cdot\mathbf{q})^2]\sigma d\Omega = 2p^4\alpha/3$. Another way is to expand $\sigma(\vartheta)$ in Legendre polynomials and use the addition theorem of spherical harmonics together with (40.11,12).

If (40.22) is inserted into (40.21) the angular integral in p-space can be carried out, leading to

$$\langle i|\gamma|i\rangle m^4 = \frac{8}{5\sqrt{\pi}}\left(\frac{m}{\beta}\right)^{3/2}\int_0^\infty g^7 e^{-g^2}\int \sigma(g,\vartheta)(1-\cos^2\vartheta)d\Omega dg \quad , \qquad (40.23)$$

where we have introduced the dimensionless momentum $g = (m/\beta)^{1/2}p$. Combining this with (39.46) and (40.7,19) one obtains the final lower limit for the transverse viscosity

$$\boxed{\begin{aligned} \eta &= mv_T/\sigma_T = (mkT)^{1/2}/\sigma_T; \quad \text{with} \\ \sigma_T &= \frac{8}{5\sqrt{\pi}}\int g^7 e^{-g^2}\int \sigma(g,\vartheta)(1-\cos^2\vartheta)d\Omega dg \quad . \end{aligned}} \qquad (40.24)$$

This is as far as a theory with a phenomenological cross section $\sigma(p,\vartheta)$ can go. The variational ansatz (40.7) reduces the calculation of the viscosity to the evaluation of a double integral (40.24). A similar procedure can be applied to the thermal conductivity (39.28). After an equally lengthy calculation one finds [40.1]

$$\boxed{\kappa = \frac{5}{2}C_v^*\eta = \frac{15k}{4m}\eta} \qquad (40.25)$$

where $C_v^* = 3k/2m$ is the specific heat per unit mass. This result, which is a direct consequence of the variational ansatz (40.7) and is independent of the cross section, is fulfilled experimentally for the noble gases to within about 5%.

An actual evaluation of (40.24) would first require a microscopic calculation of the differential cross section $\sigma(p,\vartheta)$. This is more a task for the third part of this volume, which is intended to deal with such microscopic calculations. Since, however, a classical calculation of cross sections from an interaction potential $V(r)$ is not difficult in principle (it amounts to another quadrature [40.2]) let us at least discuss a few simple results of such calculations in the context of (40.24).

The simplest example is, of course, the hard core potential, which becomes infinitely repulsive at a radius a. The total cross section for this potential is πa^2 and hence $\sigma(p, \vartheta) = a^2/4$, independent of p and ϑ. Then (40.24) can be evaluated easily, yielding

$$\eta = \frac{5}{16\pi a^2}\left(\frac{mkT}{\pi}\right)^{1/2} ; \quad \text{hard core gas} \quad . \tag{40.26}$$

The hard core potential is easy enough to allow an estimate of corrections to the simple variational ansatz. One finds [40.3] corrections to (40.26) of the order of 2%.

A somewhat more realistic potential is one with a soft repulsive core, for instance, a power law $V(r) = d/r^n$. In this case a simple dimensional analysis shows that $\sigma \propto a_e^2$ where the effective radius a_e is determined by the condition $d/a_e^n = p^2/m = g^2 kT$ leading to $\sigma \propto p^{-4/n} \propto T{-2/n}$. Particularly interesting is the case $n = 4$ (considered first by Maxwell) where $\sigma \propto 1/p \propto T^{-1/2}$. Then $\eta \propto T$ and the quantity $p\sigma(p, \vartheta)$ occurring in the collision operator (40.16) is independent of p. As a consequence of this simplification, the integral equations for φ can be solved exactly and *the variational ansatz* $|\varphi\rangle = |i\rangle$ *is exact*. The viscosity turns out to be

$$\eta = \frac{1}{0.654}\left(\frac{2m}{\pi d}\right)^{1/2} kT ; \quad \text{"Maxwell" gas } (V = d/r^4) \quad . \tag{40.27}$$

Hence for the two cases $n = 4$ and $n = \infty$ (hard core) the variational solution is either exact or correct up to about 2%. This supports the idea that also for intermediate n or more realistic potentials such as the Lennard-Jones potential $V(r) = V_0[(r/a)^{12} - (r/a)^6]$ the variational ansatz (40.7) leads to a good approximation for η and κ. Figure 40.1 shows results for the "Maxwell", hard core and Lennard-Jones potentials.

40.3 The Electron–Phonon System

For the coupling of the electrons to the phonons we assume a short-range potential u between the electrons and nuclei (either a screened Coulomb potential or a deformation potential [40.5]). The interaction Hamiltonian then contains essentially a local coupling between the density $n(r)$ of electrons and $\varrho_i(r)$ of nuclei. There are other interaction Hamiltonians, in which the stress tensor of the electrons couples locally to the velocity gradients of the nuclear motion, which are mainly used for the discussion of ultrasonic attenuation [40.6]). Both Hamiltonians are assumed to be equivalent, although, so far, the equivalence has not been demonstrated in all details.

If the scattering is treated in the Born approximation one can use the result (9.5) for the double differential cross section. The spectral function $s(k, \omega)$

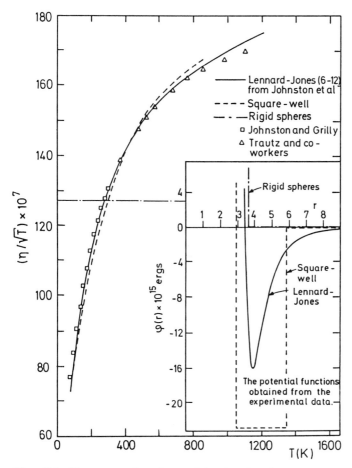

Fig. 40.1. Temperature dependence of the viscosity of argon gas: comparison of theoretical results for different molecular potentials with experiment [40.4]

relevant for electron scattering then is essentially the same as the one for neutron scattering. One only has to subtract out the elastic part leading to von Laue scattering. This part occurs only for the scattering of plane waves. The electrons, however, are in Bloch states which are not scattered any more by the periodic potential. Let us for this section introduce the spectral function $s_i(\boldsymbol{k},\omega)$ of ionic density flutuations, which should not be confused with the incoherent spectral function of Chap. 9 but is defined as

$$s_i(\boldsymbol{k},\omega) = s_{\varrho_i}(\boldsymbol{k},\omega) - 2\pi\delta(\omega)|\langle\varrho_i(\boldsymbol{k})\rangle|^2 \quad . \tag{40.28}$$

The collision term, on the one hand, is now simplified, since one has only single-particle scattering. On the other hand one has to take into account the Fermi statistics of the electrons as indicated in (32.17). For comparison with (32.17) and Chap. 44 we keep discrete momentum variables in a periodic box and write

$$[\dot{n}_p]_c = \sum [r_{pq} n_q (1 - n_p) - r_{qp} n_p (1 - n_q)] \qquad (40.29)$$

with

$$r_{pq} = 2\pi |u(\mathbf{k})|^2 \int s_i(\mathbf{k}, \omega) \delta(\varepsilon_p - \varepsilon_q - \hbar\omega) d\omega / \hbar V \quad . \qquad (40.30)$$

Momentum conservation requires

$$\mathbf{p} = \mathbf{q} + \hbar \mathbf{k} \qquad (40.31)$$

Before linearization it is convenient to make use of detailed balance and start from (32.20). This leads to

$$\gamma \varphi_p = \beta \sum r_{pq} n_q (1 - n_p)(\varphi_q - \varphi_p) \quad , \qquad (40.32)$$

which defines the collision operator. The occupation numbers in this equation may be replaced by their equilibrium values.

The relation between the normalizations for discrete and continuous momenta is given by $\sum v_p n_p / V = \int v_p n(p) d^3 p$. Hence one has

$$\langle i|\chi|i\rangle = \sum \varphi_{ip}^2 \chi_p / V \quad ; \quad \langle i|\gamma|i\rangle = \sum \varphi_{ip} \gamma_{pq} \varphi_{iq} / V \quad . \qquad (40.33)$$

Let us start with the mobility, for which $\varphi_{ip} = p_x/m$. Using $\chi_p = \partial n_p / \partial \mu$ one finds

$$\langle i|\chi|i\rangle = n/m \qquad (40.34)$$

and

$$\langle i|\gamma|i\rangle = \frac{\beta}{2m^2} \sum r_{pq} n_q (1 - n_p)(\hbar k_x)^2 / V \quad , \qquad (40.35)$$

where we have used momentum conservation.

After inserting (40.30) we introduce the spectral function of electronic excitations

$$s(\mathbf{k}, \omega) = \pi \hbar \sum_p n_{p-\hbar k}(1 - n_p) \delta(\varepsilon_{p-\hbar k} - \varepsilon_p - \hbar\omega)/V \quad . \qquad (40.36)$$

This takes care of the p sum in (40.35). The remaining q sum can be replaced by a k sum. Collecting everything together, and keeping (39.26) in mind one may write the final result as

$$\sigma = nBe^2 \quad ; \quad B = \tau_{\rm el}/m \qquad (40.37)$$

with

$$1/\tau_{\rm el} = \sum |k_x u(\mathbf{k})|^2 \int s_i(\mathbf{k}, \omega) s(\mathbf{k}, -\omega) \beta d\omega / \pi m N \quad . \qquad (40.38)$$

For further evaluation of this integral we refer to Chap. 44, where the same result is derived again by a different method. The essential ingredient is an approximation for $s(\mathbf{k}, -\omega)$, namely

$$s(\mathbf{k}, -\omega) = \pi \hbar \omega n(\omega)(dn/d\mu)\Theta(2k_{\rm f} - k)/2v_{\rm f} k \quad . \qquad (40.39)$$

Here $n(\omega) = [\exp(\beta\hbar\omega) - 1]^{-1}$ is the Bose function. Inserting (40.39) into (40.38) one obtains the final result

$$\frac{1}{\tau_{\text{el}}} = \frac{m}{12(\hbar\pi)^3 Z} \int |u(k)|^2 k^3 s_i(k,\omega)\beta\hbar\omega n(\omega) dk d\omega/\pi \quad . \tag{40.40}$$

Here $Z = n/n_i$ is the number of conduction electrons per ion. Hence for the electron–phonon system again the variational principle reduces the calculation of the mobility to a double integral, although the quantities $u(k)$ and $s_i(k,\omega)$ occurring in this integral are more difficult to obtain than the quantities for the dilute Boltzmann gas. An estimate for (40.40) can be found in Chap. 44.

For the calculation of the thermal conductivity one has to consider the second of the variational functions φ_i in (40.1). The calculation is more involved than for the electrical conductivity. Besides the function $s(k,\omega)$, more complicated quantities occur. We do not want to go into the details, but for the sake of completeness we quote the result [40.5]

$$\kappa = c_v n v_f^2 \tau_{\text{th}}/3 \tag{40.41}$$

where c_v is the specific heat per particle and

$$\frac{1}{\tau_{\text{th}}} = \frac{m}{12(\hbar\pi)^3 Z} \int |u(k)|^2 [k^3 + (\beta\hbar\omega/\pi)^2 (3kk_f^2 - k^3/2)] \\ \times s_i(k,\omega)\beta\hbar\omega n(\omega) dk d\omega/\pi \quad . \tag{40.42}$$

At high temperatures $T \gg \theta_D$ (Θ_D: the Debye temperature) the $(\beta\hbar\omega)^2$ term is negligible and the two collision times for electrical and thermal conduction are equal, which is the basis of the *Wiedemann–Franz law*.

Another transport property of metals is ultrasonic attenuation. Here one has to distinguish two regimes [40.7]: for wavelengths $1/k$ large compared to the electron mean free path ℓ the electrons are in local equilibrium with the ultrasonic wave. To first order there is only a centre-of-gravity motion of the combined electron–ion system and no relative motion as in electrical conduction. There is approximate relative momentum conservation and only the velocity gradients relax according to a viscosity. The attenuation is $\propto (k\ell)^2$, see (29.38). At shorter wavelengths there is no local equilibrium and the attenuation mechanism is different [40.7], leading to an attenuation $\propto (k\ell)$.

The viscosity can again be calculated from the variational ansatz (40.7) using the third variational function of (40.1). The calculation is very lengthy, and we again only quote the result [40.5]

$$\eta = mnv_f^2 \tau_v/5 \tag{40.43}$$

with

$$\boxed{\frac{1}{\tau_v} = \frac{m}{12(\hbar\pi)^3 Z} \int |u(k)|^2 (3k^3 - 3k^5/4k_f^2) \\ \times s_i(k,\omega)\beta\hbar\omega n(\omega) dk\,d\omega/\pi \quad .}$$
(40.44)

The leading temperature dependence of the collision frequencies is estimated in Chap. 44 and found to be $\propto T^5$ at low temperatures (this is the *Bloch–Grüneisen law*) and $\propto T$ at high temperatures.

40.4 Fermi Liquids

The collision term for Fermi liquids describes two-particle (or quasiparticle) scattering processes. The Pauli principle leads to four Fermi functions in the collision rate. The "loss" term of (40.29), for instance, now contains the factor $n_{p_1} n_{p_2}(1 - n_{q_1})(1 - n_{q_2})$. At first sight this seems to complicate the situation but, in fact, it leads to great simplifications at low temperatures. Since the initial-state excitation energy $\varepsilon_1 - \mu = \varepsilon \approx kT$ is small compared to the Fermi energy μ, the Pauli principle together with energy conservation in the scattering event severely restricts the available phase space. The initial state of excitation energy ε can produce a particle–hole excitation with an energy of order ε and leave the particle at an energy of at most ε. Thus the phase space is of order $\varepsilon^2 \approx (kT)^2$. This fixes the leading T dependence for all collision rates at low temperatures. From dimensional considerations one expects

$$1/\tau \approx m(kT)^2 \langle\sigma\rangle/\hbar^3 \quad , \tag{40.45}$$

where $\langle\sigma\rangle$ is some average cross section.

Furthermore, since the absolute magnitude of all four momenta occurring in the collision probability is fixed at about the Fermi momentum p_f the cross section depends only on angles. The average in (40.45) is an angular average. A convenient choice for two such angles — taking momentum conservation into account — is (i) the angle Θ between p_1 and p_2 (which is equal to the angle between q_1 and q_2) and (ii) the angle φ between the two planes spanned by the vectors p_1, p_2 and q_1, q_2. Hence one has $\sigma = \sigma(\Theta, \varphi)$.

As a consequence of all these geometrical simplifications it turns out that the Boltzmann equation for Fermi liquids at low temperatures can be solved exactly in terms of a rapidly converging series [40.8]. The variational principle has also been applied [40.9] to liquid ^{3}He. Its results could be checked by means of the exact solutions and turned out to be correct to within a few per cent. The variational calculations yield relatively simple expressions for the appropriate angular averages of $\sigma(\Theta, \varphi)$ to be used for the heat conductivity, viscosity and spin diffusion [40.9]. The calculations are again somewhat lengthy and hence will not be discussed here.

PROBLEMS

40.1 Determine the Euler equations of the variational problem

$$\delta\mu_i = \delta(|\langle i|\chi|\varphi\rangle|^2/\langle\varphi|\gamma\varphi\rangle) = 0$$

and demonstrate that they lead to the exact result $\mu_i = \langle i|(1/\gamma_{11})|i\rangle$ for the transport coefficient.

40.2 The function $s(k,\omega)$ introduced in (40.36) is related to the (imaginary part of) the Lindhard function (34.5). Determine the relation and use it to derive (40.39) independently of Chaps. 43 and 44.

Part III

Calculation of Kinetic Coefficients

41. Approximation Methods

As in [41.1], the third part of this volume will be devoted to examples of more microscopic calculations. Whereas equilibrium theory deals with partition functions and the corresponding thermodynamic functions, we will deal with the determination of the parameters in correlation functions, such as kinetic coefficients, transport coefficients and rate constants.

In this context, microscopic calculation means a calculation that starts directly from a Hamiltonian, or (for dynamic critical phenomena) from a Ginzburg–Landau functional, or (for Fermi liquids) from a Landau functional. The input parameters of such a calculation are interaction parameters and (effective) masses, etc. Usually there are several steps, or levels, between the Hamiltonian and the corresponding Schrödinger equation and a correlation function or a phenomenological kinetic coefficient. Let us, for example, consider the frictional motion of a macroscopic object in a dilute gas described by the kinetic equation $dP/dt = -\gamma P$. "Above" this purely phenomenological equation there is usually a hydrodynamic level described in terms of the Navier–Stokes equations (with a viscosity η) for the hydrodynamic motion of the gas. For a spherical object with a radius a this leads to Stokes' formula $\gamma = 6\pi\eta a/M$, M being the mass of the object. The next level above the Navier–Stokes equation would be the Boltzmann equation, containing the collision rates as parameters, and finally one would come to the truly microscopic level described in terms of the Liouville equation with the interaction potentials as parameters. It is, of course, just a matter of definition to call only the step from the Liouville level to the Boltzmann level a microscopic calculation. Enskog's solution of the Boltzmann equation, for instance, amounts to a determination of hydrodynamic kinetic coefficients. We have considered this step as "phenomenological" and allocated it to the second part of this volume.

None of the four levels in the scheme Liouville equation → Boltzmann equation → Navier–Stokes equation→ "relaxator" equation can be derived with complete rigor from its corresponding "higher" level. The Liouville equation can be derived from the Schödinger equation (or von Neumann equation) only in the classical limit; the Boltzmann equation is valid only in the low-density limit, the Navier–Stokes equation can be derived from the Boltzmann equation (via the Chapman–Enskog scheme) in the low-frequency, long-wavelength limit and Stokes law is valid only for low Reynolds numbers [41.2]. The corresponding approximations are contained in the scheme of Table 41.1.

Table 41.1. Approximations in the calculation of the friction constant

von Neumann eq.	
⇓	classical approx.
Liouville eq.	
⇓	low density
Boltzmann eq.	
⇓	low frequency etc.
Navier–Stokes eq.	
⇓	low Reynolds no.
Stokes' formula	

Some of the approximations are known already from equilibrium theory, some of them are new. Since the correlation functions of nonequilibrium theory (at least in linear response approximation) are equilibrium expectation values one might think that practically all approximations of equilibrium theory can be carried over to nonequilibrium theory. To some extent this is true, but in most cases there are additional problems and complications. Quite often the evaluation of higher-order terms beyond the low-order limits of the approximations is very difficult, if possible at all. The zero-order limits of many approximations therefore remain uncontrolled. The regions of their validity are often not known very precisely and outside these regions qualitatively new phenomena occur. Outside the classical regime typical quantum transport phenomena occur, such as the so-called weak localization [41.3]. See also Chap. 46. The low-density limit, on the other hand, cannot be considered as the first-order term of a simple power expansion such as the virial expansion of equilibrium theory: there are logarithmic terms in the density in higher order [41.4]. Outside the hydrodynamic regime one finds typical nonlocal effects, in particular anomalous boundary conditions, or high-frequency effects, such as zero sound.

In the limit of low Reynolds numbers the nonlinear, convective acceleration term $(v\,\text{grad})v$ in the Navier–Stokes equations can be dropped. This corresponds to the usual linear response approximation. Outside the linear regime typical nonlinear effects such as turbulence may occur.

Because of the enormous difficulties in treating the higher-order corrections mentioned above analytically, numerical "simulation" methods have become increasingly important. One considers simplified models (time-dependent Ising models, classical models, etc.) and solves the differential equations for their time evolution numerically. Approximations in these calculations result from the fact that one can consider only finite times and finite sizes. The approach to equilibrium and the thermodynamic limit can be studied only by extrapolation and sometimes erroneous conclusions are drawn due to uncertainties in these extrapolations.

In the third part of this volume we consider some typical examples of microscopic calculations in lowest-order perturbation theory and some higher-order effects which can be obtained from partial summations of higher-order perturbation terms, so-called mode coupling theories, and from expansions around quasi-static equilibrium, rather than total equilibrium.

42. Correlation Functions for Single-Particle Problems

An important field for the application of transport theories has always been electrical conduction in solids. Particularly simple, at least with respect to many-body effects, is the residual resistivity due to the scattering of electrons by impurities and imperfections. Normally this mechanism governs the conduction at low temperatures, where contributions from electron–phonon and electron–electron interactions are frozen out and one can describe the situation approximately by a single-particle Hamiltonian

$$\underline{H} = \sum \hbar \omega_\mu c_\mu^* c_\mu \quad . \tag{42.1}$$

Here $c^*(c)$ are creation (annihilation) operators for electrons with the usual Fermion commutation rules

$$c_\mu c_\nu + c_\nu c_\mu = 0 \quad , \quad c_\mu^* c_\nu + c_\nu c_\mu^* = \delta_{\nu\mu} \quad . \tag{42.2}$$

42.1 General

For our further considerations it will be useful to derive the special form of the correlation functions of linear response theory for single-particle problems. Let us introduce, besides the Hamiltonian $\underline{H}$, the operators

$$\underline{a} = \sum a_{\mu\nu} c_\mu^* c_\nu \quad , \quad \underline{b} = \sum b_{\mu\nu} c_\mu^* c_\nu \tag{42.3}$$

and the corresponding single-particle counterparts ($|\mu\rangle = c_\mu^*|0\rangle$, where $|0\rangle$ is the vacuum state)

$$a = \sum a_{\mu\nu} |\mu\rangle\langle\nu| \quad , \quad b \text{ analogously} \tag{42.4}$$

and

$$h = \sum \hbar \omega_\mu |\mu\rangle\langle\nu| \quad . \tag{42.5}$$

In the calculation of the susceptibilities one encounters the expectation values of commutators

$$\langle [\underline{a}, \underline{b}] \rangle = \sum a_{\kappa\lambda} b_{\mu\nu} \langle [c_\kappa^* c_\lambda, c_\mu^* c_\nu] \rangle \quad , \tag{42.6}$$

which can be evaluated using (42.2) and is found to be

$$\langle [c_\kappa^* c_\lambda, c_\mu^* c_\nu] \rangle = \langle c_\kappa^* c_\nu \rangle \delta_{\lambda\mu} - \langle c_\mu^* c_\lambda \rangle \delta_{\kappa\nu}$$
$$= [f(\varepsilon_\kappa) - f(\varepsilon_\mu)] \delta_{\kappa\nu} \delta_{\lambda\mu} \tag{42.7}$$

with the Fermi function ($\varepsilon_\kappa = \hbar\omega_\kappa$)

$$f(\varepsilon) = 1/\{\exp[\beta(\varepsilon - \mu)] + 1\} \quad . \tag{42.8}$$

Inserting (42.7) into (42.6) one obtains

$$\langle [\underline{a}, \underline{b}] \rangle = \sum a_{\kappa\mu} b_{\mu\kappa} [f(\varepsilon_\kappa) - f(\varepsilon_\mu)] = \sum \langle \kappa | [a, b] | \kappa \rangle f(\varepsilon_\kappa) \tag{42.9}$$

or, independent of a special representation in Hilbert space,

$$\mathrm{Tr}\{[\underline{a}, \underline{b}]\varrho(\underline{H})\} = \mathrm{Tr}\{[a, b]f(h)\} \quad , \tag{42.10}$$

which is easy to remember.

Now, if one wants to calculate the susceptibility for the response of a to an external force acting on b one needs the commutator of $a(t)$ with $b = b(0)$, where $a(t)$ is an operator such as defined in (42.4) with

$$a_{\mu\nu}(t) = a_{\mu\nu}\exp[i(\omega_\mu - \omega_\nu)t] \quad . \tag{42.11}$$

Thus the susceptibility of single-particle operators for a single-particle problem with the Hamiltonian h can be written as

$$\boxed{\chi_{ab}(t) = i\mathrm{Tr}\{[a(t), b]f(h)\}\Theta(t)/\hbar} \tag{42.12}$$

and its Fourier transform as

$$\boxed{\chi_{ab}(z) = \frac{1}{\hbar}\sum a_{\kappa\mu} b_{\mu\kappa} \frac{f(\varepsilon_\kappa) - f(\varepsilon_\mu)}{\omega_\mu - \omega_\kappa - z}} \quad . \tag{42.13}$$

The spectral function takes the form

$$\boxed{\chi''_{ab}(\omega) = \frac{\pi}{\hbar}\sum a_{\kappa\mu} b_{\mu\kappa} [f(\varepsilon_\kappa) - f(\varepsilon_\mu)]\delta(\omega_\mu - \omega_\kappa - \omega)} \quad . \tag{42.14}$$

Because of the delta function the energy ε_κ in the Fermi function can be replaced by $\varepsilon_\mu - \hbar\omega$. At low frequencies ω and low temperatures T the difference of the Fermi functions in (42.14) is practically zero except in a narrow region of width $\hbar\omega + kT$ near the Fermi energy μ. If in addition $\hbar\omega$ is small compared to kT one can expand (42.14) and obtains

$$\chi''_{ab}(\omega) = -\pi\omega \sum a_{\kappa\mu} b_{\mu\kappa} f'(\varepsilon_\mu)\delta(\omega_\mu - \omega_\kappa - \omega) + \cdots \quad . \tag{42.15}$$

Here $-f'(\varepsilon) = df(\varepsilon)/d\mu$ is a function peaked near the Fermi energy with a width kT. In particular, if kT is small compared to the Fermi energy μ one can replace $-f'(\varepsilon)$ approximately by $\delta(\varepsilon - \mu)$. Then, because of the second delta function in (42.15) both energies ε_κ and ε_μ are fixed near the Fermi energy:

The behaviour of the absorptive part of the susceptibility at low frequencies and low temperatures is completely determined by the properties of the single-particle states near the Fermi level.

Let us now carry these results over to the relaxation function

$$\Phi_{ab}(z) = [\chi_{ab}(z) - \chi_{ab}^T]/z \tag{42.16}$$

and consequently

$$\Phi_{ab}''(\omega) = \chi_{ab}''(\omega)/\omega \quad . \tag{42.17}$$

If again one is interested in the low-frequency behaviour of $\Phi_{ab}(\omega + i0)$ one can use (42.15) to obtain

$$\Phi_{ab}''(\omega) = -\pi \sum a_{\kappa\mu} b_{\mu\kappa} f'(\varepsilon_\mu) \delta(\omega_\mu - \omega_\kappa - \omega) + \cdots \quad . \tag{42.18}$$

For the nonabsorptive part of Φ one can use (42.13) with (42.16) to obtain

$$\Phi_{ab}'(\omega) = \Phi_{ab}'(0) + O(\omega) \quad . \tag{42.19}$$

Here $\Phi_{ab}'(0)$ differs from $\chi_{ab}(0)$, compare (42.13), only by having a denominator $(\omega_\mu - \omega_\kappa)^2$ instead of $(\omega_\mu - \omega_\kappa)$. Thus the diagonal part $a = b$ of the relaxation function vanishes at $\omega = 0$ because of the antisymmetry of the terms in the sum for Φ under interchange of μ and κ. The diagonal part of Φ at $\omega = 0$ is therefore equal to $\Phi_{aa}''(0)$ as given by (42.18):

The diagonal part of the relaxation function (at low frequencies) is completely determined by the properties of the single-particle states near the Fermi level.

42.2 Impurity Conduction (Greenwood Formula)

The electrical conductivity can be derived (Chap. 31) by considering the response of the electrical current density env_m to an electric field E_n [42.1]. If we restrict ourselves to homogeneous fields their effect can be described by a term $-\sum eE_n x_n$ in the Hamiltonian. Linear response theory then tells us that the expectation value $en\langle v_m \rangle = enV_m$ ($n = N_e/V$ the unperturbed particle density) obeys Ohm's law

$$enV_m(\omega) = \sum \sigma_{mn}(\omega) E_n(\omega) \tag{42.20}$$

and the conductivity $\sigma(\omega)$ is given, apart from a trivial factor e^2/V, as the Fourier transform of the susceptibility

$$\hat{\chi}_{mn}(t) = i\langle [v_m(t), x_n(0)]\rangle \Theta(t)/\hbar \quad . \tag{42.21}$$

Taking into account the relation (3.25) between commutators and bracket symbols (or dynamic susceptibilities and relaxation functions) one can write this as (remember $\dot{x}_n = v_n$)

$$\hat{\chi}_{mn}(t) = \beta \langle v_m(t); v_n(0) \rangle = -\mathrm{i}\Phi_{mn}(t) \quad, \tag{42.22}$$

where Φ_{mn} is the velocity–velocity relaxation function. Thus the conductivity is given by

$$\boxed{\sigma_{mn}(z) = -\mathrm{i}e^2 \Phi_{mn}(z)/V} \quad . \tag{42.23}$$

One can now combine this relation with the Fourier transform (3.32), or more generally (6.15), of (3.25), in our case

$$\Phi_{mn}(z) = [\chi_{mn}(z) - \chi_{mn}^T]/z \quad, \tag{42.24}$$

where χ is the Fourier transform of the retarded velocity commutator

$$\chi_{mn}(z) = \frac{1}{\hbar} \sum v_{m\kappa\mu} v_{n\mu\kappa} \frac{f(\varepsilon_\kappa) - f(\varepsilon_\mu)}{\omega_\mu - \omega_\kappa - \omega} \tag{42.25}$$

and, using (3.25) in reverse,

$$\begin{aligned}\chi_{mn}^T &= \beta \langle v_m; v_n \rangle = \frac{\mathrm{i}}{\hbar} \langle [v_m, x_n] \rangle \\ &= N_e \langle \partial v_m(p)/\partial p_n \rangle = \langle (N_e/m)_{mn} \rangle \quad .\end{aligned} \tag{42.26}$$

Here we have taken into account that v in general (in particular in the presence of periodic potentials) is not just proportional to the momentum p but may be some function of p. Equation (42.26) then defines the (inverse) mass tensor. Furthermore, of course, we have used the commutation relation $\mathrm{i}[p_m, x_n]/\hbar = \delta_{mn}$.

If one finally considers $\chi_{mn}(0)$ according to (42.25) and uses the equation of motion $v_{m\mu\kappa} = \mathrm{i}(\omega_\mu - \omega_\kappa) x_{m\mu\kappa}$, one sees that $\chi_{mn}(0)$, too, is given by the equal time commutator of velocity and position:

$$\chi_{mn}(0) = \mathrm{i}\langle [v_m, x_n] \rangle / \hbar = \chi_{mn}^T \quad . \tag{42.27}$$

Here the second equation is obtained by looking at (42.26) and guarantees ergodicity. The general results of the foregoing section, therefore, can be applied to the conductivity. In particular:

The diagonal part of the conductivity tensor at low frequencies and low temperatures is completely determined by the single-particle states near the Fermi level. In particular, the static diagonal conductivity at zero temperature has to vanish, if the Fermi energy lies in an energy gap or in a region of localized states (since such states cannot carry a current).

PROBLEMS

42.1 Determine the Lindhard function [42.2]. Use (42.13) with $a = \varrho(k) = \exp(-\mathrm{i}k \cdot r)/V^{1/2}$, $b = \varrho(-k)$. Compare the result for small k with (34.5,6).

43. Perturbation Theory for Impurity Conduction

After the preparatory work of the previous chapter and the general results of the formalism in Chaps. 12 and 13, a microscopic calculation of the residual resistivity due to the scattering of electrons by impurities and imperfections is quite straightforward. The starting point is the equation

$$\Phi(z) = \frac{1}{N(z) - z}\chi^T \tag{12.1}$$

for the relaxation function defining the memory kernel $N(z)$.

Let us restrict ourselves to the diagonal terms of the conductivity, for instance σ_{xx}. Furthermore, for simplicity, we consider parabolic energy momentum relations with $p = mv$ (p and v now being the x components of momentum and velocity). The generalized Langevin equation (13.14) for the memory kernel $M = N\chi^T = NN_e/m$ then reduces to

$$M(t) = -\mathrm{i}\beta\langle F(t); \dot{v}\rangle\Theta(t) \quad . \tag{43.1}$$

Note that the first term on the r.h.s. of (13.14) *always* vanishes for $k = l$. The stochastic force F(t) according to (13.8) reduces to

$$F(t) = \dot{v}(t) + \int_{t'}^{t} N(t - t'')v(t'')dt'' \quad . \tag{43.2}$$

Our aim is to evaluate the memory kernel (43.1) by perturbation theory (power expansion in the random potential U of the impurities and/or imperfections) given by

$$U(r) = \sum U(k)\exp(-\mathrm{i}k \cdot r)/\sqrt{V} = \sum U(k)\varrho(k) \quad . \tag{43.3}$$

The acceleration $\dot{v}(t)$ can be obtained from this potential by

$$m\dot{v}(t) = \frac{-\partial U(r(t))}{\partial x(t)} = \sum \mathrm{i}k_x U(k)\varrho(k, t) \quad . \tag{43.4}$$

It is therefore of order $O(U)$. Looking at (43.1, 2) one sees that the leading term in N is $O(U^2)$. The leading term in the stochastic force F, therefore, is the first term on the r.h.s. of (43.2), whereas the second term can be neglected in lowest order. The memory kernel in this order is identical to the acceleration relaxation function.

After inserting (43.4) into (43.1) it is useful to carry out an average over an ensemble of the random potential fluctuations. In doing so we approximate the average of products of the potentials and electron densities by the product of averages (factorization of correlations) in the form

$$N(z) = mM(z)/N_e = -\sum \langle |k_x U(k)|^2 \rangle_d \Phi(k,z)/mN_e \quad . \tag{43.5}$$

Here $\langle \rangle_d$ indicates the "disorder" average (over the random potential distribution), and $\Phi(k,z)$ is the Fourier transform of the density–density relaxation function

$$\Phi(k,t) = -i\beta \langle \langle \varrho(k,t); \varrho(-k,0) \rangle \rangle_d \Theta(t) \quad . \tag{43.6}$$

A simple example of the disorder average occurring in the first factor on the r.h.s. of (43.5) is given by a sum of impurity potentials located at random sites r_i:

$$U(k) = u(k) \sum \exp(i k \cdot r_i)/\sqrt{V} = u(k)\varrho_i(-k) \quad . \tag{43.7}$$

This leads to

$$\langle |kU(k)|^2 \rangle_d = |k_x u(k)|^2 s_i(k) \quad . \tag{43.8}$$

Here $s_i(k) = \langle |\varrho_i(k)|^2 \rangle_d$ is the Fourier transform of the (static) impurity density–density correlation function. For uncorrelated random sites r_i one has $s_i(k) = n_i = N_i/V$, which is the density of impurities.

The result (43.5) together with (43.8) can then be represented graphically in analogy to a Feynman diagram (Fig. 43.1). Since the impurity density fluctuation is static, the scattering is elastic. The energy is conserved. But the momentum is changed, and this is sufficient for a macroscopic momentum relaxation and a nonzero resistivity.

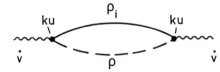

Fig. 43.1. Second-order "self-energy" (i.e. memory function) diagram describing the "virtual decay" of an acceleration line into a (static) impurity density and an electron density line

The factorization approximation (43.5) is correct only if $\Phi(k,z)$ is treated in zeroth order in the interaction U. Then $r(t)$ is independent of the random potential and thus has no correlations with the random force $kU(k)$. We thus have a justification of (43.5) for the present chapter, where we are going to work in lowest order in U.

Later on we are going to use (43.5) beyond this order. It is then a more or less uncontrolled approximation. It corresponds to an infinite "partial summation" of the complete perturbation series, which has certain analogies in field theory and the many-body problems, but no straightforward justification in terms of a systematic expansion procedure.

The diagram in Fig. 43.1, however, has a simple physical interpretation, which gives additional physical plausibility to the factorization approximation (43.5). The conductivity $\sigma(\omega)$ is related to the absorption of electromagnetic energy of a field (or light quantum) of frequency ω. The absorption leads to an excitation of the electron system. Energy conservation requires the excitation energy to be $\hbar\omega$. The momentum of the photon, however, is too small to be absorbed by the electron system directly. It is transferred to the impurity system instead. The system photon + impurities can then be considered in analogy to, say, an electron (or neutron) scattered by the electron system. It is known that the scattering intensity of such processes is proportional to the imaginary part of the electron density propagator $\Phi(k,\omega)$, measuring the "recoil spectral density".

If there exist two well-separated time scales (in this case of the conductivity and the memory kernel N) the frequency dependence of the conductivity at low frequencies is determined by the memory kernel at zero frequency. Since we are considering only the diagonal part of the conductivity the results of the Chap. 42 apply: one has only to consider the absorptive part $N''(0)$ of the memory kernel, see (43.5),

$$N(0) = iN''(0) = -\frac{i}{3}\sum \langle |kU(k)|^2\rangle_d \Phi''(k,0)/N_e m \quad . \tag{43.9}$$

Here we have replaced the k_x^2 in the k sum by $k^2/3$, assuming rotational invariance after the disorder averaging. For the correlation function $\Phi''(k,0)$ one can now use (42.18), in the present case with $a = b^* = \varrho(k)$.

Since in this chapter we are interested only in the zero-order (in U) approximation for Φ, one can use the free-electron eigenstates $|p\rangle$, $|q\rangle$ of H_0 for the general eigenstates $|\mu\rangle$, $|\nu\rangle$ of H in (42.18). Then using

$$\langle p|\varrho(k)|q\rangle = \delta_{p+\hbar k,q}/\sqrt{V} \tag{43.10}$$

one obtains from (42.18)

$$\Phi_0''(k,0) = -\frac{\pi}{V}\sum_p f'(\varepsilon(p))\delta(\omega(p+\hbar k) - \omega(p)) \quad . \tag{43.11}$$

To evaluate this sum it is convenient to replace the delta function by its angular average over the angles of p. Since the f' function restricts the values of p to the Fermi momentum p_f, this average can be written as

$$\langle \delta \rangle = \int \delta \frac{v_f k c - \hbar k^2}{2m} dc \bigg/ \int dc = \Theta \frac{2k_f - k}{2v_f k} \quad , \tag{43.12}$$

where the integration over $c = \cos\vartheta$ (ϑ the angle between p and k) has to be taken from -1 to $+1$. The angular average can then be taken out of the sum and the remaining sum over f' yields $\sum df/d\mu = dN_e/d\mu$. In writing this a possible factor of 2 from the spin degree of freedom is automatically taken into account (although it has been omitted so far) by assuming N_e to be the total number of electrons. The existence of spin then manifests itself only in the relation between the Fermi energy $\mu = \varepsilon_f = p_f^2/2m$ and the particle number N_e. For parabolic energy–momentum relations one has in any case

$$N_e \propto \mu^{3/2} \quad \text{and therefore} \quad \frac{dN_e}{d\mu} = \frac{3N_e}{2\mu} \quad . \tag{43.13}$$

Collecting everything together one finds (k now being used for $|\boldsymbol{k}|$)

$$\Phi_0''(\boldsymbol{k},0) = \pi \frac{dN_e}{d\mu} \Theta \frac{2k_f - k}{2v_f k} \quad . \tag{43.14}$$

We now insert (43.14) with (43.13) into (43.9) and restrict ourselves to the case of random impurity scattering (43.8). After replacing the k sum by the corresponding integral and carrying out the angular integration in k-space the result can be written as

$$-N_0''(0) = \frac{1}{\tau_{el}} = \frac{m}{4\pi p_f^3} \int_0^{2k_f} k^3 |u(k)|^2 s_i(k) dk \quad . \tag{43.15}$$

This final result[1] has a simple "gas kinetic" interpretation. It can be found after introducing the angle θ between $\boldsymbol{p}$ and $\boldsymbol{q}$ instead of $\boldsymbol{k}$ as the integration variable. The transformation can be determined from

$$(\hbar k)^2 = (\boldsymbol{p} - \boldsymbol{q})^2 = 2p_f^2(1 - \cos\theta) \quad . \tag{43.16}$$

Introducing $d\Omega = 2\pi d\cos\theta$ one finds for uncorrelated impurities [$s_i(k) = n_i$]

$$\frac{1}{\tau_{el}} = v_f n_i \left(\frac{m}{2\pi\hbar^2}\right)^2 \int |u(\theta)|^2 (1 - \cos\theta) d\Omega \quad . \tag{43.17}$$

Now the quantity

$$d\sigma = \left(\frac{m}{2\pi\hbar^2}\right)^2 |u(\theta)|^2 d\Omega \tag{43.18}$$

is nothing but the differential cross section for electron impurity scattering in the Born approximation. Thus (43.17) may be written as

$$\frac{1}{\tau_{el}} = v_f n_i \sigma_{tr} = \frac{v_f}{\ell_{el}} \quad . \tag{43.19}$$

Here σ_{tr} is what is known as a transport cross section: a total cross section weighted in a special way. The weighting takes into account that different parts of the differential cross section contribute differently to the transport process under consideration. In our case the electric resistivity is related to momentum

---

[1]There are numerous derivations of (43.15) via the Boltzmann equation. See for instance [43.1]. A derivation of (43.15) using (43.9) with $\Phi = \Phi_0$ can be found in [43.2] and, with $s_i(k) = n_i$, [43.3].

relaxation, to which the forward scattering contributes less than backward scattering. This is taken into account by the weighting factor $(1 - \cos\theta)$, which is directly related to the k^2 in (43.9) and thus to the fact that the change of momentum in a scattering event is proportional to the force (i.e. the gradient of the potential) and not the potential itself. The factor $\ell_{el} = 1/n_i \sigma_{tr}$ is the corresponding mean free path.

Let us now come to finite temperature effects. In our derivation so far we have tacitly assumed $T = 0$ and replaced $-f'(\varepsilon)$ by a delta function at the Fermi energy. For nonvanishing temperatures $f'(\varepsilon)$ has a finite width of order kT and τ_{el} becomes dependent on T. If the procedure of this chapter were blindly extended to nonzero T one would have to replace $\tau_{el}(\varepsilon_f)$ by $\tau_{el}(\varepsilon)$ [obtained from (43.15) by replacing p_f by $p = (2m\varepsilon)^{1/2}$] and then calculate $\langle 1/\tau_{el}(\varepsilon)\rangle$ [by integration with $-f'(\varepsilon)$]. The static conductivity then would be given by $\sigma(0) = (e^2 n_e/m)\langle 1/\tau_{el}\rangle^{-1}$. It turns out that the correct result, however, is given by $\sigma(0) = (e^2 n_e/m)\langle \tau_{el}\rangle$. The difference between the two results is negligible at low temperatures, where the leading temperature-dependent correction (of order T) to the zero-temperature result is the same for both expressions. At higher temperatures, however, (as occurs in semiconductors, for instance) the two expressions differ.

The failure of the present procedure to produce the correct result at nonzero temperatures has its origin in the fact that the momentum alone is not the only slow variable in the problem. A sufficient number of slow variables are contained in the Boltzmann transport equation, which works with the phase space function $f(\boldsymbol{p}, \boldsymbol{r}, t)$. A derivation of the Boltzmann collision rate will be given later (Chap. 52).

A simplified procedure starting from something close to (12.1) for the momentum correlation function works with (12.1) "on the energy shell", i.e. with (42.10) where $f(h)$ is replaced by $\delta(\varepsilon - h)$. This leads to a memory kernel depending on the energy ε, which then finally is inserted into (42.10) and then averaged over the Fermi function.

Finally, let us discuss the frequency dependence of the memory kernel. This will lead us to a first microscopic estimate of the time scale $\delta\tau$ mentioned several times in earlier chapters (Chaps. 4, 15, 18). We will then be able to check the conditions under which it is small compared to τ_{el} so that a Markovian description such as the Drude theory for $\sigma(\omega)$ is valid.

The dominant reason for a frequency dependence of the memory kernel in our present approximation is the sharp drop of $\Phi_0''(k,\omega)$ for frequencies of the order of $\omega \approx v_f k$ (see the imaginary part of the Lindhard function in Chap. 34). If this is inserted into the k sum in (43.5) one has to distinguish two cases. (1) The potential is sufficiently long ranged (with ranges $a \gg 2\pi/2k_f$). Then the cutoff in the k sum for $N''(k,\omega)$ is provided by $u(k)$, leading to a characteristic frequency of the order of $\omega = 2\pi/\delta\tau = v_f 2\pi/a$. The physical meaning of $\delta\tau$ in this case is the duration of the collision (time which the electron spends in the range of the potential).

(2) The potential is short ranged. The cutoff in the k sum is provided by the upper limit $2p_f$, leading to a characteristic frequency given by $\hbar\omega = 4\varepsilon_f$. In terms of our "recoil interpretation" in Fig. 43.1, this cutoff also corresponds to an "effective range cutoff": even a point scatterer looks smeared out over the Fermi wavelength $\lambda_f = \hbar/p_f$. Only in the classical limit $\hbar \to 0$ would the upper limit of the k sum be replaced by infinity and the only cutoff would be provided by the actual range of the potential.

Usually one has $a \geq \pi/2k_f$ so that the first case prevails. In both cases the characteristic frequency is too high to describe the corresponding memory effects by means of the simplified free-electron model used in this chapter. For energies of the order of ε_f usually band structure effects such as interband transitions become important. Cases of other observable memory effects will be discussed in subsequent chapters.

PROBLEMS

43.1 Determine the relaxation time (43.15) for a screened Coulomb potential

$$u(k) = \frac{4\pi e^2}{k^2 + \kappa^2}$$

and uncorrelated impurities ($s_i = n_i$). The result

$$\frac{1}{\tau_{el}} = \frac{2\pi e^4 m n_i}{p_f^3} \ln(1+\alpha) - \frac{\alpha}{1+\alpha}$$

with $\alpha = (2k_f/\kappa)^2$ is in rather good agreement with experimental data for alkali metals.

43.2 Determine the zero-frequency recoil spectrum (43.11) for the dimensions $d = 2$ and 1. Compare it with (43.14) for $d = 3$.
Result: $\Phi_0''(k,0) = [\pi(\partial n/\partial \mu)/2v_f]\{\Theta(2k_f-k)/k, 2/[k(1-k/2k_f)^{1/2}, [\delta(k-2k_f) + \delta(k)]\}$ for $d = 3, 2, 1$.

44. Electron–Phonon Conduction

The factorization approximation (43.5) can easily be generalized to many-body problems, such as electron–phonon or electron–electron interactions. It is convenient to start out from t-space and generalize (43.4), taking into account that the positions r_i of the scattering centres now are functions of t. Equation (43.7) then takes the form

$$U(\boldsymbol{k},t) = \frac{u(\boldsymbol{k})\exp[i\boldsymbol{k}\cdot\boldsymbol{r}_i(t)]}{\sqrt{V}} = u(\boldsymbol{k})\varrho_i(-\boldsymbol{k},t) \quad . \tag{44.1}$$

Inserting this into (43.4) one can derive an expression for the acceleration correlation function, which, apart from a factor mN_e, is the spectral function $s(t)$ of the self-energy N. Factorizing the correlations of the ionic and electronic densities as in (43.5) one obtains

$$s(t) = -\sum \frac{|k_x u(\boldsymbol{k})|^2 s_i(\boldsymbol{k},t) s(\boldsymbol{k},t)}{mN_e} \tag{44.2}$$

with

$$s_i(\boldsymbol{k},t) = \langle \varrho_i(-\boldsymbol{k},t)\varrho_i(\boldsymbol{k},0)\rangle \quad \text{and} \quad s(\boldsymbol{k},t) = \langle \varrho(\boldsymbol{k},t)\varrho(-\boldsymbol{k},0)\rangle \quad . \tag{44.3}$$

Here the averages are thermal averages. In Fourier space, according to (6.10) (44.3) leads to a convolution

$$s(\omega) = -\sum |k_x u(\boldsymbol{k})|^2 \int \frac{s_i(\boldsymbol{k},\omega')s(\boldsymbol{k},\omega-\omega')d\omega'}{\pi m N_e} \quad . \tag{44.4}$$

The spectral function s_i in this formula is practically identical with the neutron scattering spectral function s_n of (9.10) and thus directly measurable. The only difference is the elastic scattering contribution to s_n (the Laue peaks), which is absent in s_i. The elastic scattering occurs for incident plane waves, whereas the electron states are Bloch waves, which are not scattered by the periodic potential any more. Only the *deviations* from periodicity due to the thermal motion of the ions contribute to the resistivity.

We now introduce the dissipative spectral functions $N''(\omega)$ and $\Phi''(\omega)$, making use of the detailed balance relations (8.3). In our case

$$s(\boldsymbol{k},\omega) = \hbar\omega[1+n(\omega)]\Phi''(\boldsymbol{k},\omega) \tag{44.5}$$

and

$$s(\omega) = \hbar\omega[1+n(\omega)]N''(\omega) \quad , \tag{44.6}$$

which has to be inserted into (44.4).

If one uses again the free-electron approximation for Φ'', the frequency scale of Φ'' is set by the Fermi frequency. The frequency scale of s_i, on the other hand, is given by the Debye frequency. As long as the relaxation frequency is small compared to both frequencies it is again sufficient to consider $N''(0)$ in order to calculate the low-frequency conductivity. The prefactor of $N''(0)$ occurring on the r.h.s. of (44.6) then reduces to $k_B T$, and on the r.h.s. of (44.4) we make use of $1 + n(-\omega') = -n(\omega')$. The result then is (using ω instead of ω' as the integration variable)

$$\frac{1}{\tau_{\text{el}}} = \sum |k_x u(\mathbf{k})|^2 \int \frac{s_i(\mathbf{k},\omega)\beta\hbar n(\omega)\Phi''(\mathbf{k},\omega)d\omega}{\pi m N_e} \quad . \tag{44.7}$$

If one uses for Φ'' the same approximation as in (43.14) one can write the final result in a form completely analogous to (43.15), namely (k is now used for the absolute magnitude of $\mathbf{k}$, and dimension $d = 3$ is assumed)

$$\frac{1}{\tau_{\text{el}}} = \frac{m}{4\pi p_f^3} \int_0^{2k_f} k^3 |u(k)|^2 s_i(k,T) dk \tag{44.8}$$

and $s_i(k,T)$ is given by

$$s_i(k,T) = \int s_i(k,\omega) \frac{\hbar \omega n(\omega)}{\pi k_B T} d\omega \quad . \tag{44.9}$$

Another form of (44.8) is obtained if one takes into account that s_i is proportional to the density n_i of ions. Introducing the density n_e of conduction electrons and the number $Z = n_e/n_i$ of conduction electrons per ion one can write (44.8) as

$$\frac{1}{\tau_{\text{el}}} = \frac{m}{12(\hbar\pi)^3 Z} \int_0^{2k_f} \frac{k^3 |u(k)|^2 s_i(k,T)}{n_i} dk \quad . \tag{44.10}$$

Equations (44.8,10) can also be derived from a generalized Boltzmann equation using a simple variational ansatz [44.1,2]. The result contains multiphonon effects, Debye-Waller factors, umklapp processes, etc. In the single-phonon approximation $s_i(k,\omega)$ takes the form

$$s_i(k,\omega) = \frac{\hbar\pi n_i}{m_i} \sum_\lambda (\mathbf{k}\cdot\mathbf{e}_{\mathbf{k}\lambda})^2 \delta(\omega^2 - \omega_{\mathbf{k}\lambda}^2) \frac{\omega}{\omega_{\mathbf{k}\lambda}} [1 + n(\omega)] \quad . \tag{44.11}$$

Here $\mathbf{e}_{\mathbf{k}\lambda}$ is the polarization unit vector of a phonon with polarization index λ. For an elastic continuum only longitudinal waves with $\omega_k = ck$ contribute, and $s_i(k,T)$ according to (44.9,11) is given by

$$s_i(k,T) = \frac{n_i}{m_i} \beta(\hbar k)^2 n(\omega_k)[1 + n(\omega_k)] \tag{44.12}$$

with $\beta = 1/k_B T$. If this is inserted into (44.10) and the k dependence of $u(k)$ is neglected, the final result can be written as

$$\frac{1}{\tau_{\text{el}}} = \frac{m}{m_{\text{i}}} \frac{|u(0)|^2 (k_{\text{B}}T)^5}{12\hbar \pi^3 Z(\hbar c)^6} J_5\left(\frac{2\hbar c k_{\text{f}}}{k_{\text{B}}T}\right) \quad . \tag{44.13}$$

Here $J_5(y)$ is one of the Debye functions

$$J_5(y) = \int_0^y \frac{x^5 e^x}{(e^x - 1)^2} dx = \begin{cases} y^4/4, & y \text{ small}, \\ 5!\zeta(5) \approx 124.4, & y \text{ large} \end{cases} . \tag{44.14}$$

Equation (44.13) is (a simplified version of) the Bloch–Grüneisen formula [44.3,4]. The approximations used in the derivation of (44.13) [omission of k dependence of $u(k)$ and single-phonon continuum approximation] are not good enough for accurate quantitative purposes. For such purposes one would have to go deeper into several details of lattice dynamics. But the essentials come out right: the resistivity according to (44.13, 14) increases proportional to T^5 at low temperatures and to T at high temperatures.

At the beginning of this chapter we mentioned that the factorization approximation (44.2) also allows electron–electron interactions to be taken into account. They, of course, lead to a modification of the density relaxation function $\Phi(k,\omega)$. A simple approximation for this modification is the so-called "random phase approximation" (Problem 33.1). In the static limit $\Phi(k,0)$ this leads to a screening of the bare interaction $u(k)$. This screening has tacitly been taken into account anyway (Problem 43.1). The dominant modification of the frequency dependence of $\Phi(k,\omega)$ is the occurrence of plasmon poles. In three-dimensional systems at normal metallic electron densities the plasmon frequencies are very high and do not lead to any modifications of the low-frequency conductivity. The situation is different, however, in two-dimensional semiconductors. In this case the plasmon frequencies have a dispersion, vanishing proportional $k^{1/2}$ for small k. This leads to strong memory effects and deviations from Drude behaviour of the low-frequency conductivity [44.5,6].

PROBLEMS

44.1 Use an ansatz $s_i(k,\omega) = a\delta(\omega - \omega_k) + b\delta(\omega + \omega_k)$ with $\omega_k = ck$ and determine the two coefficients a and b from the two conditions: 1. detailed balance (6.17) $s_i(k,-\omega) = s_i(k,\omega)\exp(-\beta\hbar\omega)$ and 2. the "compressibility sum rule" (11.5), in our case, $2 \int s_i(k,\omega)d\omega/\pi\hbar = n_i/m_i c^2$. Compare the result with the continuum approximation of (44.11).

44.2 Determine the high T behaviour of (44.10) by direct expansion of the Bose function $n(\omega)$ in (44.9). Compare the result for a screened Coulomb potential with the result of Problem 43.1.

45. Mode Coupling Theory for Impurity Conduction

45.1 Particle Diffusion and Current Relaxation

Equation (43.5) can be used as a starting point for the discussion of the localization of electrons by (arbitrary) random potentials in one and two dimensions and by sufficiently strong random potentials in three dimensions. An important point in this discussion is a "self-consistent feedback" contained in (43.5): the random potential modifies the density "propagator" Φ occurring in (43.5) from free-particle propagation Φ_0 to diffusive-type propagation. This slowing down of the particle propagation leads to a slowing down of the random force fluctuations via (43.5) manifesting itself as a low-frequency increase of $N(z)$, and a corresponding low-frequency decrease of $\sigma(z)$ via (42.23) and (12.1) (see the first equation of the previous chapter).

Mathematically speaking, the density propagator $\Phi(k,z)$ can be expressed approximately in terms of $N(z)$ itself. Equation (43.5) then reduces to a transcendental equation for $N(z)$. The dependence of Φ on N can be determined from a generalization of the equations for the Brownian oscillator (Chap. 5) and for hydrodynamics (Chap. 27). The starting point is the continuity equation (27.6,22) describing particle number conservation. If we write the particle number density as a sum of the equilibrium density $n_e = n$ and the deviation from it, and introduce the longitudinal current density j, the continuity equation in Fourier space takes the form (where, since we do not have to consider single components of the vector $\boldsymbol{k}$ any more, we use k for the absolute magnitude of $\boldsymbol{k}$)

$$-\omega \varrho + kj = 0 \quad . \tag{45.1}$$

This equation may be considered as a generalization of $dx/dt = p/m$ for the Brownian oscillator. The second equation $dp/dt + m\omega_0^2 x + \gamma p = 0$ is generalized by introducing a pressure term $\mathrm{grad} P = (\partial P / \partial n) \mathrm{grad} \varrho$ as the restoring force, a k-dependent susceptibility $\chi(k) = \beta \langle \varrho(\boldsymbol{k}); \varrho(-\boldsymbol{k}) \rangle$ approaching $(\partial n/\partial \mu) = n(\partial n/\partial P)$ for $k = 0$, and a wave number and frequency dependent friction coefficient $N(k,\omega)$, approaching $-i\gamma$ for $k = \omega = 0$. The second equation for our generalized Brownian oscillator then takes the form

$$(N(k,\omega) - \omega)j + \frac{kn}{m\chi(k)}\varrho = 0 \quad . \tag{45.2}$$

We then use Onsager's regression ansatz and obtain for the correlation functions Φ between density and density and Φ_j between longitudinal current density and density the two equations

$$-z\Phi(k,z) + k\Phi_j(k,z) = \chi(k) \quad , \tag{45.3a}$$

$$\frac{kn}{m\chi(k)}\Phi(k,z) + [N(k,z) - z]\Phi_j(k,z) = 0 \quad . \tag{45.3b}$$

The r.h.s. of the second equation would be the corresponding static susceptibility, say $\chi_j(k)$, which, however, is zero since the static susceptibility between a quantity (in this case ϱ) and its time derivative (in this case j) always vanishes. We now eliminate Φ_j from (45.3a) by inserting its value determined from (45.3b) and obtain an equation for Φ alone:

$$-\left(nk^2/\{m\chi(k)[N(k,z)-z]\} + z\right)\Phi(k,z) = \chi(k) \quad . \tag{45.4}$$

This equation is completely general. It may be considered as a definition of $N(k,z)$ in terms of $\Phi(k,z)$ (or vice versa). A simple approximation can be obtained by using the $k = z = 0$ limit $-i(\partial n/\partial \mu)m\gamma$ for the denominator on the l.h.s. of (45.4). Then one has

$$-(iDk^2 + z)\Phi(k,z) = \chi(k) \quad . \tag{45.5}$$

So Φ has a simple diffusion pole with a diffusion constant

$$D = n\frac{\partial \mu/\partial n)}{\gamma m} \quad , \tag{45.6}$$

which is nothing but Einstein's relation between the diffusion constant and mobility.

The simple diffusion approximation can be improved by considering only small (rather than vanishing) k and z. Now for k going to zero the current density j becomes proportional to the total momentum p. Thus (45.2) becomes the momentum relaxation equation $[N(\omega)-\omega]p(\omega) = 0$. Hence we have $N(0,z) = N(z)$. If we write $N(k,z) = N(z)+N_1(k,z)$, N_1 has to approach zero for k going to zero. Since we are interested in a low-frequency increase in $N(z)$ one can consider N_1 as small compared to $N(z)$ for small k. Another obviously small quantity compared to $N(z)$ is z itself. Our improved diffusion approximation — let's call it the hydrodynamic approximation Φ_h — for Φ then can be obtained from (45.4) by replacing $N(k,z) - z$ by $N(k,z)$, multiplying both sides by $N(k,z)\chi m/n$ and neglecting the small term $N_1 z$ on the l.h.s.

$$-\frac{k^2 + zN(z)\chi m}{n}\Phi_h(k,z) = \frac{N(k,z)\chi^2 m}{n} \quad . \tag{45.7}$$

45.2 Backscattering Effects

There is another low-frequency anomaly in Φ that is not contained in (45.7) since it is not related to the small wave number behaviour of $\Phi(k,z)$. It had been overlooked for a long time until attention was drawn to it [45.1–4]. It is related to an important symmetry property which can be expressed in terms of the phase space correlation function. To formulate it we first express the density operator $\varrho(k) = \exp(-i k \cdot r)/V^{1/2}$ in terms of the the complete set $|p\rangle$ of momentum eigenstates. Making use of $\langle p|\exp(-i k \cdot r)|q\rangle = \delta_{p+\hbar k, q}$, one can write

$$\varrho(k) = \sum \frac{|p - \hbar k/2\rangle\langle p + \hbar k/2|}{V^{1/2}} = \sum \varrho_p(k) \quad . \tag{45.8}$$

This may be considered as the decomposition of the (Fourier transform of the) density into the (Fourier transform of the) phase space density $\varrho_p(k)$. Similarly the density correlation function can be decomposed into a sum of phase space density correlation functions

$$\Phi(k, z) = \sum \Phi_{pq}(k, z) \tag{45.9}$$

describing the correlations between $\varrho_p(k)$ and $\varrho_q(-k)$.

Now the phase space correlation function obeys an important symmetry relation which may be called "detailed time reversal symmetry". It can be derived from the representation of Φ_{pq}, or χ_{pq}, in terms of the single-particle eigenstates $|\kappa\rangle, |\mu\rangle$ of h using (42.14). According to (45.8) this representation then contains the terms

$$a_{\kappa\mu} b_{\mu\kappa} = \langle \kappa|p - \hbar k/2\rangle\langle p + \hbar k/2|\mu\rangle\langle \mu|q + \hbar k/2\rangle\langle q - \hbar k/2|\kappa\rangle \quad . \tag{45.10}$$

Now we assume the single-particle Hamiltonian h to be time reversal invariant. Then its eigenstates may be chosen as real. Hence

$$\langle \kappa|p\rangle = \langle -p|\kappa\rangle \quad . \tag{45.11}$$

If one applies this relation to the two terms involving κ in (45.10) one sees that (45.10) is invariant under the exchange of $p - \hbar k/2$ with $-q + \hbar k/2$. If the phase space correlation function $\Phi_{pq}(k)$ is represented graphically by the four pole with the four ends labelled by the four momenta occurring in (45.10) the symmetry operation can be represented as in Fig. 45.1.

Introducing the proper incoming and outgoing momenta, as well as the momentum transfer after the transformation, one finds the symmetry relation

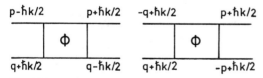

Fig. 45.1. Diagrammatic representation of the phase space correlation function and the symmetry operation of detailed time reversal: the momenta of one pair of legs are exchanged and turned into their negatives

$$\Phi_{pq}(k,z) = \Phi_{(p-q+\hbar k)/2,(q-p+\hbar k)/2}((p+q)/\hbar, z) \quad . \tag{45.12}$$

The term *detailed* time reversal is intended to indicate that the time reversal operation is applied only to one pair of legs of the four pole. The full time reversal operation with both pairs of legs is a symmetry operation for all Hamiltonians that are invariant under time reversal. Detailed time reversal invariance requires in addition the absence of many-body effects. It can be violated either by time reversal noninvariant terms, such as magnetic fields in the Hamiltonian, or by many-body terms, such as electron–phonon, and/or electron–electron interactions. We shall come back to this point in the next chapter.

Our next step will be to find an approximation for the phase space correlation function in accord with the hydrodynamic approximation (45.7). This can be done by approximating the phase space density $\varrho_p(k,t)$ by its projection [in the sense of (13.19)] onto the density $\varrho(k,t)$ itself

$$\varrho_p(k,t) = \frac{\chi_p(k)}{\chi(k)} \varrho(k,t) + \cdots \quad . \tag{45.13}$$

We are going to approximate the phase space susceptibility

$$\chi_p(k) = \beta \langle \varrho_p(k,t); \varrho(-k,t) \rangle = \beta \langle \varrho_p(k); \varrho_p(-k) \rangle \tag{45.14}$$

by its value for the free-particle system [using (42.13) and (45.8)]

$$\chi_p(k) = \frac{f(p - \hbar k/2) - f(p + \hbar k/2)}{V \hbar p \cdot k/m} \quad . \tag{45.15}$$

In the same approximation one can calculate $\chi(k)$ using (45.15) and

$$\chi(k) = \sum \chi_p(k) \quad . \tag{45.16}$$

We expect (45.13) to be a good approximation for small k and ω where the only relevant mode of the system is hydrodynamic diffusion. Hence we expect a good approximation for the phase space correlation at small k after inserting (45.13) into the phase space correlation function and replacing the density correlation function by its hydrodynamic approximation (45.7)

$$\Phi_{pq}(k,z) = \frac{\chi_p(k)\chi_q(k)}{\chi^2(k)} \Phi_h(k,z) + \cdots \quad ; \quad k, z \text{ small} . \tag{45.17}$$

The symmetry (45.12) is, of course, not fulfilled for this approximation, but it can be installed by applying the symmetry operation to the r.h.s. of (45.17) and adding the result to the r.h.s.. Because of (45.12) the wave number occurring in Φ_h of the additional term is now $(p+q)/\hbar$. While the low-frequency enhancement of the r.h.s. of (45.17) occurs for small k it occurs in the additional term for small $p+q$, i.e. for approximately opposite momenta p and q. Hence the term "backscattering effects" introduced for this section.

The symmetry transformation generates a diffusion pole in $p+q$ from the original pole in k. Since both contributions are large in different regions of phase space one can expect that they contribute additively to (45.17) in lowest order.

Another, in a sense more systematic, approach to the two additive contributions to the phase space correlation function uses infinite partial summations of the perturbation series for Φ in the random potential. Without going into details [45.3], let us just mention that the diffusion contribution (45.17) to the phase space correlation function can be represented as a partial summation of particle–hole "ladder" diagrams (Fig. 45.2.). Application of the symmetry operation turns these diagrams into the maximally crossed particle–hole "fan" diagrams. (If the arrow on the r.h.s. of such diagrams is reversed they become the topologically equivalent particle–particle "ladder" diagrams.) The approximation corresponding to (45.17) plus the symmetric addition then can be graphically represented as in Fig. 45.2.

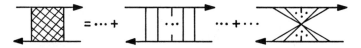

Fig. 45.2. An approximation for the phase space correlation function obeying detailed time reversal invariance is obtained by summing the particle–hole ladder plus fan diagrams

45.3 Self-Consistency Relations

Both terms, finally, have to be inserted into the r.h.s. of (43.5) and the p,q sum of (45.9) as well as the k sum of (43.5) have to be carried out, in order to obtain the desired self-consistency equation for $N(z)$. For the first term we use (45.17) directly. For the second, symmetry generated, term we introduce the two subscripts on the r.h.s. of (45.12) and $(p+q)/\hbar$ as new summation variables. Then the two terms can be collected together as

$$N(z) = -\sum \langle |U(k)k|^2 + |U((p+q)/\hbar)(p+q)/\hbar|^2 \rangle_d$$
$$\times [\chi_p(k)\chi_q(k)] \Phi_h(k,z)/(dmN_e) \quad . \tag{45.18}$$

Here we have introduced the dimension d of the system in order to generalize (43.9) from $d = 3$ to arbitrary dimensions. The p,q sum of the first term involving $U(k)$ can easily be carried out using (45.16). The same is true for the special case of a Coulomb potential in $d = 2$. In this case $U(p+q)$ is proportional to $1/|p+q|$ and cancels the $(p+q)$ in the numerator of (45.18). On the other hand, for a short-range potential, U becomes constant in Fourier space. In this case the p,q sum can again be carried out. Writing $(p+q)^2 = p^2 + q^2 + 2p \cdot q$ the $p \cdot q$ term vanishes after summing over p,q. The first two terms can, in principle, be summed up exactly using (45.15). We shall approximate the sums, taking into account that the dominant contributions to (45.18) come from small k where $\chi_p(k)$ reduces to a delta function at the Fermi energy according to (45.15). This suggests the approximation

$$\sum p^2 \chi_p(k) = (\hbar k_f)^2 \chi(k) + \cdots \quad . \tag{45.19}$$

Finally we collect together everything in terms of a simplifying ansatz interpolating between long-range and short-range potentials

$$N(z) = -\langle U^2 \rangle \sum \frac{(k^2 + 2k_f^2)\Phi_h(k,z)}{dmN_e} \quad . \tag{45.20}$$

If one uses

$$\langle U^2 \rangle = \langle |U(k)|^2 \rangle_d \quad \text{with} \quad k^2 = 2k_f^2 \tag{45.21}$$

one represents most of the limiting cases discussed above correctly and has a reasonable interpolation for practically all other cases. What is not described by (45.20) is a possible suppression of the "$2k_f^2$" contribution for long-range potentials with a much sharper drop off for large k than the Coulomb potential [45.5]. Note that the k^2 contribution in (45.20) is directly due to the r.h.s. of (45.17) whereas the $2k_f^2$ contribution is due to the fan diagrams.

We now insert Φ_h from (45.7) into (45.20), decompose $N(k,z) = N(z) + N_1(k,z)$ and neglect the small term $zN(z)$ in the numerator of the N_1 term, because N is already a small correction compared to $N(z)$ for small k. Then the r.h.s. of (45.20) yields a term

$$N_0(z) = \langle U^2 \rangle \sum \frac{(k^2 + 2k_f^2)N_1(k,z)}{dnN_e} \quad . \tag{45.22}$$

The only property of this term we know is that to lowest order in U [note that to lowest order N_1 is $O(U^0)$] has to be equal to (43.15) since all other terms on the r.h.s. of (45.20) are at least $O(U^4)$. We assume that the possible low-frequency anomalies of $N(z)$ are due only to the terms on the r.h.s. of (45.20) that depend *explicitly* on $N(z)$. Hence we drop the z dependence of $N_0(z)$ completely and replace it by a constant $-i\nu$ which, to lowest order, may be identified with $-i/\tau$ — the lowest order result of Chap. 44. Then (45.20) takes the simple form

$$N(z) = -i\nu + \frac{\langle U^2 \rangle}{dnN_e} N(z) \sum \chi^2 \frac{k^2 + 2k_f^2}{k^2 + zN(z)\chi m/n} \quad . \tag{45.23}$$

This equation is the final result of the self-consistent current relaxation theory including backscattering effects. It is a transcendental equation for $N(z)$ that describes the feedback mechanism mentioned at the beginning of this section.

PROBLEMS

45.1 Determine the recoil spectrum $\Phi''(k,0)$ for the diffusion approximation (45.5) and compare its k dependence with the free-particle recoil spectrum (43.14).

45.2 Calculate the free-particle static susceptibilities $\chi(k)$ from (45.15,16). Show that they are given by $\chi(k)(\partial\mu/\partial n) = (1/2)+(1-\kappa^2)\ln|(1+\kappa)/(1-\kappa)|/4\kappa$, $1-\Theta(\kappa^2-1)(\kappa^2-1)^{1/2}/\kappa$, $\ln|(1+\kappa)/(1-\kappa)|/2\kappa$ for $d = 3, 2, 1$, respectively, with $\kappa = k/2k_{\rm f}$.

46. Electron Localization

The discovery by *Anderson* [46.1] of the localization of electrons in sufficiently strong random potentials in 1958 inspired numerous investigations of disordered systems. The early development is reviewd by *Mott* and *Davis* [46.2]. Renormalization group and scaling arguments [46.3–5] were introduced into the theory and have led to scaling laws. Scaling functions were calculated in certain limiting cases using partial summations of the perturbation series, in particular including the backscattering effects discussed in Chap. 45 [46.5,6]. The results led to the conclusion that not only in one but also in two dimensions electrons are always localized in a random potential, regardless of its strength.

The self-consistency arguments presented in Chap. 45 [46.7] offer another way to deal with the localization transition. They lead not only to scaling laws but also to explicit forms of the scaling functions. If the backscattering effects are included in the self-consistency equation [46.8–10] self-consistency and scaling arguments lead to essentially the same results. In this chapter we shall pursue the self-consistent theory, in particular the consequences of (44.23) derived at the end of the previous chapter.

Replacing the sum in this equation by an integral leads to

$$N(z) = -i\nu + \left(\frac{\langle U^2 \rangle}{n^2} \frac{\Omega_d}{d(2\pi)^d}\right) N(z) \int_0^\infty \chi^2 \frac{k^2 + 2k_f^2}{k^2 + zN(z)\chi m/n} k^{d-1} dk \quad . \tag{46.1}$$

Here d is the dimension of the system and ω_d the surface of the unit "sphere" in d dimensions ($\Omega_d = 4\pi, 2\pi, 2$ in $d = 3, 2, 1$). The susceptibility χ drops off sharply above $k = 2k_f$ (Problem 41.2). We therefore introduce a cutoff k_0 of the order of $2k_f$ by $\chi = (\partial n/\partial \mu)\Theta(k_0 - k)$.

In order to express the prefactor of the integral in (46.1) in terms of physically more transparent quantities we evaluate the expression (46.5) for the relaxation time $N(0) = -i/\tau$ in d dimensions with $\Phi'' = \Phi_0''$ taken from the results of Problem 46.2. Making use of the same approximations as discussed in the context of (46.20,21) one obtains

$$\frac{1}{\tau} = 2\pi \frac{\langle U^2 \rangle}{\hbar n} \frac{\Omega_d}{d(2\pi)^d} \frac{\partial n}{\partial \mu} k_f^d \quad . \tag{46.2}$$

Finally we express ν by its lowest order expression $1/\tau$ [compare the discussion in the context of (44.22)] and introduce the unperturbed and "renormalized" diffusion constants

$$D_0 = n\frac{\partial \mu}{\partial n}\frac{\tau}{m} \quad \text{and} \quad D(z) = \frac{n(\partial \mu \partial n)}{iN(z)m} \quad . \tag{46.3}$$

Then, after some rearrangement, (46.1) takes the more transparent form

$$D(z) = D_0 - \frac{\hbar k_f^{-d}}{2\pi m} \int_0^{k_0} \frac{k^2 + 2k_f^2}{k^2 - iz/D(z)} k^{d-1} dk \quad . \tag{46.4}$$

Since we will mainly be interested in situations where $\kappa^2 = -iz/D(z)$ is small we expand the integral in the small quantity κ and keep only the first two leading terms of the expansion. Inspection of the integral then shows that for the k^2 term in the numerator the (small) z/D in the denominator can be completely neglected. Equation (46.4) can then be written as

$$D(z) = D_0 - \frac{\hbar}{\pi m d}\left(\frac{k_0}{k_f}\right)^d - \frac{2\hbar}{\pi m}\left(\frac{k_0}{k_f}\right)^{d-2} I(\kappa) \tag{46.5}$$

with $\kappa^2 = -iz/D(z)$ and with the integral

$$I(\kappa) = (k_0)^{2-d} \int_0^{k_0} \frac{k^{d-1}}{k^2 + \kappa^2} dk \quad . \tag{46.6}$$

It has become fashionable in the context of scaling laws to consider the dimension d as a continuous variable. The expansion of the integral (46.6) in the small quantity κ then runs differently for d greater or smaller than 2. The final result for the first two leading terms in κ, however, can be collected in one simple formula valid for all $d < 4$. Making use of

$$\int_0^\infty \frac{x^{2\mu-1}}{1+x^2}dx = \frac{\Gamma(1+\mu)\Gamma(1-\mu)}{2\mu} = \frac{\pi}{2}\sin(\mu\pi) \quad , \tag{46.7}$$

one finds after some intermediate steps

$$I(\kappa) = \left[1 - \Gamma\frac{d}{2}\Gamma\left(2 - \frac{d}{2}\right)\left(\frac{\kappa}{k_0}\right)^{d-2}\right]/(d-2) \tag{46.8}$$

in general and

$$I = 1 - \frac{\pi\kappa}{2k_0}, \quad \ln\left(\frac{k_0}{\kappa}\right), \quad \frac{\pi k_0}{2\kappa} - 1; \quad \text{for} \quad d = 3, 2, 1. \tag{46.9}$$

Here the case $d = 2$ has been obtained from (46.8) by expanding in the small quantity $d - 2$.

46.1 Breakdown of Perturbation Theory

For a first orientation concerning the solution of the transcendental equation (46.5) with (46.8) we consider the *perturbation regime*. In this regime one has $D(z) = D_0$ in lowest order with small corrections due to the additional terms on the r.h.s. of (46.5). In this case one can use $\kappa^2 = -iz/D_0$ in (46.8) and obtains with $\omega_0 = k_0^2 D_0$

$$\operatorname{Re}\{I(\omega)\} = 1 - \frac{\pi}{4}\left(\frac{2\omega}{\omega_0}\right)^{1/2}, \quad \ln\frac{\omega_0}{\omega}, \quad \frac{\pi}{4}\left(\frac{2\omega_0}{\omega}\right)^{1/2} - 1 \qquad (46.10)$$

again for $d = 3, 2, 1$ respectively. From this, the corrections to $\operatorname{Re}\{D(\omega)\}$ and to the conductivity $\operatorname{Re}\{\sigma(\omega)\} = e^2(\partial n/\partial \mu)\operatorname{Re}\{D(\omega)\}$ (for $\omega\tau \ll 1$) can easily be calculated by inserting (46.10) into (46.5). Some typical results are shown in Fig. 46.1.

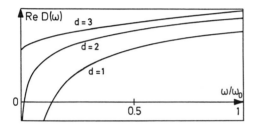

Fig. 46.1. Frequency-dependent corrections to the conductivity in the perturbation regime according to (46.10)

The correction to the diffusion coefficient is negative for all d and its magnitude increases with decreasing frequency, producing a deviation from *Drude* type behaviour, which would mean a decrease of $D(\omega)$ with increasing frequency in contrast to Fig. 46.1. For $d = 3$ (generally speaking for $4 > d > 2$) and sufficiently large D_0 the static conductivity remains positive and the perturbative solution of (46.5) converges. With decreasing D_0, however, the perturbative solution at $z = 0$ turns negative below a certain critical value D_c. This clearly indicates a breakdown of the perturbation approximation (46.10).

More interestingly the approximation (46.10), for *arbitrary* D_0, leads to negative (even negatively infinite) static conductivities for $d = 1, 2$ (generally speaking for $d \leq 2$): the perturbation approximation *always* breaks down for $d \leq 2$ at low frequencies. Note that this breakdown is generated by the $2k_f^2$ contribution of the integral in the self-consistency equation (46.4), and hence by the backscattering terms.

46.2 Localization and Nonergodicity

Whenever the perturbative solution would lead to a negative $D(0)$ the correct $D(0)$ has to vanish [since otherwise (46.5) would predict a negative $D(0)$]. Thus the l.h.s. of (46.5) vanishes for $z = 0$; the r.h.s. then requires a nonzero limit of κ for $z = 0$. Let us denote this limit by $1/\xi$, then one has

$$D(z) = -iz\xi^2 \quad \text{for small } z \text{ and} \tag{46.11}$$

$$0 = \begin{cases} \delta + \Delta(k_0\xi)^{2-d} & \text{for } d \neq 2 \text{ and} \\ \delta - \Delta \ln(k_0\xi) & \text{for } d = 2 \end{cases} \tag{46.12}$$

with

$$\delta = \begin{cases} D_0 - \frac{\hbar}{\pi m d}(\frac{k_0}{k_f})^d - \frac{2\hbar}{\pi m(d-2)}(\frac{k_0}{k_f})^{d-2} & \text{for } d \neq 2 \text{ and} \\ D_0 - \frac{\hbar}{\pi m d}(\frac{k_0}{k_f})^d & \text{for } d = 2 \end{cases} \tag{46.13}$$

$$\Delta = \begin{cases} \frac{2\hbar}{\pi m d(d-2)}(\frac{k_0}{k_f})^{d-2}\Gamma(d/2)\Gamma(2-d/2) & \text{for } d \neq 2 \text{ and} \\ \frac{2\hbar}{\pi m d}(\frac{k_0}{k_f})^{d-2}\Gamma(d/2)\Gamma(2-d/2) & \text{for } d = 2 \end{cases} \tag{46.14}$$

The solution of (46.12) is

$$\xi k_0 = \begin{cases} |\Delta/\delta|^{1/(d-2)} & \text{for } d \neq 2 \text{ and} \\ e^{\delta/\Delta} & \text{for } d = 2 \end{cases} \tag{46.15}$$

For $4 > d > 2$ the critical point where the perturbation solution breaks down is given by $\delta = 0$. Below this critical point, i.e. for negative δ, (46.15) yields a divergent length $\xi \propto 1/\delta^{1/(d-2)}$ for δ approaching the critical point. For $d = 2$ the length ξ is exponentially large for large δ (i.e. for large D_0). For $d < 2$ one usually has ξ of order $1/k_0$ or smaller. The physical meaning of ξ can be found by inserting (46.11) into the density relaxation function $\Phi = -\chi/(iDk^2+z)$. This leads to

$$\Phi = -\frac{1}{z}\frac{\chi}{[1+(k\xi)^2]} \quad \text{for small } z. \tag{46.16}$$

Since, in general, $z\Phi(z) = \chi^T - \chi(z)$, the factor multiplying $1/z$ in this equation is nothing but $\chi(0) - \chi^T$: there is a difference between the "isolated" and thermodynamic static susceptibilities. Since, at $T = 0$, $\chi^T = \chi^S = \chi$, the isolated and adiabatic susceptibility are different: *the system is non-ergodic* in the sense of our discussion in the context of (3.15). This is one of the many fascinating aspects of disordered systems. Using (46.16) one finds

$$\chi(k,0) = \frac{(k\xi)^2}{1+(k\xi)^2}\chi. \tag{46.17}$$

If one applies a localized perturbation (i.e. independent of k) the static response according to (46.17) is equal to the thermodynamic value χ for large k (i.e. short distances in coordinate space) but drops to zero for $k \ll 1/\xi$ (i.e. for distances large compared to χ). Whenever (46.11) holds, the electrons are localized and ξ is the "localization length". Fourier transformation of (46.17) implies exponential localization in coordinate space. See Problem 46.1.

46.3 Critical Behaviour and Scaling Laws

Using the notation (46.13,14) one can write the self-consistency equation (46.5) as

$$D = \delta + \Delta \left(\frac{-iz}{Dk_0^2} \right)^{(d-2)/2} = \delta + a(-iz/D)^{(d-2)/2} \ . \tag{46.18}$$

A particularly useful way of writing the solution of this equation is obtained from the scaling idea. The physics behind this idea can be described as follows. In a diffusion process, to each time t there corresponds a length ℓ given by $\ell^2 = Dt$, or else in frequency space $\ell^2 = D(z)/(-iz)$. Inside a volume ℓ^d one has essentially free-particle motion, outside the collective diffusion mode [consider the discussion in the context of (4.15)]. Hence one might expect for the conductance g (conductivity times ℓ^{d-2}) of a volume ℓ^d a smooth dependence on frequency. This would imply that $g = D[D/(-iz)]^{(d-2)/2}$ has a smooth dependence on z and suggests a representation for $D(z)$ of the form $D = (-iz)^{(d-2)/d} g^{2/d}$. Furthermore, the natural unit δ for D leads to a natural scaling for the frequency by the factor $\delta^{d/2} z^{(d-2)/2}$. Altogether one expects a representation for D of the form

$$d(z) = (-iz)^{(d-2)/d} g^{2/d} z |\delta|^x \quad \text{with } x = d/(2-d) \ . \tag{46.19}$$

An equation for g can be obtained by inserting (46.19) into (46.18) yielding

$$g^{2/d} - ag^{(2-d)/d} = \delta(-iz)^{(2-d)/d} \ . \tag{46.20}$$

Equation (46.19) is a typical scaling expression for the three small quantities D, z, δ near the localization transition familiar from the theory of critical phenomena. For $4 > d > 2$ one can derive from it three power laws

$$D(z) \propto z^{(d-2)/d} \quad \text{for } \delta = 0 \ , \tag{46.21}$$

$$D(0) \propto z^{(d-2)/d} (z\delta^x)^{(2-d)/d} = \delta^s \quad \text{for } \delta \geq 0 \tag{46.22}$$

and furthermore $z\xi^2 \propto z^{(d-2)/d} (z|\delta|^x)^{2/d}$ or

$$\xi \propto |\delta|^{x/d} = |\delta|^{-\nu} \quad \text{for } \delta \leq 0 \ . \tag{46.23}$$

Elimination of x from the two critical exponents s and ν in the last two equations leads to the scaling law

$$s = (d-2)\nu \ , \qquad (46.24)$$

which can be derived more generally from renormalization group considerations [46.4]. Note that in our case one has in addition $x = d/(2-d)$ and thus [from (46.22)] $s = 1$. Combined with (46.22–24) this leads to

$$D(0) \propto \delta \quad \text{and} \quad \xi \propto |\delta|^{-1/(d-2)} \ . \qquad (46.25)$$

The second of these equations had, of course, been derived already in (46.15). The first one could have been obtained also directly from (46.18) by putting $z = 0$. Equations (46.21–23) fix the scaling function g determined by (46.20) in three limiting cases by direct inspection.

Instead of (46.20) one often characterizes the scaling function only by a differential equation $d\ln g/d\ln z = -\beta(g)/2$, where $\beta(g)$ can easily be determined from (46.20) by taking the logarithmic derivative. This yields

$$-2\frac{d\ln g}{d\ln z} = \beta(g) = 2\frac{(d-2)g + (d-2)a}{2g + (d-2)a} = d - 2 - \frac{(4-d)(d-2)a}{2g + (d-2)a} \ . \qquad (46.26)$$

In writing the eqs. (46.18–20) one should, in principle, again treat the case $d = 2$ separately. Since for the scaling relation we were mainly interested in $4 > d > 2$ we did not write down the corresponding equations. It turns out, however, that (46.26) is valid for $d = 2$ as well.

Looking at (46.14,18) one sees that $(d-2)a$ is a positive (nonvanishing) quantity. This suggests incorporating the $(d-2)a$ in the definition of g and replacing g by $(d-2)ag$. The $(d-2)a$ in (46.24) is then replaced by 1. The factor -2 was introduced on the l.h.s. of (46.24) in order to define $\beta(g)$ in close analogy to a corresponding function introduced in the context of "length scaling". In this approach [46.3,5] one considers the dependence of the static conductance $g(L)(\propto D(0,L)L^{(d-2)})$ of a system with volume L^d on the length L. Then asymptotically one has $d\ln g/d\ln L = \beta(g) = d-2$ as in (46.24).

In the localized regime (i.e. for small g), on the other hand, one expects an exponential dependence $g \propto \exp(-L/\xi)$. This implies $d\ln g/d\ln L = \ln g$. For intermediate g the β function for the length scaling approach can be calculated [46.8]. It has the same asymptotic behaviour as the z-scaling function for large g but starts to deviate with decreasing g.

PROBLEMS

46.1 Calculate the Fourier transform of the static isolated susceptibility $\chi(k,0)$ according to (46.17).
For $d = 2$ one finds, for instance, $\chi(r) = \chi[\delta(r) - (2\xi)^{-1}\exp(-r/\xi)]$.

46.2 Consider the limiting case $d = 2$ in (46.18–20); determine the scaling function $\beta(g)$ according to (46.24) and convince yourself that it is obtained from

(46.24) by inserting $d = 2$ into the r.h.s. [Note that $(d-2)a$ is nonzero for $d = 2$].

46.3 Using the result of (46.1) calculate the scaling function $\beta(g)$ for the length scaling problem. For the solution see [46.8].

47. Localization and Quantum Interference*

In this chapter we shall try to give a simple physical interpretation of the backscattering terms that proved to be so important in the theory of localization as discussed in the foregoing chapters.

Since the correction terms to D_0 described by the integral on the r.h.s. of (46.4) contain quantum objects (such as $\hbar$ and the de Broglie wave number k_f of electrons at the Fermi energy), we look for differences between classical and quantum mechanical diffusion. Now classically the electrons move as point particles along infinitesimally narrow trajectories, but quantum mechanically as waves along ray tubes with a width $1/k$. In most cases the difference between these two descriptions is small because the interference terms of the quantum waves average out for the random trajectories occurring in the diffusion process.

There is one exception, however: if a trajectory, or a ray tube, returns to the origin in a system with time reversal invariance. Then this trajectory can be traversed forward and backward with equal probability. The two corresponding waves would then interfere destructively and thus not contribute to diffusion. Wave interference, generally speaking, slows down diffusion.

To estimate this effect let us assume that the reduction $\delta D/D_0$ of diffusion is proportional to the fraction of particles in self-intersecting ray tubes. Now the fraction of particles in a ray tube near the origin in a time interval dt is given by $n(t)dV$ where $n(t) = (D_0 t)^{-d/2}$ is the density of particles near the origin [since they have moved a distance $(D_0 t)^{1/2}$ after time t] and $dV = k_f^{1-d} v_f dt$ is the volume of a ray tube of cross section k_f^{1-d} and length $v_f dt$. After integration over t one expects

$$\frac{\delta D}{D_0} \approx - \int k_f^{1-d} v_f (D_0 t)^{-d/2} dt \quad . \tag{47.1}$$

As the upper limit of the integration we chose the time $t_{\max} \approx 1/\omega$, and as the lower limit the time $t_{\min}$ for which the length $(D_0 t_{\min})^{1/2}$ is the de Broglie wave length $1/k_f$. The final result can then be written as

$$D \approx D_0 - \frac{\hbar}{m} \frac{[1 - (\omega/\omega_0)^{(d-2)/2}]}{(d-2)} \quad , \tag{47.2}$$

where we have introduced $\omega_0 = D_0 k_f^2$ in analogy to (46.10). Comparing (47.2) with (46.10) one sees that the simple heuristic discussion described so far is in pretty good agreement with the accurate perturbation result (46.10). We now continue this heuristic line of reasoning to take into account deviations from detailed time reversal invariance. The two ingredients of this invariance were: the absence of many body effects, and time reversal invariance.

a) In the presence of many-body effects, such as electron–phonon and electron–electron interactions, the phase coherence between the two interfering self-intersecting waves is destroyed by inelastic scattering processes if the time t is larger than the inelastic relaxation time τ_{in}. This relaxation time, therefore, occurs as a new upper cutoff of the integral (47.1). The static diffusion coefficient D then can be obtained from (47.2), replacing ω_0/ω by $\omega_0\tau_{in}$. Usually $1/\tau_{in}$ has some power dependence $\propto T^p$ on temperature T (for instance electron–phonon scattering in $d = 3$ has $1/\tau_{in} \propto T^5$). Equation (47.2) then leads to typical T dependences such as

$$\delta D \propto \begin{cases} -T^{p/2}, & d = 1 \\ (p/2)\ln(T_0/T), & d = 2 \\ T^{-p/2}, & d = 3 \end{cases} \quad , \qquad (47.3)$$

in the perturbation regime. Figure 47.1 exhibits some experimental results for a two-dimensional system.

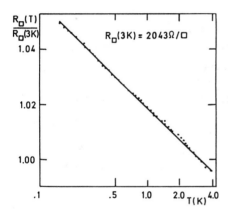

Fig. 47.1. Experimental results on the conductivity of thin palladium films [47.1] plotted on a logarithmic scale

b) The presence of *magnetic fields* violates time reversal invariance. It modifies the interference between the two self-intersecting waves by introducing a phase factor $\exp(ie\int A \cdot dl/\hbar c)$ to each of them. The line integral over the (closed) paths in opposite directions can be replaced by the flux BS of the magnetic field B, where the area S can be approximated by $D_0 t$. The relative phase of the two waves is then of order $\exp(2ieBD_0 t/\hbar c)$. Thus instead of the complete destructive interference without magnetic field [factor -1 in the integral (47.1)] one now has a reduction of D of the order of

$$\delta D = -\int \cos(2eBD_0 t/\hbar c) k_f^{1-d} v_f (D_0 t)^{-d/2} D_0 dt \quad . \qquad (47.4)$$

We now consider the difference between the integral (47.1) for $B = 0$ and (47.4) and concentrate on $d = 2$. The result can be written as

$$\Delta D(B) = D(B) - D(0) = \frac{\hbar}{mk_f^2} \int^{eBD_0\tau_{in}/\hbar c} 2\sin^2 x \, dx/x \quad . \quad (47.5)$$

The lower cutoff of the integral is now provided by the $\sin^2 x$ itself and is thus of order $x_{min} \approx 1$. The final result is

$$\Delta D(B) \approx \frac{\hbar}{mk_f^2} \begin{cases} x_{in}^2 & x_{in} \text{ small} \\ \ln x_{in} & x_{in} \text{ large} \end{cases}, \quad (47.6)$$

with $x_{in} = eBD_0\tau_{in}/\hbar c$.

Two things should be pointed out in connection with this formula. First of all the sign of ΔD. It is positive: a magnetic field *increases* diffusion, and hence conduction, since it weakens the destructive quantum interference leading to localization. The corresponding anomalous magnetoresistance (*decreasing with B*) had been a puzzle for many years and finds a natural and simple explanation. Secondly the magnitude of the effect: the change from the B^2 behaviour to the $\ln B$ behaviour occurring at $x_{in} \approx 1$ depends on temperature, and for temperatures normally used occurs for a field of the order of $B \approx 100$–500 Gauss (10–50 mT), which is quite small. The normal magnetoconductivity $\approx -(\omega_c\tau)^2$ (ω_c: the cyclotron frequency) has the opposite sign and is usually much smaller.

c) For the sake of completeness, we finally mention spin flip scattering processes, which have, so far, not found simple treatments as above, but have been investigated by field-theoretical methods [47.2,3]. They lead to divergences in perturbation theory only in higher order. Generally speaking this has the consequence that localization tendencies are weaker than for normal impurities.

Secondly, for impurities with high nuclear charge numbers Z, spin–orbit scattering has to be taken into account. This leads to corrections to the diffusion coefficient of a similar form to normal scattering, but of opposite sign: the conductivity *increases* [47.3–6].

48. Scaling Laws for Dynamic Critical Phenomena

In this chapter we introduce the concept of dynamical scaling and consider its application to the lambda transition of liquid helium as a first example.

48.1 General

When a system is at or close to a critical point of a continous phase transition, anomalies occur not only in static but also in dynamic phenomena. The physical origin of all these critical anomalies is the increase of the correlation length of the order parameter on approaching the critical point. The increase of the range of correlated regions leads to a corresponding increase of time scales for dynamic processes ("critical slowing down").

A number of theoretical concepts have been applied to "critical dynamics": (i) the conventional theory of critical slowing down, (ii) the dynamic scaling hypothesis, (iii) mode-coupling theories and (iv) the renormalization group approach.

The conventional theory, due to van Hove (1954) and Landau and Khalatnikov (1954), assumes that the transport coefficients and kinetic coefficients remain finite at the critical point. Since the resonance frequencies and relaxation rates are generally determined by the ratio of transport or kinetic coefficients to static susceptibilities and since the order parameter susceptibility diverges at the transition, the conventional theory gives a simple explanation of the occurrence of slow (or "soft") modes at critical points.

The original argument behind the assumption of finite transport and kinetic coefficients is the assertion that they are essentially determined by the interactions between particles on the length scale of interatomic separations that are not affected by the increase of the coherence length of the order parameter. It turns out, however, that the conventional theory is incorrect in most cases. In many systems kinetic coefficients are influenced by interactions between the slow modes on a length scale comparable to the coherence length. These interactions are treated in detail in the mode-coupling theory and, more generally, in the renormalization group approach.

The fact that close to a critical point all length scales — except for the one diverging correlation length — become irrelevant leads to the so-called scaling laws. They have been very successful for static phenomena [48.1,2] and have been extended to dynamic processes [48.3]. In this chapter we consider several typical examples of the application of dynamic scaling laws.

Close to the transition the microscopic details become irrelevant and the structure of the dynamics depends essentially only on the symmetry of the system. There are two basic characteristics which follow solely from symmetry. Firstly, the order parameter may or may not be a conserved quantity. Secondly, the symmetry, which is broken in the ordered phase, may be continuous or discrete. Table 48.1 exhibits a few examples.

Table 48.1. Examples of the four basic symmetry characteristics in dynamic processes

Broken symmetry is continuous
Nonconserved order parameter
Superfluid He, isotropic antiferromagnet
Conserved order parameter
Isotropic ferromagnet
Broken symmetry is discrete
Nonconserved order parameter
Uniaxial antiferromagnet, SrTiO$_3$
Conserved order parameter
Liquid–gas transition, uniaxial ferromagnet

A general formulation of dynamic scaling can be expressed in terms of the dynamic susceptibility $\chi(k,\omega,\tau)$, where τ in this and the next two chapters will be used as an abbreviation for $T/T_c - 1$. First of all, in accord with static scaling [48.1,2], the static susceptibility $\chi(k,\tau)$ close to the critical point can be written as

$$\chi(k,\tau) = k^{-2+\eta} X(\kappa/k) \quad , \tag{48.1}$$

where κ is the inverse of the correlation length ξ and obeys a power law

$$\kappa = 1/\xi \propto |\tau|^\nu \quad . \tag{48.2}$$

The idea behind (48.1) is that for wavelengths small compared to the correlation length (but still large compared to atomic lengths), i.e. for $k \gg \kappa$, the argument of the function X can be replaced by zero, corresponding to $\tau = 0$, i.e. the critical point: the length dependences at length scales small compared to the coherence length are already the same as at the critical point. Since at the critical point the coherence length is infinite, there is no length scale available anymore. The k dependence of χ is a scale-invariant power law. At temperatures slightly below, or above, the critical point the coherence length is the *only relevant length scale*. So, besides the scale-invariant power law, there may be an additional dependence on the single dimensionless variable κ/k. In the limit $k \ll \kappa$ one usually has again simple power laws for $X(\kappa/k)$, which, in the "matching regime" $k \approx \kappa$, can be joined smoothly to the the "critical" power laws.

The terminology of the various regimes is not quite unambiguous. In critical dynamics one usually calls $k \ll \kappa$ the "hydrodynamic regime" while $k \gg \kappa$ is called the "critical regime". In static critical phenomena, sometimes the long-wavelength region is called the "critical regime", and the short-wavelength region the "homogeneous regime". See Fig. 48.1.

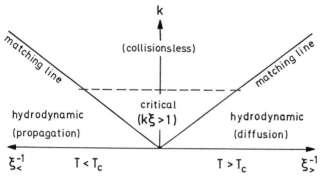

Fig. 48.1. The five regimes in the context of scaling laws

In analogy to (41.1) the dynamic scaling hypothesis assumes a similar form for the characteristic frequencies $\omega(k, \tau)$ (resonance and relaxation frequencies). Namely

$$\omega(k, \tau) = k^z \Omega(\kappa/k) \quad . \tag{48.3}$$

The arguments for this form are precisely the same as described above for $\chi(k, \tau)$. The dynamic susceptibility then is assumed to depend on ω only via the dimensionless combination $\omega/\omega(k, \tau)$, or, equivalently in the form

$$\chi(k, \omega; \tau) = \chi(k, \tau) Y(\omega/k^z, \kappa/k) \quad . \tag{48.4}$$

In these expressions z occurs as a new dynamic critical exponent besides the two independent static exponents η and ν. We shall see that in systems where the order parameter has a propagating (instead of a purely dissipative) mode, dynamic scaling (48.3) allows z to be related to static exponents.

48.2 The Lambda Transition in Liquid Helium

We begin our discussion of examples with the lambda transition from normal to superfluid He. The broken gauge invariance in the ordered phase is continuous. The order parameter $\psi(r)$ — the condensate wave function — is not conserved. Nevertheless, the order parameter has a propagating mode, the so-called second sound, corresponding to an entropy or temperature wave with a frequency $\omega_k = c_2 k$ in the hydrodynamic regime. This looks very similar to ordinary sound, corresponding to the propagating mode of the (conserved) density. The reason for the vanishing frequency in the long-wavelength limit in the case of second sound is not a conservation law but the fact that the corresponding susceptibility $\chi(k, \tau)$ diverges for small k even for $\tau \neq 0$. The free energy density does not depend on the phase $\varphi(r)$ of the complex order parameter directly, i.e. there is no restoring force for a change of a constant phase. There is only a term corresponding to the kinetic energy density of the superfluid component of the two-fluid model of superfluid He. This model is valid in the hydrodynamic regime and contains the *gradient* of the phase in the form

$$F_{s,\text{kin}} = \int \varrho_s v_s^2/2 \, d^d r \int \varrho_s \hbar^2 [\text{grad}\varphi(r)]^2/2m^2 d^d r \quad , \qquad (48.5)$$

where ϱ_s is the mass density of the superfluid component and

$$v_s = \hbar(\text{grad}\varphi)/m \qquad (48.6)$$

its velocity.

In Fourier space (48.5) takes the form

$$F_{s,\text{kin}} = \frac{\varrho_s \hbar^2}{2m^2} \sum k^2 |\varphi_k|^2 = \sum |\varphi_k|^2 / 2\chi_\varphi(k) \qquad (48.7)$$

with the "phase susceptibility"

$$\chi_\varphi(k, \tau) = m^2/\hbar^2 \varrho_s(\tau) k^2 \quad . \qquad (48.8)$$

For $\tau \neq 0$ the order parameter susceptibility can be obtained from (48.8) if one takes into account that there is a nonzero restoring force for the *amplitude* fluctuations of the order parameter. Thus the long-wavelength fluctuations of the order parameter are completely determined by its phase fluctuations. Expanding $\Delta\psi = \psi i \Delta\varphi + \cdots$ one obtains

$$\chi(k, \tau) = \langle |\psi| \rangle^2 \chi_\varphi(k, \tau) \quad \text{for} \quad k < \kappa \, . \qquad (48.9)$$

In the critical regime, on the other hand, there is no amplitude restoring force and (48.1) leads to the "similarity law"

$$\chi(k, \tau) = k^{-2+\eta} X(0) \quad \text{for} \quad k > \kappa \, . \qquad (48.10)$$

Combining (48.8) with (48.9) and matching it to (48.10) for $k \approx \kappa$ yields

$$\varrho_s(\tau) = \langle|\psi|\rangle^2 \lim_{k\to 0} \frac{m^2}{\hbar^2 k^2 \chi(k,\tau)} \propto \langle|\psi|\rangle^2 \kappa^{-\eta} . \qquad (48.11)$$

If one furthermore takes into account the temperature dependence of the order parameter [48.1,2], near the critical point

$$\langle|\psi|\rangle \propto |\tau|^\beta \quad \text{with } 2\beta = (d-2+\eta)\nu \qquad (48.12)$$

as well as (48.2) one finally obtains

$$\varrho_s(\tau) \propto \kappa^{d-2} \qquad (48.13)$$

as a result of the static matching procedure.

If one takes into account [48.1,2] that the free energy (for vanishing external field) is proportional to $|\tau|^{d\nu}$ or κ^d and that $\varrho_s \langle v_s^2\rangle/2$ is one contribution to it, (48.13) implies that $\langle v_s^2\rangle \propto (\hbar\kappa/m)^2$. This is very plausible in view of (48.6) and may even be taken as a simplified "proof" for (48.13).

We have, so far, treated d as arbitrary. We should, however, mention that for $d \leq 2$ the order parameter fluctuations destroy the ordered phase. The problem of superfluidity in lower dimensional systems needs special treatment [48.4].

Let us now come to the application of the scaling hypothesis (48.3). We start out again from the hydrodynamic regime where the two-fluid hydrodynamics predicts a second sound frequency (Problem 48.1)

$$\omega(k,\tau) = \sqrt{\left(\frac{Ts^2 \varrho_s}{mc_p \varrho_n}\right) k} . \qquad (48.14)$$

Here s and c_p are the entropy and specific heat per particle and $\varrho_n = \varrho - \varrho_s$ is the mass density of the normal component. The only singular quantities in (48.14) are the superfluid density (48.13) and the specific heat

$$c_p \propto |\tau|^{-\alpha} \propto \kappa^{-\alpha/\nu} \quad \text{with } \alpha = 2 - d\nu . \qquad (48.15)$$

The specific heat exponent α for the lambda transition of He is very close to zero: the specific heat singularity is very close to logarithmic. Inserting the κ dependences of ϱ_s and c_p into (48.14) one finds

$$\omega(k,\tau) \propto \kappa^{d/2+\alpha/2\nu-1} k \quad \text{for } k < \kappa . \qquad (48.16)$$

Now according to the scaling hypothesis the κ dependences turn into k dependences in the matching regime, and in the critical regime one has

$$\omega(k,\tau) \approx \omega(k,0) \propto k^z \quad \text{with } z = d/2 + \alpha/2\nu . \qquad (48.17)$$

If we consider in particular $d = 3$ and $c_p = A\ln(\kappa_c/\kappa)$, we obtain

$$\omega(k,\tau) \propto [\ln(\kappa_c/k)]^{-1/2} k^{3/2} \quad \text{for } k > \kappa . \qquad (48.18)$$

This is the prediction of the scaling hypothesis for the resonance frequency. In practice there is also an attenuation of second sound described by an imaginary part $\propto iD_2 k^2/2$ of the resonance frequency. We have not considered this attenuation in detail since in the critical region it has to have the same k dependence as (48.18). Otherwise an absolute length besides ξ could be determined from the real and imaginary parts of ω, which would be inconsistent with the scaling hypothesis.

Above T_c the thermal wave propagation turns into thermal diffusion with a purely imaginary characteristic frequency

$$\omega(k,\tau) = -iD_T k^2 \quad \text{for } k < \kappa. \tag{48.19}$$

Matching this behaviour to the critical behaviour (48.18) for $k \approx \kappa$ one finds

$$D_T \propto [\ln(\kappa_c/\kappa)]^{-1/2} \kappa^{-1/2} \quad \text{for } k < \kappa \tag{48.20}$$

and, taking into account that for He the critical exponent ν is $\approx 2/3$,

$$D_T \propto [\ln(1/|\tau|)]^{-1/2} |\tau|^{-1/2}. \tag{48.21}$$

So the scaling hypothesis with its double matching procedure from the two-fluid hydrodynamic region via the critical regime into the normal hydrodynamic region above T_c leads to the remarkable prediction of a singular $|T - T_c|$ dependence of the thermal diffusion constant. This is in sharp contrast to conventional critical slowing down.

Early experiments [48.5] seemed to verify the predictions of dynamical scaling laws for superfluid He rather well. More recently [48.6] small but systematic deviations have been detected. More detailed theoretical investigations [48.7] demonstrate that this is not an indication that, in principle, the scaling laws are not valid. Present experiments only have not yet reached the asymptotic regime, where the scaling behaviour holds.

Superfluid helium has been the first testing ground for dynamic scaling [48.3]. Concluding this section we mention that rather similar results can be obtained for isotropic antiferromagnets [48.8] and easy-plane ferromagnets [48.9]. The hydrodynamic frequencies in this example are also given by $\omega = ck - iDk^2/2$, and since the structure of hydrodynamics is rather similar to that of superfluids one also finds a critical frequency and damping $\propto k^{3/2}$.

PROBLEMS

48.1 Derive (48.14) for the velocity of second sound. There are essentially two modes in superfluid helium: (i) normal sound with a momentum density $\varrho_n v_n + \varrho_s v_s = g$, driven by a pressure gradient grad$P$, and (ii) second sound with $g = 0$, driven by a temperature gradient grad$T$. The two main ingredients for second sound are 1) the Josephson equation [48.10] for the time evolution of the phase of the order parameter $\hbar\partial\varphi/\partial t = \mu$, and 2) the continuity equation for the entropy s per particle with the condition that the entropy is carried only by the normal component: $\partial s/\partial t = c_p \partial T/\partial t = -s\,\text{div}\,\boldsymbol{v}_n$. These ingredients have to be combined with $\boldsymbol{g} = 0$, Eq. (48.6) and the Gibbs–Duhem relation (linearized in $\boldsymbol{v}_{s,n}$) $d\mu = -sdT + vdP$ with $dP = 0$. One then can derive a wave equation for T.

48.2 Compare the condition for stationary superfluid flow $\partial \boldsymbol{v}_s/\partial t = 0$ with the result (19.24) for the fountain pressure in superfluid helium. Use again the Josephson phase equation and the Gibbs–Duhem relation.

49. Applications of Dynamic Scaling Laws

In this chapter we consider more examples from Table 48.1.

49.1 Isotropic Ferromagnets

Let us begin with isotropic ferromagnets, an example of systems with a continuous broken symmetry (3d rotations in spin space) where the order parameter (for instance the z component of the total spin) is conserved. In fact, all components of the total spin are conserved for an isotropic Hamiltonian. If the interactions between the spins is short ranged one even has a differential conservation law of the form

$$\partial s_n(\mathbf{r}, t)/\partial t + \mathrm{div}\, \mathbf{j}_n(\mathbf{r}, t) = 0 \quad . \tag{49.1}$$

Here s_n is the nth component of the spin density and $\mathbf{j}_n$ is the spin current density with the components j_{mn}.

In the nonmagnetic regime — above T_c — the hydrodynamics of the spin system is described by a diffusion equation for $\langle s \rangle = S$

$$J_n = -D\,\mathrm{grad}\, S_n \quad , \tag{49.2}$$

which, together with the conservation law (49.1), after Fourier transformation leads to

$$\partial S_n(\mathbf{k}, t)/\partial t = -Dk^2 S_n(\mathbf{k}, t) \quad . \tag{49.3}$$

In the magnetic regime — below T_c — one has a nonzero order parameter: the static expectation value of s in some preferred direction, say S_z. As a consequence one now has propagating modes for the two spin components perpendicular to S_z. This can easily be seen from (15.5,6) by keeping only the reversible part of M in (15.6), i.e. the first term on the r.h.s. of (15.6). This yields

$$\frac{\partial S_m(\mathbf{r}, t)}{\partial t} = -\mathrm{i} \sum_n \int M_{mn}(\mathbf{r}, \mathbf{r}', t) h_n(\mathbf{r}', t) d^3 r' \quad , \tag{49.4}$$

where h is the field thermodynamically conjugate to the spins (proportional to the magnetic field) and

$$M_{mn}(\mathbf{r}, \mathbf{r}') = \frac{1}{\hbar} \langle [s_m(\mathbf{r}), s_n(\mathbf{r}')] \rangle \tag{49.5}$$

where we have suppressed the t dependence. Using the commutation rules for spin operators one finds

$$M_{xy}(\boldsymbol{r}, \boldsymbol{r}') = -M_{yx}(\boldsymbol{r}, \boldsymbol{r}') = iS_z \delta(\boldsymbol{r} - \boldsymbol{r}') \quad, \tag{49.6}$$

all other components of M being zero. Inserting this into (49.4) one obtains, after introducing the circular components $S_\pm = S_x \pm iS_y$ and $h_\pm = h_x \pm ih_y$ and Fourier transformation,

$$\frac{\partial S_\pm(\boldsymbol{k}, t)}{\partial t} = -iS_z h_\pm(\boldsymbol{k}, t) = -i \left[\frac{S_z}{\chi(\kappa, \tau)}\right] S_\pm(\boldsymbol{k}, t) \quad. \tag{49.7}$$

Thus the propagating modes in the perpendicular direction have a frequency

$$\boxed{\omega(k, \tau) = \frac{S_z}{\chi(k, \tau)}} \tag{49.8}$$

In order to calculate the k and τ dependence of this frequency we start out again in the hydrodynamic regime where the static transverse (or perpendicular) susceptibility $\chi(k, \tau)$ can be determined from a free-energy functional. Similar to superfluid helium there is no restoring force for a rotation of the order parameter with a polar angle ϑ which is constant in space. Again the free energy only contains a term proportional to the square of the gradient of the angle ϑ, and the "angular susceptibility" is given in complete analogy to (48.8) in superfluid helium. Of course, in this case ϱ_s cannot be interpreted as a superfluid density but just as an "order parameter bending" stiffness constant. In analogy to liquid He, (45.9) now takes the form

$$\chi(k, \tau) = S_z^2 \chi_\vartheta(k, \tau) \quad. \tag{49.9}$$

Furthermore, (48.9) will be valid for ferromagnets, since at the critical point the perpendicular susceptibility, for symmetry reasons, will be proportional to (in fact twice) the order parameter susceptibility, and thus will have the same scaling properties.

The matching procedure is then completely analogous to that for superfluid helium and leads to

$$\chi(k, \tau) \propto S_z^2 \kappa^2 - \frac{d}{k^2} \propto \frac{\kappa^\eta}{k^2} \quad. \tag{49.10}$$

One now has to insert this into (49.8) and combine it with the behaviour of the order parameter near the critical point [compare (48.12)]

$$S_z \propto |\tau|^\beta \propto \kappa^{(d-2+\eta)/2} \tag{49.11}$$

leading to

$$\omega(k, \tau) \propto \kappa^{(d-2+\eta)/2} k^2 \quad \text{for } k < \kappa \quad. \tag{49.12}$$

In the matching regime again the κ dependence is turned into a k dependence so that in the critical regime one has a critical dispersion law

$$\boxed{\omega(k,0) \propto k^{(d+2-\eta)/2}} \qquad \text{for } k > \kappa \quad . \tag{49.13}$$

We have kept the critical indices general, so far, but our considerations above always assumed an isotropic ferromagnet of dimension $d = 3$. So, apart from the correction term η, which is usually small compared to unity, one has a critical dispersion law $\omega(k,0) \propto k^{5/2}$. Again there will be a damping of spin waves, which, in the critical regime, has to have the same k dependence.

This k dependence now has to be matched to that of the diffusion behaviour above T_c in order to find the critical τ dependence of the diffusion constant D. Since the diffusion frequency has the same dependence on $k (\propto k^2)$ as the spin waves (49.12), D has to have the same τ dependence as the prefactor of k^2 in (49.12), hence

$$D \propto \kappa^{(d-2-\eta)/2} \propto |\tau|^{(d-2-\eta)\nu/2} \quad . \tag{49.14}$$

Since $\nu \approx 2/3$ for ferromagnets, one expects an approximate $|\tau|^{1/3}$ dependence of the diffusion coefficients in the hydrodynamic regime above T_c. All predictions of dynamical scaling are in approximate agreement with experimental results. There are deviations close to T_c, which can be explained in terms of dipole–dipole interactions between the spins. They are normally small compared to the exchange interactions responsible for ferromagnetism, but become relevant near T_c. Since for dipole interactions the spin is no longer conserved one expects a crossover [49.1] from diffusion behaviour with $z = 5/2 - \eta$ to a simple relaxation behaviour with $z \approx 2$.

Up to now we have applied the scaling hypothesis to the perpendicular modes, which are not the modes of the order parameter. Apparently the scaling laws work for these modes as well. One speaks of "extended scaling" in this case, in contrast to "restricted scaling" for the order-parameter modes. Above T_c, of course, the order-parameter modes are the same as the transverse modes because of symmetry. Below T_c one expects no propagating modes for the order parameter, according to (49.6), since the equilibrium expectation values of s_x and s_y vanish. There will be a relaxation behaviour of the order parameter according to an equation of motion

$$\frac{\partial \Delta S_z(\mathbf{k},t)}{\partial t} = -\omega_\mathrm{r}(k,\tau) \Delta S_z(\mathbf{k},t) \quad . \tag{49.15}$$

For a nonconserved order parameter, such as the condensate wave function of a superfluid, the relaxation frequency will be independent of k for small k. For a conserved order parameter, such as S_z, because of the differential conservation law (49.1) one expects a vanishing relaxation frequency in the limit $k \to 0$. In fact, if the relaxation of the order parameter were due to interactions with noncritical short-wavelength modes, one would expect a simple diffusion behaviour $\omega_{\mathrm{r}} = D_{\mathrm{r}} k^2$. The situation is complicated somewhat by the fact that the coupling of the order parameter to the long-wavelength perpendicular modes produces singular k dependences of the static order parameter susceptibility as well as the irreversible part of the kinetic coefficient M. If one adopts the point of view that the coupling to the long-wavelength modes is taken into account by the scaling laws, one would predict the same temperature dependence of D_{r} (by the double matching procedure from the diffusion regime above T_{c} — via the critical region — to the hydrodynamic regime just below the critical point) and D according to (49.14). Detailed mode-coupling calculations confirm these results.

49.2 Uniaxial Antiferromagnets, Structural Phase Transitions

The transition to a uniaxial antiferromagnet and to the ferroelectric state are examples for discrete broken symmetries and nonconserved order parameters. The ferroelectric transition is a special case of so-called structural phase transitions from one discrete lattice symmetry to another.

The order parameter in the ferroelectric state is the displacement coordinate S of some optical Brillouin zone boundary phonon. Let us define the corresponding zone boundary k vector as zero and the deviations from it by $\mathbf{k}$. Then the equation of motion for the displacement coordinate is that of a damped oscillator whith a frequency $\omega(k, \tau)$ whose real part for small damping is given by $\omega^2(k, \tau) = 1/m\chi(k, \tau)$. If the susceptibility is approximated by a generalized Ornstein–Zernike ansatz

$$\chi(k, \tau) \propto (\kappa^2 + k^2)^{-1+\eta/2} \tag{49.16}$$

the oscillator will (for a constant friction coefficient γ) always go over to the overdamped case near the critical point. The low-lying relaxation frequency of this overdamped oscillator then is given by [see (4.5) for the "relaxator"]

$$\boxed{\omega_r(k, \tau) = \frac{1}{m\chi(k, \tau)\gamma}} \tag{49.17}$$

The situation is similar for the uniaxial antiferromagnet, where the order parameter is the staggered magnetization, which is nonconserved, in contrast to the magnetization, which would be conserved. In this case the time dependence of the order parameter is purely relaxational with a relaxation frequency given by a kinetic coefficient divided by a static susceptibility similar to (49.17). Since the kinetic coefficient cannot be expressed in terms of equilibrium quantities, the dynamic scaling laws do not allow one to calculate the k and τ dependence of the relaxation frequency. If, in addition to the scaling laws for χ, one assumes a constant kinetic coefficient (conventional slowing down) one would expect a critical dispersion law determined by the scaling law for χ, i.e. a dynamic critical exponent $z = 2 - \eta$. Renormalization group calculations [49.2] yield

$$\boxed{z = 2 + c\eta}\qquad \text{with } c \text{ of the order of 1.} \tag{49.18}$$

Since η is usually small compared to 1, the conventional theory is approximately valid.

There has been a long debate concerning what is known as the central peak in structural phase transitions (a peak near $k = 0$, which sharpens up at the critical point, somehow similar to the Rayleigh peak in the hydrodynamics of ordinary liquids, see Chaps. 28,29). For some time the idea of a linear coupling of the order parameter [49.3] to impurities was favoured, but complete consensus about what is going on apparently has not been reached yet.

49.3 Anisotropic Ferromagnets

The general situation, so far, is that the scaling laws have been able to predict the k and τ dependences of characteristic frequencies in all other regimes, if in one regime there was a propagating mode whose frequency was determined in terms of equilibrium quantities. The uniaxial antiferromagnet and the structural phase transition of the previous section are examples of cases in which the scaling laws are insufficient to determine the characteristic frequencies, since there is no propagating mode close to the critical point.

The situation is no better if the order parameter is conserved. Let us consider, for instance, the uniaxial ferromagnet. In this case the transverse modes are nonsingular and the longitudinal order-parameter mode is purely dissipative. Since the order parameter is conserved its relaxation is diffusive with an equation of motion

$$\frac{\partial \Delta S_z(\boldsymbol{k},t)}{\partial t} = -Dk^2 \Delta S_z(\boldsymbol{k},t) \ . \tag{49.19}$$

The diffusion coefficient again is not determined by equilibrium quantities alone but is given by

$$\boxed{D = \frac{\lambda_m}{\chi(k,\tau)}} \ , \tag{49.20}$$

where λ_m is a "magnetic conductivity", the temperature dependence of which is not known a priori. The conventional theory assumes λ_m to be noncritical. Then the Ornstein–Zernike-type behaviour (49.16) of χ leads to

$$D \propto \kappa^{-2+\eta} = |\tau|^{-(2-\eta)\nu} \quad \text{for } k < \kappa \tag{49.21}$$

and

$$\boxed{\omega_r \propto k^z \quad \text{with } z = 4 - \eta} \quad \text{for } k > \kappa \quad . \qquad (49.22)$$

It turns out that this example is one of the few cases where the conventional van Hove hypothesis of noncritical kinetic coefficients is valid. This can be readily understood by mode-coupling theory (Chap. 50).

49.4 Liquid–Gas Transition

At first sight the liquid–gas transition at the critical point seems to be a good and straightforward example of the application of scaling laws. The order parameter in this case would be density (or better the difference between the density and the critical density). The density is a conserved quantity and has a propagating mode — the sound wave — which becomes soft at the critical point. The dominant term in the attenuation of sound waves near the critical point turns out to be the conduction of heat associated with the temperature modulations set up by the density modulation, which is adiabatic to first order (Chap. 27). Dynamic scaling laws would then predict the temperature dependence of the coefficient of the conductivity of heat, which then could be used to predict the behaviour of heat diffusion, which is the second critical mode.

Unfortunately these predictions disagree with with experimental results. The situation is far less trivial. The first indication that the sound wave is not the critical mode is the fact that because $C_V/C_P \propto |\tau|^{(d-2)\nu}$ the intensity ratio between the sound wave propagation component and the thermal diffusion component in the density propagator (Chap. 29) approaches zero at the critical point. The dominant critical mode therefore is the thermal diffusion and not the sound wave. Since thermal waves are diffusive, they cannot be determined from equilibrium parameters. One has to refer to mode-coupling techniques [49.4]. The theory of ultrasonic attenuation is quite nontrivial and has a long history [49.4,5].

50. Mode-Coupling Theory for Dynamic Critical Phenomena

Mode coupling calculations have found widespread applications in the theory of dynamic critical phenomena [50.1]. Here we discuss only the three special examples we mentioned in the previous chapter. Let us start with the *isotropic ferromagnet*. In this example one has, first of all, a simple geometric constraint

$$s_x^2(\mathbf{r}) + s_y^2(\mathbf{r}) + s_z^2(\mathbf{r}) = S(S+1) \quad , \tag{50.1}$$

leading to a static mode coupling and a corresponding equation for the static order-parameter susceptibility. Let us assume a nonzero order parameter S_z and solve (50.1) for $\delta s_z(\mathbf{r}) = s_z(\mathbf{r}) - S_z$. Expanding the occurring square root to first order in the small quantity $s_x^2 + s_y^2$ one obtains

$$\delta s_z(\mathbf{r}) = \pm \left([S(S+1)]^{1/2} - \frac{s_x^2(\mathbf{r}) + s_y^2(\mathbf{r})}{2[S(S+1)]^{1/2}} \right) - S_z + \cdots \quad . \tag{50.2}$$

We now consider the correlation function

$$\langle \delta s_z(\mathbf{r}) \delta s_z(0) \rangle = \frac{\langle [s_x^2(\mathbf{r}) + s_y^2(\mathbf{r})][s_x^2(0) + s_y^2(0)] \rangle}{4S(S+1)} \quad . \tag{50.3}$$

A possible constant has been averaged out to zero since for large r the correlation (50.3) has to vanish. An approximation which is used frequently is the factorization of higher-order correlations. In our case

$$\langle s_x^2(\mathbf{r}) s_x^2(0) \rangle \approx \langle s_x(\mathbf{r}) s_x(0) \rangle^2 \quad \text{etc.} \tag{50.4}$$

Taking into account that terms like $\langle s_x(\mathbf{r}) s_y(0) \rangle$ vanish because of isotropy and inversion symmetry, and using the symmetry in x and y, (50.3) takes the form

$$\langle \delta s_z(\mathbf{r}) s_z(0) \rangle \approx \frac{\langle s_x(\mathbf{r}) s_x(0) \rangle^2}{2S(S+1)} \quad . \tag{50.5}$$

Now we take into account that $\langle s_x(\mathbf{r}) s_x(0) \rangle$ is proportional to the Fourier transform of $\chi(k, \tau)$ in (49.10), which is $\propto \kappa^\eta/k^2$. Fourier back transformation leads to $\langle s_x(\mathbf{r}) s_x(0) \rangle \propto \kappa^\eta/r$ and hence $\langle \delta s_z(\mathbf{r}) \delta s_z(0) \rangle \propto \kappa^{2\eta}/r^2$. If this result is in turn Fourier transformed one obtains

$$\chi_{zz}(k) = \beta \langle |\delta s_z(\mathbf{k})|^2 \rangle \propto \kappa^{2\eta}/k \quad . \tag{50.6}$$

Thus the coupling of longitudinal and transverse modes induced by the geometric constraint (50.1) leads to a singular behaviour of the (longitudinal) order-parameter susceptibility, although the singularity is only $\propto 1/k$ instead of $1/k^2$ for the transverse susceptibilities. A corresponding $1/k$ singularity then has to occur in the Onsager mobility $\mu_{zz}(k)$ for the order parameter relaxation in order to have a nonsingular diffusion coefficient $D_r = \mu_{zz}(k)/\chi_{zz}(k)$ in accord with dynamic scaling laws (Sect. 49.1). In order to investigate the singularities of the Onsager coefficient one has to go back to the general expressions (13.8,9,14–16). The first term on the r.h.s. of, for example, (13.15) vanishes, because it is actually a commutator. See (3.25) or (15.6). In the second term the force F can be replaced by $\dot{s}_z(\underline{k},t)$ if one is interested only in the leading powers in k for small k. For a quantity that obeys a local conservation law like $s_z(\boldsymbol{k},t)$ the time derivative is proportional to k, whereas $(1/\chi^T)M \propto Dk^2$. Hence one has

$$\mu_{zz}(k) = \beta \int_0^\infty \langle \dot{s}_z(\boldsymbol{k},t)\dot{s}_z(-\boldsymbol{k},0)\rangle dt \quad . \tag{50.7}$$

For the calculation of the time derivatives we use a Heisenberg model with a Hamiltonian $\sum J(\boldsymbol{k}')s(\boldsymbol{k}')s(-\boldsymbol{k}')/2N$ leading to the equations of motion

$$\dot{s}_z(\boldsymbol{k},t) = (1/2N)\sum[J(\boldsymbol{k}') - J(\boldsymbol{k}-\boldsymbol{k}')]s_x(\boldsymbol{k}',t)s_y(\boldsymbol{k}-\boldsymbol{k}',t) \quad . \tag{50.8}$$

They describe the "decay" of the longitudinal mode into two transverse modes. Obviously there will be great differerences between uniaxial and isotropic ferromagnets. In *uniaxial ferromagnets* the two transverse modes have nonzero restoring forces for small k and hence are noncritical at the critical point. One therefore obtains a nonsingular behaviour of the Onsager coefficient (50.7) as assumed in (49.19,20). On the other hand, in isotropic ferromagnets the transverse modes have divergent susceptibilities. It is to be expected that this singular behaviour is transferred to the Onsager coefficient similar to (50.5,6). This is indeed confirmed by a detailed analysis of (50.7) together with (50.8).

As a third example we consider a simplified treatment [50.2] of the heat conductivity near the gas–liquid critical point. If one neglects the quadratical heat production processes by irreversibilities one has an approximate differential conservation law

$$T\dot{\sigma} = \dot{q} \approx -\text{div}\boldsymbol{j}_q \quad . \tag{50.9}$$

For the heat current we use $T\Delta s\boldsymbol{j}_n = \Delta i\boldsymbol{j}_n$ (assuming constant pressure). Furthermore, we neglect longitudinal current fluctuations, since they involve density fluctuations, which occur at higher frequencies than the transverse fluctuations. Then one may write

$$\boldsymbol{j}_q(\boldsymbol{r}) = \Delta i(\boldsymbol{r})n\boldsymbol{v}^t(\boldsymbol{r}) = q(\boldsymbol{r})\boldsymbol{v}^t(\boldsymbol{r}) \quad . \tag{50.10}$$

When we consider the thermal fluctuations of the heat current we assume not only $q(\boldsymbol{r})$ but also $\boldsymbol{v}^t(\boldsymbol{r})$ to fluctuate around their average values zero.

For the heat conductivity one can use arguments analogous to the ones leading to (50.7), now yielding

$$\mu_T(k) = \beta \int_0^\infty \langle \dot{q}(k,t)\dot{q}(-k,0)\rangle dt = D_T c_P k^2 \quad , \tag{50.11}$$

where c_P is the specific heat per unit volume. Using the conservation law (50.9) and isotropy one may finally write

$$D_T = \frac{\beta}{3c_P} \int_0^\infty dt \int d^3r \langle \boldsymbol{j}_q(\boldsymbol{r},t) \cdot \boldsymbol{j}_q(0,0)\rangle \quad . \tag{50.12}$$

Here we have taken the limit $k \to 0$. We now insert (50.10) and factorize the correlations. We shall see that $\langle q(\boldsymbol{r},t)q(0,0)\rangle$ is slowed down near the critical point and hence will be approximated by the $t=0$ limit. Then

$$D_T = \frac{\beta}{3c_P} \int_0^\infty dt \int d^3r \langle q(\boldsymbol{r},0)q(0,0)\rangle \langle \boldsymbol{v}^t(\boldsymbol{r},t) \cdot \boldsymbol{v}^t(0,0)\rangle \quad , \tag{50.13}$$

which describes the decay of a heat current fluctuation into a heat density and a velocity fluctuation. For the first factor in the integrand we use $\int q(\boldsymbol{r}) d^3r = \Delta I$ and $\beta \langle (\Delta I)^2 \rangle = c_P V$, see (50.10), hence

$$c_P = \beta \int \langle q(\boldsymbol{r})q(0)\rangle d^3r \quad . \tag{50.14}$$

For the r dependence we assume a decay length of the order of the correlation length ξ, for instance an Ornstein–Zernike–type form

$$\langle q(\boldsymbol{r})q(0)\rangle \propto \frac{1}{r} \exp(-r/\xi) \quad . \tag{50.15}$$

The detailed shape of the function on the r.h.s. of this equation is unimportant since the second factor in the integrand of (50.13) is slowly varying, as we shall see. The second factor is essentially the transverse momentum relaxation function, which can be obtained from hydrodynamics, see (29.25), in $(\omega, \boldsymbol{k})$ space:

$$\beta \langle g_i^t(\boldsymbol{k},t) g_j^t(-\boldsymbol{k})\rangle_\omega = \chi_{ij}^t(\boldsymbol{k}) \frac{1}{\omega - \eta k^2/nm} \tag{50.16}$$

with $\chi_{ij}^t = mn(\delta_{ij} - k_i k_j/k^2)$. In (50.13) one only needs the trace $\chi_{ii}^t = 2mn$, and the zero frequency limit. Using $\boldsymbol{g} = mn\boldsymbol{v}$ one obtains

$$\beta \int_0^\infty dt \langle \boldsymbol{v}^t(\boldsymbol{k},t) \cdot \boldsymbol{v}(-\boldsymbol{k},0)\rangle = \frac{2}{\eta k^2} \tag{50.17}$$

and after Fourier back transformation

$$\int_0^\infty dt \langle \boldsymbol{v}^t(\boldsymbol{r},t) \cdot \boldsymbol{v}^t(0,0)\rangle = \frac{kT}{2\pi\eta r} \quad . \tag{50.18}$$

Inserting this and (50.14,15) into (50.13) one finally arrives at the surpisingly simple result

$$D_T = \frac{kT}{6\pi\eta\xi} \quad . \tag{50.19}$$

This has the physically transparent interpretation that the thermal diffusion constant near the critical point is equal to kT times the mobility of droplets of radius ξ moving in the viscous fluid according to Stokes' formula [50.3].

Equation (50.19) says that the product $D_T\eta$ vanishes $\propto 1/\xi$ near the critical point. If one wants to know both factors separately one needs an independent mode-coupling calculation for η. We do not want to go into the details of such a calculation but just quote [50.1] the result that η is only weakly divergent. Thus one may say that the thermal diffusion is slowed down essentially $\propto 1/\xi$ near the critical point.

51. Broken Symmetry and Low-Frequency Modes**

We have discussed many examples of collective excitations and hydrodynamic modes with low excitation frequencies. In many cases the reason for the slowness of a mode was a conservation law. The first example was diffusion, see (20.8–10). A global conservation law (in the case of particle diffusion for the total particle number: $dN/dt = 0$) always implies the existence of a zero-frequency mode for the zero wave number variable $N(t) = n(\boldsymbol{k} = 0, t)$. For short-range forces one then expects a branch of low-frequency excitations to develop continuously out of this zero-frequency mode on increasing the wave vector k to nonzero values. In qualitative terms: a conservation law prevents the conserved quantity from relaxing to zero locally. It can only vanish from a limited region in space by means of currents through the surface of that region. The larger the region is (i.e. the longer the wavelength is), the longer such a process will take. Hence there will be slow modes for long wavelengths.

Another typical example is hydrodynamics (Chap. 29). The existence of three conservation laws guarantees the existence of three types of low-lying hydrodynamic modes (sound waves and the diffusion of heat and transverse momentum).

A second cause of the occurrence of low-frequency collective modes is the *spontaneous breaking of a continuous symmetry* and the presence of long-range order. We discussed several related examples in the previous two chapters on critical phenomena. In somewhat more general terms one may say that the breaking of a symmetry is connected with a "degeneracy" of the equilibrium state. Let us assume the symmetry operation is $U(\alpha) = \exp(i\alpha q)$, where q is the generator of the continuous symmetry and α some continuous phase. Then the state of broken symmetry is not an eigenstate of q. Application of the transformation U will not reproduce the state but lead to another state. "Spontaneous" symmetry breaking means that the Hamiltonian is invariant under U, in contrast to "external" symmetry breaking by asymmetric external fields. Since the Hamiltonian is invariant under $U(\alpha)$ the application of $U(\alpha)$ to a particular state produces a continuous manifold of degenerate states (i.e. states of the same energy). Transitions between these states will have zero energy. Again, for short-range forces one will expect a branch of low-energy excitations joining on continuously to the homogeneous zero energy excitations, considering phases α that vary slowly in space.

Typical examples of broken symmetries encountered in Chaps. 48 and 49 were spin rotations about a fixed axis n with $U(\alpha) = \exp(i\alpha n \cdot s)$ in ferro- and antiferromagnets, as well as gauge transformations with $U(\alpha) = \exp(i\alpha N)$ in superfluids and superconductors. Another example is translational symmetry with the translation operator $U(a) = \exp(ia \cdot p)$. Translational symmetry can be broken in two ways:

(i) The symmetry operation is translation of matter in space. Condensed matter in the limit of large mass has a fixed centre of gravity. $U(a)$ shifts this centre of gravity by the vector a. Such a state is associated with a breaking of Galilean invariance. The low-lying excitations occurring in connection with this symmetry breaking are the long-wavelength phonons.

(ii) The symmetry operation is translation of excitations relative to matter. This symmetry is usually called "homogeneity". The generator p may be called "pseudo-momentum". Homogeneity may be broken in crystals, or in so-called "topological defects" (phase boundaries, solitons, Bloch walls, vortex lines, etc.). Possible low energy excitations are "umklapp" phonons or slowly moving topological defects.

The approximate degeneracy of long-wavelength excitations is associated with vanishing restoring forces for modulations of the degenerate ground state at long wavelengths and manifests itself in a divergence of the corresponding susceptibilities for small k. This finds a quantitative expression in a $1/k^2$ theorem first discussed in the context of superfluidity [51.1,2] and then more generally [51.3]. The proof of this theorem uses Schwartz's inequality for the bracket symbol (3.11) following from its positivity (see Problem 13.2). For the two variables in this inequality one choses (i) the Fourier transform $a(k)$ of the quantity for which one wants to know the susceptibility, (ii) the Fourier transform of the time derivative $\dot{q}(k)$ of the density $q(r)$ of the conserved quantity q considered above (the generator of the symmetry operation). The inequality then takes the form

$$\langle a^*; a \rangle \geq \frac{|\langle a^*; \dot{q} \rangle|^2}{\langle \dot{q}^*; \dot{q} \rangle} \quad . \tag{51.1}$$

Here all quantities are taken at the same time, say $t = 0$. We multiply both sides by β and use (3.25) for the numerator on the r.h.s. of this inequality. For the denominator we use the differential conservation law

$$\dot{q}(k) + ik j_q(k) = 0 \quad , \tag{51.2}$$

where j_q is the longitudinal component of the current density of q. Using (3.13) we then introduce the static susceptibilities

$$\chi_a(k) = \beta \langle a^*(k); a(k) \rangle \quad \text{and} \quad \chi_j(k) = \beta \langle j_q^*(k); j_q(k) \rangle \tag{51.3}$$

and [after the use of (3.25)] the expectation value of the commutator

$$\frac{i}{\hbar} \langle [a^*(k), q(k)] \rangle = B(k) \quad , \tag{51.4}$$

which is nonzero because of the broken symmetry, if $a(\mathbf{k})$ is chosen properly. For instance, in ferromagnetic systems $a(\mathbf{k})$ may be chosen as $s_x(\mathbf{k})$ and $q(\mathbf{k}) = s_y(\mathbf{k})$. Then B is the expectation value S_z of the order parameter. In superfluid systems $a(\mathbf{k})$ may be taken to be the order-parameter operator $\psi(\mathbf{k})$ and $q(\mathbf{k}) = n(\mathbf{k})$ the particle density. Then B is again the expectation value of the order parameter, etc. Inserting everything, (53.1) takes the form

$$\chi_a(k) \geq \frac{|B(\mathbf{k})|^2}{k^2 \chi_j(k)} \quad . \tag{51.5}$$

It turns out that B is in most cases independent of $\mathbf{k}$. Let us assume that it at least approaches a nonzero limit for vanishing k. Often $\chi_j(k)$ can also be evaluated explicitly and is found to be independent of k. For instance, if q is the particle number, the f sum rule yields $\chi_j = n/m$. Let us again assume that it at least approaches a nonzero limit for vanishing k. Then (51.5) says that $\chi_a(k)$ diverges at least as $1/k^2$ for $k \to 0$. The conservation law (51.2) together with the broken symmetry implies vanishing restoring forces for modulations of the quantity $a(\mathbf{r})$ in the long-wavelength limit, as discussed in qualitative terms above.

In Chaps. 48 and 49 we had examples where χ_a diverged precisely as $1/k^2$

$$\chi_a(k) = \frac{B^2}{Rk^2} \tag{51.6}$$

with some "stiffness constant" R. Equation (51.5) then implies

$$R \leq \chi_j(0) \quad . \tag{51.7}$$

In the case of superconductivity with $R = \varrho_s/m^2$, see (48.11), Eq. (51.7) would imply the trivial inequality $\varrho_s \leq \varrho$.

Equation (51.5 or 6) in turn will lead to low-lying modes for $a(\mathbf{k})$ even if it is not a conserved quantity. An example is antiferromagetism, where $a(\mathbf{k})$ corresponds to the the staggered magnetization or the Fourier component $s_x(\mathbf{k} + \mathbf{K})$ of the magnetization, where $\mathbf{K}$ is some reciprocal lattice vector that is not conserved, in contrast to the total magnetization.

Rather than considering specific examples as in the foregoing chapters, let us give some general plausibility arguments for the existence of low-lying modes. The arguments are based on the reversible part of the mobility matrix (15.6) for the two slow variables $a(\mathbf{r}, t)$ and $q(\mathbf{r}, t)$. According to (51.4) this antisymmetric matrix has the off-diagonal elements iB and $-iB$. In order to calculate the frequency matrix one has to divide by the susceptibility matrix. Let us consider the case (which normally prevails) that this matrix, for reasons of symmetry (in particular space and time reversal symmetry), is diagonal with the diagonal elements $\chi_a(k)$ and $\chi_q(k)$. Then one may distinguish three different cases:

(i) *The isotropic ferromagnet*
In this case a and q may be identified as s_x and s_y. One has

$$\chi_a(k) = \chi_q(k) = \frac{B^2}{Rk^2} \quad . \tag{51.8}$$

The essential point is that not only $q(\mathbf{k})$ but also $a(\mathbf{k})$ is a conserved quantity. Hence both diagonal elements of the static susceptibility diverge quadratically. The reversible part of the frequency matrix is then given by

$$\nu = \frac{\mu}{\chi} = \frac{Rk^2}{B}\begin{pmatrix} 0 & i \\ -i & 0 \end{pmatrix} \tag{51.9}$$

and the frequencies are

$$\omega_{1,2} = \pm Rk^2/B \quad . \tag{51.10}$$

Due to two quadratically diverging susceptibilities one has a quadratic dispersion of the frequencies of the propagating modes.

(ii) *Other systems with short-range interactions*

The usual situation is that $a(\mathbf{k})$ is not conserved and $\chi_q(0)$ is finite. For instance, in antiferromagnets a is the staggered magnetization, in superfluids and superconductors a may be identified with the phase of the condensate, etc. The frequency matrix then takes the form

$$\nu = \mu/\chi = \begin{pmatrix} 0 & iRk^2/B \\ -iB/\chi_q(0) & 0 \end{pmatrix} \tag{51.11}$$

and the frequencies

$$\omega_{1,2} = \pm\sqrt{R/\chi_q(0)}k \tag{51.12}$$

are now linearly dependent on k.

Superfluids and superconductors are somewhat special since actually they are described by the two-fluid model. This model has the usual longitudinal hydrodynamic variables n, g^ℓ, s and one additional variable, the superfluid phase φ or velocity $\boldsymbol{v}_s = \hbar\,\mathrm{grad}\,\varphi/m$. As a consequence, there are two low-lying excitations with frequencies like (51.12): first and second sound. In principle the frequency matrix in this case has 4×4 elements. But in a simplified treatment one can use two separate calculations with 2×2 matrices as above: one with the pair of variables φ and s (Problem 48.1), where $g^\ell = 0$ and $\chi_q(0) \propto (\partial s/\partial T)_\varrho$, leading to second sound; and another one with the pair of variables φ and n, where $v_\mathrm{s} = v_\mathrm{n} = v$ and $\chi_q(0) \propto (\partial n/\partial P)_s$ leading to first sound. First sound is an elementary excitation existing down to zero temperature, whereas second sound is a hydrodynamic mode working with thermodynamic variables s and T and the concept of local equilibrium. Nevertheless, both excitations can be considered to have a low frequency $\omega \propto k$ because of the $1/k^2$ divergence of the long-ranged superfluid phase fluctuations (48.8).

(iii) *Systems with long-range Coulomb forces*

In systems with Coulomb interactions, long-wavelength density fluctuations are plasma vibrations with a nonzero frequency ω_p. Superconductors are no exceptions in this respect, despite their long-ranged phase fluctuations. In the scheme used above one has to consider a susceptibility

$$\chi_q(k) = \frac{\chi(k)}{1 + 4\pi e^2 \chi(k)/k^2} \to \frac{k^2}{4\pi e^2} \ . \tag{51.13}$$

This k^2 behaviour cancels the k^{-2} behaviour of χ_a and leads to a nonzero frequency for the long-wavelength excitations [51.4].

A complete treatment would have to include a discussion of the irreversible part of the frequency matrix. It turns out [51.5] that [e.g. in case (ii)], partly due to conservation laws, partly to $1/k^2$ singularities, it is proportional to k^2. The denominators of relaxation functions therefore show a $z^2 - c^2 k^2 + izk^2\gamma$ behaviour. The imaginary parts of the poles therefore vanish proportional to k^2, and hence in the long-wavelength limit the excitations derived above become undamped. An exception is the behaviour at the critical point, as discussed in the previous two chapters.

In relativistic particle physics the connection between broken symmetry, degeneracy (of the vacuum in this case) and low-lying excitations has found a parallel in the celebrated *Goldstone* theorem [51.6]. Statements about the k dependence are simplified enormously by relativistic invariance. It is sufficient to prove the existence of a zero-energy particle at $k = 0$ to prove the existence of a zero-mass particle with $\omega = ck$. The fact that long-range forces [case (iii)] push up the energy to nonzero values has its parallel in the Kibble–Higgs mechanism [51.7] of relativistic field theory. The broken symmetry is the electroweak gauge symmetry. The condensate wave function corresponds to the Higgs field. The long-range forces are mediated by the gauge fields of the vector bosons W, Z and γ. Both W and Z acquire nonzero masses by the analogue of the Meissner effect of superconductivity (the London penetration depth corresponding to the Compton wavelength of W and Z). The plasma frequency then corresponds to the mass of the Higgs boson, which is pushed up in energy by the long-range forces.

52. Collision Rates

In this chapter we apply the general formulae (15.6) or (15.14) to the calculation of collision rates. We first evaluate these formulae using perturbation theory and then apply the result to impurity scattering and chemical reactions.

52.1 Perturbation Theory

The variables we consider are occupation numbers $n_p = \psi_p^* \psi_p$ with the thermodynamically conjugate "forces" φ_p, see (32.1,2). The label p in our first example below will denote momenta but, for the time being, may be any quantum number.

Since the equal time commutator of the occupation numbers vanishes,

$$[n_p, n_q] = 0 \quad , \tag{52.1}$$

the reversible part in (15.6) (the first term on the r.h.s.) vanishes. If one then uses (15.14) one has to calculate the time derivatives in the stochastic force F and the commutator with the enthalpy (15.11). In both cases, if one writes

$$H = \sum \varepsilon_p n_p + U \quad ; \quad I = \sum (\varepsilon_p - \mu - \varphi_p) n_p + U \tag{52.2}$$

with an interaction Hamiltonian U one has

$$[H, n_p] = [I, n_p] = [U, n_p] = \frac{\hbar}{i} \dot{n}_p \quad . \tag{52.3}$$

As in Chap. 43, we want to use perturbation theory in the interaction U. The time derivatives (52.3) are then of first order in U, and M [according to (15.14)] of order U^2. Hence for the stochastic force (15.7) one can write

$$F_p = \dot{n}_p + O(U^2) \tag{52.4}$$

and neglect the correction term on the r.h.s. One then obtains for the memory kernel

$$M_{pq}(t, t') = -i\beta \langle \dot{n}_p(t); \dot{n}_q(t') \rangle \Theta(t - t') \quad . \tag{52.5}$$

The averages have to be taken with the statistical operator

$$\varrho = \exp[-\beta(I_0 - K)] \quad , \tag{52.6}$$

which is obtained from (52.2) by neglecting U.

We now assume a separation of time scales and interest ourselves only in the low-frequency limit of $M(t,0)$, which is nothing but the collision rate. (We specialize the formula for $t' = 0$, but in principle t' could be any time.) Then using (3.11) for the bracket symbol with the generalization (15.8) we finally obtain

$$\gamma_{pq} = \int_0^\infty \int_0^\beta \langle \dot{n}_p(t) \dot{n}_q(\alpha) \rangle d\alpha dt \quad ; \quad p \neq q \quad . \tag{52.7}$$

In the examples below we are going to consider only models with particle number conservation. Then according to (32.12,16) we only need to know the off-diagonal elements of the collision matrix γ.

52.2 Impurity Scattering

To fix our attention we consider a dilute Fermi gas and specify U as a static impurity potential

$$U = \sum U_{pq} \psi_p^* \psi_q \quad . \tag{52.8}$$

This is, of course, simpler than a two-body interaction, but once one has understood the formalism in this case, the corresponding generalization is not difficult in principle.

First we have to calculate the time derivatives (52.3). One finds, using the commutation relations $\psi_p^* \psi_q + \psi_q \psi_p^* = \delta_{pq}$ and $\psi_p \psi_q + \psi_q \psi_p = 0$,

$$\dot{n}_p = \frac{i}{\hbar} \sum (U_{p'p} \psi_{p'}^* \psi_p - \text{h.c.}) \quad . \tag{52.9}$$

Here the Hermitian conjugate can also be obtained by interchanging p and p'. If this is now inserted into (52.7) one encounters terms like

$$\langle \psi_{p'}^* \psi_p \psi_{q'}^* \psi_q \rangle = \langle \psi_{p'}^* \psi_q \rangle \langle \psi_p \psi_{q'}^* \rangle + \langle \psi_{p'}^* \psi_p \rangle \langle \psi_{q'}^* \psi_q \rangle \quad . \tag{52.10}$$

This factorization holds since the average is taken with the unperturbed statistical operator (52.6). After insertion into (52.9) the second term on the r.h.s. of (52.10) cancels out because $\langle \dot{n}_p \rangle = 0$. In the remaining term we introduce the time dependences of (52.7)

$$\psi_p(t) = \psi_p(0) \exp(-i\varepsilon_p t/\hbar) \quad ; \quad \psi_{p'}^*(t) = \psi_{p'}^*(0) \exp(i\varepsilon_{p'} t/\hbar) \tag{52.11}$$

and, see (15.8),

$$\psi_q(\alpha) = \exp(-\alpha I_0) \psi_q(0) \exp(\alpha I_0) = \psi_q(0) \exp[\alpha(\varepsilon_q - \mu - \varphi_q)] \quad ,$$
$$\psi_{q'}^*(\alpha) = \exp(-\alpha I_0) \psi_{q'}^*(0) \exp(\alpha I_0) = \psi_{q'}^*(0) \exp[-\alpha(\varepsilon_{q'} - \mu - \varphi_{q'})] \quad .$$
$$\tag{52.12}$$

Furthermore, we take into account that the expectation values on the r.h.s. in the unperturbed state (52.6) are diagonal

$$\langle \psi_{p'}^* \psi_q \rangle = n_q \delta_{p'q} \exp[i\varepsilon_q t/\hbar + \alpha(\varepsilon_q - \mu - \varphi_q)] \quad \text{etc.} \tag{52.13}$$

with

$$n_q = \{\exp[\beta(\varepsilon_q - \mu - \varphi_q)] + 1\}^{-1} \ . \tag{52.14}$$

Taking into account that we are only interested in $p \neq q$, there are only two remaining terms, which, after carrying out the α integration, take the form

$$\gamma_{pq} = (|U_{pq}|/\hbar)^2 \int_0^\infty dt\, e^{i(\varepsilon_q - \varepsilon_p)t/\hbar}\, n_q(1 - n_p)$$
$$\times \frac{\exp[\beta(\varepsilon_q - \varphi_q - \varepsilon_p + \varphi_p)] - 1}{\varepsilon_q - \varphi_q - \varepsilon_p + \varphi_p} + \{p \Leftrightarrow q\} \ . \tag{52.15}$$

After taking the symmetry into account, carrying out the t integral, and making some rearrangements, one obtains for statistically independent impurity positions with a density n_i the final result

$$\gamma_{pq} = \frac{2\pi}{\hbar V} n_i |u(\boldsymbol{p} - \boldsymbol{q})|^2 \delta(\varepsilon_p - \varepsilon_q) n_p n_q e^{\beta(\varepsilon_p - \mu)} \frac{e^{-\beta\varphi_q} - e^{-\beta\varphi_p}}{\varphi_p - \varphi_q} \ . \tag{52.16}$$

This then is in complete agreement with (32.20, 27). Thus we have finally reached a derivation of the more heuristic results of Chap. 32 from the formalism of Chap. 15.

For our simple model with only elastic scattering the nonlinearities of the rate equation (32.20) are spurious: (32.17) and (32.16) can easily be seen to be equivalent (see Problem 52.1). Nevertheless, our derivation allows us to see how the exponential factors of the form $[\exp(\beta \Delta \varphi) - 1]/\Delta \varphi$ arise which occur in the kinetic coefficients γ if one forces nonlinear rate equations to take the Onsager form $dn/dt = \gamma \Delta \varphi$: they result straightforwardly from the α integrals.

52.3 Chemical Reaction Rates

A rather similar situation occurs for chemical reactions. Let us consider as an example the reaction

$$A + B \leftrightarrow C + D \tag{52.17}$$

with the rate equation

$$\dot{n}_a = \dot{n}_b = -\dot{n}_c = -\dot{n}_d = \dot{n} = k n_c n_d - k' n_a n_b$$
$$= k n_c n_d \left[1 - \frac{n_a}{n_c} \frac{n_b}{n_d} \left(\frac{n_c}{n_a} \frac{n_d}{n_b} \right)_0 \right] = \lambda \Delta \alpha_c \tag{52.18}$$

where α_c is the chemical affinity

$$\alpha_c = \mu_a + \mu_b - \mu_c - \mu_d \tag{52.19}$$

and λ the rate "constant"

$$\lambda = k n_c n_d (1 - e^{\beta \alpha_c})/\alpha_c \quad . \tag{52.20}$$

In analogy to (52.7) one can now derive the perturbation expression [52.1, 2]

$$\lambda = \int_0^\infty \int_0^\beta \langle \dot{n}(t) \dot{n}(\alpha) \rangle d\alpha \, dt \tag{52.21}$$

where $\dot{n}$ is now an operator obtained from

$$\dot{n} = \frac{i}{\hbar}[U, n_a] \quad . \tag{52.22}$$

Here U is the interaction responsible for the chemical reaction. In order to keep the calculation simple let us consider only a caricature of such an interaction, namely

$$U = u(a^* b^* dc + \text{h.c.}) \quad , \tag{52.23}$$

where $a, b, c, d, (a^*, b^*, c^*, d^*)$ are the annihilation (creation) operators of the chemical species A, B, C, D. We take them to be fermions, although in the dilute limit, which is what we shall consider, this is irrelevant.

Inserting (52.23) into (52.22) and using the fermion commutation relations one obtains

$$\dot{n} = \frac{i}{\hbar} u(c^* d^* ba - \text{h.c.}) \quad . \tag{52.24}$$

The time dependences now take the form

$$a(t) = a(0) \exp(-i\varepsilon_a t/\hbar) \quad \text{etc.} \tag{52.25}$$

and the α dependencies

$$c(\alpha) = c(0) \exp[\alpha(\varepsilon_c - \mu_c)] \quad ; \quad c^*(\alpha) = c^*(0) \exp[-\alpha(\varepsilon_c - \mu_c)] \quad . \tag{52.26}$$

Collecting everything together, one obtains after some intermediate steps quite similar to those in Sect. 52.2

$$\lambda = (|u|/\hbar)^2 \int_0^\infty dt \, e^{i \Delta \varepsilon t/\hbar} n_c n_d (1 - n_a)(1 - n_b) \frac{1 - e^{\beta(\Delta \varepsilon + \alpha_c)}}{\Delta \varepsilon - \alpha_c} + \{a, b \Leftrightarrow c, d\}$$
with
$$\tag{52.27}$$

$$\Delta \varepsilon = \varepsilon_c + \varepsilon_d - \varepsilon_a - \varepsilon_b \quad . \tag{52.28}$$

Finally, after carrying out the t integration and using energy conservation one arrives at

$$\lambda = k n_c n_d (1 - n_a)(1 - n_b)(1 - e^{\beta \alpha_c})/\alpha_c \tag{52.29}$$

with

$$k = (2\pi/\hbar)|u|^2 \delta(\varepsilon_a + \varepsilon_b - \varepsilon_c - \varepsilon_d) \quad . \tag{52.30}$$

This is equal to (52.20) in the dilute limit $n_a, n_b \ll 1$.

Equation (52.30) says that a nonzero λ will occur only if the energy deficit $\Delta\varepsilon$ can be passed on to some other degrees of freedom, for instance the translational energy of the reaction partners, or the energy of some catalyst. We have not taken these additional effects into account because they would mask the simple calculations above.

Concluding this section let us mention that two-body collisions can be considered in close analogy to the binary reaction (52.17) if one identifies them as reactions of the form $A(\boldsymbol{p}_1) + A(\boldsymbol{p}_2) \leftrightarrow A(\boldsymbol{q}_1) + A(\boldsymbol{q}_2)$.

PROBLEMS

52.1 Convince yourself that (32.16) and (32.17) are equivalent (the quadratic terms in the occupation numbers drop out) for elastic scattering.

52.2 Generalize (52.16) to the case of electron–phonon scattering and derive (40.30).

53. Many-Body Effects in Collision Rates

Electrons in solids are scattered not only by impurities and phonons but also by other electrons. Usually the scattering by phonons is more important since they have a higher density of states at low energies, but in transition metals, for instance, one has a rather high density of d-electron states and the scattering of conduction (s-)electrons by d-electrons can lead to a momentum relaxation of s-electrons and hence contribute to the conductivity. One of the problems in treating electron–electron scattering is the divergence of the two-particle scattering cross section $\propto 1/k^4$ at small momentum transfer k. This problem is resolved by taking into account that the electrons are actually scattered not by other single electrons but by the many-body system of all other electrons. Such a scattering event can be described by a formula such as (9.5) where $s(k,\omega)$ is the spectral function of electronic charge density fluctuations. To lowest order, $s(k,\omega)$ is given by the dilute limit result (40.36), but it will be modified by the self-consistent field as, e.g. in (35.2): the spectral function can be expressed in terms of the imaginary part of the susceptibility and detailed balance (6.13,17), leading to

$$s^e(k,\omega) = \hbar[1+n(\omega)]\chi^{e''}(k,\omega) \quad . \tag{53.1}$$

Here, as in (35.2), we use the label "e" for the quantities renormalized by the self-consistent field, describing the response of the system to external fields as compared to the "bare" quantities of the noninteracting system $s(k,\omega)$ and $\chi(k,\omega)$. Using

$$\chi^e(k,\omega) = \frac{\chi(k,\omega)}{1+v(k)\chi(k,\omega)} \tag{53.2}$$

and (53.1) one obtains

$$s^e(k,\omega) = \frac{s(k,\omega)}{[1+v(k)\chi'(k,\omega)]^2 + [v(k)\chi''(k,\omega)]^2} \quad . \tag{53.3}$$

For the loss term in the kinetic equation one expects in analogy to (40.29,30)

$$[\dot{n}_p]_{\text{loss}} = -\frac{2\pi}{\hbar V}\sum |v(k)|^2 \int s^e(k,\omega)\delta(\varepsilon_p - \hbar\mathbf{k} - \hbar\omega)n_p(1-n_p-\hbar\mathbf{k})d\omega \quad . \tag{53.4}$$

Now for Coulomb interactions $v(k) = 4\pi e^2/k^2$. Equation (53.4) then tells us that the divergence in the cross section is cancelled by the divergence in the denominator of (53.3).

Thus the many-body effects in the electronic system cancel the divergence in the bare interaction. Another way of looking at the same effect is to consider the screened interaction $v^s(k,\omega)$. This is the bare interaction potential $v(k)$ plus the change in the Coulomb potential of the electronic system $-v(k)\delta n(k,\omega)$ [see (35.1)] produced by the change δn in electron density, which in turn is induced by the scattering electron corresponding to an "external field" $v(k)$. That is, $\delta n(k,\omega) = \chi^e(k,\omega)v(k)$. Collecting everything together one may write

$$[\dot{n}_p]_{\text{loss}} = -\frac{2\pi}{\hbar V} \sum |v^s(k,\omega)|^2 \int s(k,\omega)\delta(\varepsilon_p - \varepsilon_{p'} - \hbar\omega)n_p(1-n_{p'})d\omega \quad (53.5)$$

with $p' = p - \hbar k$ and

$$v^s(k,\omega) = \frac{v(k)}{1 + v(k)\chi(k,\omega)} \quad . \quad (53.6)$$

So, instead of saying the electron is scattered by the interacting electron system with the bare potential one may also say it is scattered by the noninteracting system with the screened potential.

Since the the energy transfers occurring in Fermi liquids at low temperatures are of the order of kT, they can be neglected to first order and the $\chi(k,\omega)$ can be replaced by the static limit $\chi(k,0)$, which is real since $\chi''(k,\omega)$ is proportional to ω for small ω. In particular, for forward scattering ($k=0$) one can even use the homogeneous static susceptibility $\chi(k \to 0, 0) = \partial n/\partial \mu$ and the screened potential in the Thomas–Fermi limit

$$v^s(k) = \frac{v(k)}{1 + v(k)\partial n/\partial \mu} \quad . \quad (53.7)$$

For short-range forces (for instance in liquid ^{3}He) there is no divergence in the bare cross section but the scattering, nevertheless, is by the *interacting* system. Hence (53.6) is still valid for the screened interaction to be used in the collision term. Using (35.4) the screened forward scattering amplitude is now given by [53.1]

$$f_0^s = \frac{f_0}{1 + f_0} \quad . \quad (53.8)$$

Since f_0 in liquid He is of the order of 11 (Chap. 35) there is a reduction of the scattering cross section proportional to $(f_0^s)^2$ to be used in the kinetic equation of about two orders of magnitude. Thus there can be marked screening effects even in systems with short-range interactions.

On the other hand there may be corresponding enhancements, if the sign of the bare interaction changes. The f_0 in liquid He is positive, but there are spin flip contributions $v(0)\partial n/\partial \mu = f_0 + i_0 \boldsymbol{\sigma} \cdot \boldsymbol{\sigma}'$ which are of the order of $i_0 \approx -0.8$ in liquid He. This leads to an enhancement of the spin flip scattering cross section by a factor of about 25 and a pronounced contribution of spin flip scattering processes to transport coefficients [53.2]. Such effects also occur in solids, whenever they are close to a ferromagnetic instability. In this case the real part of the dynamic magnetic susceptibility is enhanced and the spectral function $s_m^e(k,\omega)$ of magnetic excitations contains a low-energy peak corresponding to "paramagnons", which scatter electrons very efficiently and lead to large T^2 contributions in electronic transport properties [53.3].

PROBLEMS

53.1 Discuss and plot the imaginary part of the magnetic susceptibility

$$\chi_m^e(k,\omega) = \frac{\chi(k,\omega)}{1 - |i_0|\chi(k,\omega)}$$

where $|i_0|$ is close to unity. Use the small k, small ω approximation (34.6,8) for the susceptibility of the noninteracting system.

References

Chapter 1

1.1 Brenig, W.: *Statistische Theorie der Wärme I. Gleichgewicht*, 2nd ed. (Springer, Berlin, Heidelberg 1983)
1.2 Boltzmann, L.: Wien. Ber. **66**, 213, 275 (1872); *Lectures on Gas Theory*, translated by G. Brush (Berkeley, CA 1964)
1.3 Einstein, A.: Ann. Phys. (Leipzig) **17**, 549 (1905); **19**, 371 (1906)
1.4 Langevin, P.: C.R. Acad. Sci. **146**, 530 (1908)
1.5 Onsager, L.: Phys. Rev. **37**, 405 (1931); **38**, 2265 (1931)
1.6 Schottky, W.: Ann. Phys. (Leipzig) **57**, 541 (1918)
1.7 Nyquist, H.: Phys. Rev. **32**, 110 (1928)
1.8 Callen, H.B., Welton T.A.: Phys. Rev. B **83**, 34 (1951)

Chapter 2

2.1 Kadanoff, L.P., Baym, G.: *Quantum Statistical Mechanics* (Benjamin, New York 1962)
2.2 Neumann, J. von: Z. Phys. **57**, 30 (1929)
2.3 Liouville, J.: J. de Math. **3**, 348 (1838)
2.4 Brenig, W.: *Statistische Theorie der Wärme I. Gleichgewicht*, 2nd ed. (Springer, Berlin, Heidelberg 1983)
2.5 Kadanoff, L.P., Martin, P.C.: Ann. Phys. **24**, 419 (1963). A slightly different but similar initial condition is used by Zubarev, D.N.: *Nonequilibrium Thermodynamics* (Consultants Bureau, New York 1974). Translation of Russian original from 1971

Chapter 3

3.1 Brenig, W.: Statistische Theorie der Wärme I. 2nd ed. (Springer, Berlin, Heidelberg 1983)
3.2 Kubo, R.: J. Phys. Soc. Jpn. **12**, 570 (1957); *Lectures in Theoretical Physics*, Vol. 1 (Interscience, New York 1959)
3.3 Matsubara, T.: Prog. Theor. Phys. **14**, 351 (1955)
3.4 Callen, H.B., Welton, T.A.: Phys. Rev. B **83**, 34 (1951)

Chapter 5

5.1 Martin, P.C.: *Measurements and Correlation Functions* (Gordon and Breach, New York 1968) Chap. 1
5.2 Kubo, R., Toda, M., Hashitsume, N.: *Statistical Physics II*, Springer Ser. Solid-State Sci., Vol. 31 (Springer, Berlin, Heidelberg 1985) Chap. 1
5.3 Chandrasekhar, S.: Rev. Mod. Phys. **15**, 1 (1943)

Chapter 6

6.1 Kramers, H.A.: Estratto dagli Atti del Congresso Internazionale di Fisica, Como (Nicolo Zanichelli, Bologna 1927)
     Kronig, R. de: Ned. Tijdschr. Natuurkd. **9**, 402 (1942); Physica **12**, 543 (1946)

Chapter 7

7.1 For information about time reversal invariance see monographs on quantum mechanics, e.g. Gottfried, K.: *Quantum Mechanics I* (Benjamin, New York 1966) Chap. 39
7.2 Wigner, E.P.: Göttinger Nachr. **31**, 546 (1932)

Chapter 8

8.1 Brenig, W.: *Statistische Theorie der Wärme I. Gleichgewicht*, 2nd ed. (Springer, Berlin, Heidelberg 1983)

Chapter 9

9.1 Brenig, W.: Z. Phys. B **48**, 127 (1982)

Chapter 13

13.1 Langevin, P.: C.R. Acad. Sci. **146**, 530 (1908)
13.2 Zwanzig, R.: J. Chem. Phys. **33**, 1338 (1960)
13.3 Mori, H.: Prog. Theor. Phys. **33**, 423 (1965)

Chapter 14

14.1 Langevin, P.: C.R. Acad. Sci. **146**, 530 (1908)

Chapter 16

16.1 Brenig, W.: *Statistische Theorie der Wärme I. Gleichgewicht*, 2nd ed. (Springer, Berlin, Heidelberg 1983)
16.2 Prigogine, I.: *Introduction to Thermodynamics of Irreversible Processes* (Interscience, New York 1969) Chaps. III, IV

Chapter 17

17.1 Brenig, W.: *Statistische Theorie der Wärme I. Gleichgewicht*, 2nd ed. (Springer, Berlin, Heidelberg 1983)
17.2 Becker, R.: *Theorie der Wärme*, Heidelberger Taschenbücher, Vol. 10 (Springer, Berlin, Heidelberg 1955) Chap. 32d

Chapter 18

18.1 Schottky, W.: Ann. Phys. (Leipzig) **57**, 541 (1918)
18.2 Nyquist, H.: Phys. Rev. **32**, 110 (1928)

Chapter 19

19.1 Becker, R.: *Theorie der Wärme*, Heidelberger Taschenbücher, Vol. 10 (Springer, Berlin, Heidelberg 1955) Chap. 7
19.2 Tisza, L.: J. Phys. Radium **8**, 1, 165, 350 (1940)
    Landau, L.D.: Zh. Eksp. Teor. Fiz. **11**, 592 (1941)
19.3 Allen, J.F., Jones, H.: Nature **141**, 243 (1938)

Chapter 21

21.1 Callen, H.B.: Phys. Rev. **73**, 1349 (1948)

Chapter 22

22.1 Brenig, W.: *Statistische Theorie der Wärme I. Gleichgewicht*, 2nd ed. (Springer, Berlin, Heidelberg 1983)

Chapter 24

24.1 Haken, H.: *Synergetics*, 3rd ed., Springer Ser. Synergetics, Vol. 1 (Springer, Berlin, Heidelberg 1983) Chap. 9
24.2 Schlögl, F.: Z. Phys. **253**, 147 (1972)
24.3 Lotka, A.J.: J. Phys. Chem. **14**, 271 (1910)
    Volterra, V.: *Leçons sur la théorie mathématique de la lutte pour la vie* (Gauthier-Villars, Paris 1931)
24.4 Glansdoff, P., Prigogine, I.: *Thermodynamic Theory of Structure, Stability and Fluctuations* (Wiley, New York 1971)

24.5   Belousov, B.P.: Sb. Ref. Radats Med., Moscow (1959)
       Zhabotinsky, A.M., Zaikin, A.N.: J. Theor. Biol. **40**, 45 (1973)
24.6   Imbihl, R., Cox, M.P., Ertl, G., Müller, H., Brenig, W.: J. Chem. Phys. **83**, 1578 (1985)
24.7   Schieve, W.C., Allen, P.M. (eds.): *Self-Organization and Dissipative Structures* (University of Texas Press, Austin, TX 1982)
       Riste, T. (ed.): *Nonlinear Phenomena at Phase Transitions and Instabilities* (Plenum, New York 1982)
24.8   Benard, H.: Rev. Gen. Sci. Pures Appl. **11**, 1261 (1900)
24.9   Eigen, M., Schuster, P.: Naturwissenschaften **64**, 541 (1977); ibid. **65**, 7, 341 (1978)
24.10 Grossmann, S., Thomae, S.: Z. Naturforsch. A **32**, 1353 (1977)
24.11 Feigenbaum, M.J.: J. Stat. Phys. **19**, 25 (1978)
24.12 Lorenz, E.N.: J. Atmos. Sci. **20**, 130, 448 (1963)
24.13 Rössler, O.E.: Z. Naturforsch. A **31**, 259 (1976); Phys. Lett. **57A**, 397 (1976)
24.14 Hao Bai-Lin: *Chaos* (World Scientific, Singapore 1984)

Chapter 25

25.1   Brenig, W.: *Statistische Theorie der Wärme I. Gleichgewicht*, 2nd ed. (Springer, Berlin, Heidelberg 1983)
25.2   Feher, G., Weissman, M.: Proc. Natl. Acad. Sci. USA **70**, 870 (1973)

Chapter 26

26.1   Brenig, W.: *Statistische Theorie der Wärme I. Gleichgewicht*, 2nd ed. (Springer, Berlin, Heidelberg 1983)
26.2   Pelzer, H., Wigner, E.: Z. Phys. Chem. B **15**, 445 (1932)
       Eyring, H., Lin, S.H., Lin, S.M.: *Basic Chemical Kinetics* (Wiley, New York 1980)

Chapter 27

27.1   Becker, R., Döring, W.: Ann. Phys. (Leipzig) **24**, 719 (1935)
27.2   Brenig, W.: *Statistische Theorie der Wärme I. Gleichgewicht*, 2nd ed. (Springer, Berlin, Heidelberg 1983)
27.3   Binder, K., Stauffer, D.: Adv. Phys. **25**, 343 (1976)
27.4   Volmer, M., Flood, R.: Z. Phys. Chem. **170**, 273 (1934)
27.5   Fisher, M.E.: Rep. Prog. Phys. **30**, 615 (1967)

Chapter 28

28.1   Brenig, W.: *Statistische Theorie der Wärme I. Gleichgewicht*, 2nd ed. (Springer, Berlin, Heidelberg 1983)
28.2   Landau, L.D., Placzek, G.: Phys. Z. Sowjetunion **5**, 172 (1934)

Chapter 29

29.1   Euler, L.: *Principes généraux du mouvement des fluids*. Hist. Acad. Berlin (1755)
29.2   Martin, P.C., Parodi, O., Pershan, P.S.: Phys. Rev. A **6**, 2401 (1972) Appendix A
29.3   Groot, S.R. de, Mazur, P.: *Non-equilibrium Thermodynamics* (North-Holland, Amsterdam 1969) Chap. 12
29.4   Kreuzer, H.J.: *Nonequilibrium Thermodynamics and Its Statistical Foundations* (Clarendon, Oxford 1981) Chap. 2
29.5   Forster, D.: *Hydrodynamic Fluctuations, Broken Symmetry, and Correlation Functions* (W.A. Benjamin, Reading, MA 1974) Chap. 10
29.6   Stokes, G.: Trans. Cambridge Philos. Soc. **8**, 287 (1845). Incompressible fluids were already treated by Navier, C.L.M.H.: Mem. Acad. Sci. Inst. Fr. (2) **6**, 289 (1827)
29.7   Landau, L.D., Lifshitz, E.M.: *Fluid Mechanics*, Course of Theoretical Physics, Vol. 6 (Pergamon, Oxford 1959)

Chapter 30

30.1   Adler, B.J., Wainwright, T.E.: Phys. Rev. Lett. **18**, 988 (1967)
30.2   Ernst, M.H., Hauge, E.H., Leeuven, J.M.J. van: Phys. Rev. A **4**, 2055 (1971)

30.3 Landau, L.D., Lifshitz, E.M.: *Fluid Mechanics*, Course of Theoretical Physics, Vol. 6 (Pergamon, Oxford 1959)
30.4 Pomeau, Y., Resibois, P.: Phys. Rep. **19C**, 64 (1975)

Chapter 31

31.1 Abrikosov, A.A., Gorkov, L.P., Dzyaloshinski, I.E.: *Methods of Quantum Field Theory in Statistical Physics* (Prentice-Hall, Englewood Cliffs, NJ 1963)
31.2 Forster, D.: *Hydrodynamic Fluctuations, Broken Symmetry, and Correlation Functions* (W.A. Benjamin, Reading, MA 1975) Sect. 12.4

Chapter 32

32.1 Brenig, W.: *Statistische Theorie der Wärme I. Gleichgewicht*, 2nd ed. (Springer, Berlin, Heidelberg 1983)
32.2 Boltzmann, L.: Wien. Ber. **66**, 213, 275 (1872); *Lectures on Gas Theory*, translated by G. Brush (Berkeley, CA 1964)

Chapter 33

33.1 Boltzmann, L.: Wien. Ber. **66**, 213, 275 (1872); *Lectures on Gas Theory*, translated by G. Brush (Berkeley, CA 1964)
33.2 Vlasov, A.A.: Sov. Phys. – JETP **8**, 291 (1938)
33.3 Sommerfeld, A.: Z. Phys. **47**, 1 (1928)
33.4 Peierls, R.: Ann. Phys. (Leipzig) **3**, 1055 (1929)
33.5 Landau, L.D., Khalatnikov, I.M.: Zh. Eksp. Teor. Fiz. **19**, 637, 709 (1949)
33.6 Landau, L.D.: Sov. Phys. – JETP **5**, 101 (1957)
33.7 Brenig, W.: *Statistische Theorie der Wärme I. Gleichgewicht*, 2nd ed. (Springer, Berlin, Heidelberg 1983)
33.8 Wigner, E.: Phys. Rev. **40**, 749 (1932)
33.9 Bogolubov, N.N.: J. Phys. (USSR) **10**, 256 (1946) [English transl. in *Studies in Statistical Mechanics*, Vol. 1, ed. by J. de Boer, G.E. Uhlenbeck (North-Holland, Amsterdam 1962)]
3.10 Kadanoff, L.P., Baym, G.: *Quantum Statistical Mechanics* (Benjamin, New York 1962)

Chapter 34

34.1 Lindhard, J.: K. Dansk. Vidensk. Selsk., Mat.-Fys. Medd. **28**, No. 8 (1954)
34.2 Landau, L.D.: Z. Phys. **64**, 629 (1930)

Chapter 35

35.1 Landau, L.D.: Sov. Phys. – JETP **32**, 59 (1957)
35.2 Abel, W.R., Anderson, A.C., Wheatly, J.C.: Phys. Rev. Lett. **17**, 74 (1966)

Chapter 36

36.1 Fokker, A.D.: Ann. Phys. (Leipzig) **43**, 810 (1914)
Planck, M.: Sitzungsber. Preuss. Akad. Wiss. p. 324 (1917)
36.2 Haken, H.: Rev. Mod. Phys. **47**, 67 (1975)
36.3 Chandrasekhar, S.: Rev. Mod. Phys. **15**, 1 (1943)
36.4 Risken, H.: *The Fokker-Planck Equation*, 2nd ed., Springer Ser. Synergetics, Vol. 18 (Springer, Berlin, Heidelberg 1989)
36.5 Kramers, H.A.: Physica **7**, 284 (1940)
36.6 Moyal, J.E.: J. R. Stat. Soc. London B **11**, 150 (1949)
36.7 Pawula, R.F.: Phys. Rev. **162**, 186 (1967)
36.8 Risken, H., Vollmer, H.D.: Z. Phys. B **35**, 313 (1979)

Chapter 37

37.1 Einstein, A.: Ann. Phys. (Leipzig) **17**, 549 (1905); **19**, 289, 371 (1906); **33**, 1096 (1910)
37.2 Smoluchowski, M. von: Ann. Phys. (Leipzig) **48**, 1103 (1915)
37.3 Drude, P.: Ann. Phys. (Leipzig) **14**, 566 (1900); **3**, 369 (1900)

Chapter 39

39.1 Enskog, D.: Dissertation, Uppsala University (1917)
39.2 Chapman, S., Cowling, T.G.: *The Mathematical Theory of Nonuniform Gases*, 3rd ed. (Cambridge University Press, Cambridge 1970)
39.3 Abrikosov, A.A., Khalatnikov, I.M.: Rep. Prog. Phys. **22**, 329 (1959)
39.4 Brenig, W.: *Statistische Theorie der Wärme I. Gleichgewicht*, 2nd ed. (Springer, Berlin, Heidelberg 1983)

Chapter 40

40.1 Chapman, S., Cowling, T.G.: *The Mathematical Theory of Non-Uniform Gases* (Cambridge University Press, Cambridge 1952) Chaps. 9, 10
40.2 Landau, L.D., Lifshitz, E.M.: *Mechanics*, Course of Theoretical Physics, Vol. 1 (Pergamon, Oxford 1959)
40.3 Résibois, P., De Leener, M.: *Classical Kinetic Theory of Fluids* (Wiley, New York 1977)
40.4 Hirschfelder, J.O., Curtiss, C.F., Bird, R.B.: *Molecular Theory of Gases and Liquids* (Wiley, New York 1954)
40.5 Mannari, I.: Prog. Theor. Phys. **26**, 51 (1961)
Baym, G.: Phys. Rev. **135A**, 1691 (1964)
40.6 Kadanoff, L.P., Falko, I.I.: Phys. Rev. **136A**, 1170 (1964)
Schmid, A.: Z. Phys. **259**, 421 (1973)
40.7 Pippard, A.B.: Philos. Mag. **46**, 1104 (1955)
40.8 Jensen, H.H., Smith, H., Wilkins, J.W.: Phys. Rev. **185**, 323 (1969)
40.9 Baym, G., Ebner, C.: Phys. Rev. **170**, 346 (1968)
Emery, V.J.: Phys. Rev. **175**, 251 (1968)

Chapter 41

41.1 Brenig, W.: *Statistische Theorie der Wärme I. Gleichgewicht*, 2nd ed. (Springer, Berlin, Heidelberg 1983)
41.2 Landau, L.D., Lifshitz, E.M.: *Fluid Mechanics*, Course of Theoretical Physics, Vol. 6 (Pergamon, Oxford 1959)
41.3 Abrahams, E., Anderson, P.W., Licciardello, D.C., Ramakrishnan, T.V.: Phys. Rev. Lett. **42**, 673 (1979)
41.4 Dorfmann, J.R., Cohen, E.G.D.: Phys. Lett. **16**, 124 (1965)
Sengers, J.V.: Phys. Rev. Lett. **15**, 515 (1965)
Leeuwen, J.M.J. van, Wejland, A.: Physica **36**, 457 (1967)

Chapter 42

42.1 Greenwood, D.A.: Proc. R. Soc. London **87**, 775 (1966)
42.2 Lindhard, J.: K. Dansk. Vidensk. Selsk., Mat.-Fys. Medd. **28**, No. 8 (1954)

Chapter 43

43.1 Ziman, J.M.: *Principles of the Theory of Solids* (Cambridge University Press, Cambridge 1969) Chap. 7
43.2 Brenig, W.: "Transporttheorie", Vorlesungsskriptum TU München (1969)
43.3 Götze, W., Wölfle, P.: Phys. Rev. B **6**, 1226 (1972)

Chapter 44

44.1 Mannari, I.: Prog. Theor. Phys. **26**, 51 (1961)
44.2 Baym, G.: Phys. Rev. **135A**, 1691 (1964)
44.3 Bloch, F.: Z. Phys. **59**, 208 (1930)
44.4 Ziman, J.M.: *Electrons and Phonons* (Oxford University Press, Oxford 1963) Chap. 9
44.5 Gold, A., Götze, W., Mazuré, C., Koch, F.: Solid State Commun. **49**, 1085 (1984)
44.6 Gold, A., Götze, W.: Phys. Rev. B **33**, 2495 (1986)

Chapter 45

45.1 Abrahams, E., Anderson, P.W., Licciardello, D.C., Ramakrishnan, T.V.: Phys. Rev. Lett. **42**, 673 (1979)

45.2 Gorkov, L.P., Larkin, A.I., Khmelnitzkii, D.E.: JETP Lett. **30**, 228 (1979)
45.3 Vollhardt, D., Wölfle, P.: Phys. Rev. Lett. **45**, 842 (1980); Phys. Rev. B **22**, 4666 (1980)
45.4 Prelovšek, P.: In *Recent Developments in Condensed Matter Physics*, Vol. II, ed. by J.T. Devreese (Plenum, New York 1981); Phys. Rev. B **23**, 1304 (1981)
45.5 Belitz, D., Gold, A., Götze, W.: Z. Phys. B **44**, 273 (1981)

Chapter 46

46.1 Anderson, P.W.: Phys. Rev. **109**, 1492 (1958)
46.2 Mott, N.F., Davis, E.A.: *Electronic Processes in Non-Crystalline Materials* (Clarendon, Oxford 1979)
46.3 Licciardello, D.C., Thouless, D.J.: Phys. Rev. Lett. **35**, 1475 (1975)
46.4 Wegner, F.: Z. Phys.. B **25**, 327 (1976)
46.5 Abrahams, E., Anderson, P.W., Licciardello, D.C., Ramakrishnan, T.V.: Phys. Rev. Lett. **42**, 673 (1979)
46.6 Gorkov, L.P., Larkin, A.I., Khmelnitzkii, D.E.: JETP Lett. **30**, 228 (1979)
46.7 Götze, W.: Solid State Commun. **27**, 1393 (1978)
46.8 Vollhardt, D., Wölfle, P.: Phys. Rev. Lett. **48**, 699 (1982); In *Anderson Localization*, ed. by Y. Nagaoka, H. Fukuyama, Springer Ser. Solid-State Sci., Vol. 39 (Springer, Berlin, Heidelberg 1982) p. 26

Chapter 47

47.1 McGinnis, W.C., Burns, M.J., Simon, R.W., Deutscher, G., Chaikin, P.M.: Physica **107B**, 5 (1981)
Dolan, G.J., Osheroff, D.D.: Phys. Rev. Lett. **43**, 721 (1979)
47.2 Wegner, F.: Z. Phys. B **35**, 207 (1979)
47.3 Hikami, S., Larkin, A.I., Nagaoka, Y.: Prog. Theor. Phys. **63**, 707 (1980)
47.4 Lee, P.A.: J. Noncryst. Solids **35**, 21 (1980)
47.5 Oppermann, R., Jüngling, K.: Phys. Lett. **76A**, 449 (1980)
47.6 Lee, P.A., Ramakrishnan, T.V.: Rev. Mod. Phys. **57**, 287 (1985)

Chapter 48

48.1 Widom, B.: J. Chem. Phys. **43**, 3892, 3898 (1965)
Kadanoff, L.P.: Physics **2**, 263 (1966)
Wilson, K.G., Kogut, J.: Phys. Rep. **12C**, 75 (1974)
48.2 Brenig, W.: *Statistische Theorie der Wärme I. Gleichgewicht*, 2nd ed. (Springer, Berlin, Heidelberg 1983) Sects. 47,48
48.3 Ferrell, R.A., Menyhárd, N., Schmidt, H., Schwabl, F., Szépfalusy, P.: Phys. Rev. Lett. **18**, 891 (1967)
48.4 Mikeska, H.J., Schmidt, H.: J. Low Temp. Phys. **2**, 371 (1970)
Nelson, D.R., Kosterlitz, J.M.: Phys. Rev. Lett **39**, 1201 (1977)
48.5 Ahlers, G.: Phys. Rev. Lett. **21**, 1159 (1968)
48.6 Ahlers, G.: Phys. Rev. Lett. **43**, 1417 (1979)
48.7 Dohm, V., Folk, R.: Z. Phys. B **45**, 129 (1981)
48.8 Halperin, B.I., Hohenberg, P.C.: Phys. Rev. Lett. **19**, 700 (1967)
48.9 Hohenberg, P.C., Halperin, B.I.: Rev. Mod. Phys. **49**, 435 (1977)
48.10 Anderson, P.W.: Rev. Mod. Phys. **38**, 298 (1966)

Chapter 49

49.1 Frey, E., Schwabl, F.: Phys. Lett. **123A**, 49 (1987)
49.2 Halperin, B.I., Hohenberg, P.: Phys. Rev. **177**, 952 (1969)
49.3 Schmidt, H., Schwabl, F.: Phys. Lett. **61A**, 476 (1977)
49.4 Fixman, M.: J. Chem. Phys. **36**, 199 (1962)
Kawasaki, K.: Phys. Lett. **30**, 325A (1969)
Ferrell, R.A.: Phys. Rev. Lett. **24**, 1169 (1970)
49.5 Ferrell, R.A., Bhattacharjee, J.K.: Phys. Rev. **31A**, 1788 (1985)
Dengler, R., Schwabl, F.: Europhys. Lett. **4**, 1233 (1987)

Chapter 50

50.1 Hohenberg, P.C., Halperin, B.I.: Rev. Mod. Phys. **49**, 435 (1977)
50.2 Ferrell, R.A.: Phys. Rev. Lett. **24**, 1169 (1970)
50.3 Kawasaki, K.: Ann. Phys. **61**, 1 (1970)

Chapter 51

51.1 Bogolubov, N.N.: J. Phys. (USSR) **2**, 23 (1947); Physica **26** (Suppl.), 1 (1960)
51.2 Hugenholtz, N., Pines, D.: Phys. Rev. **116**, 489 (1959)
51.3 Wagner, H.: Z. Phys. **195**, 273 (1966)
51.4 Lange, R.V.: Phys. Rev. Lett. **14**, 3 (1965)
51.5 Forster, D.: *Hydrodynamic Fluctuations, Broken Symmetry, and Correlation Functions* (W.A. Benjamin, Reading, MA 1975) Chap. 7
51.6 Goldstone, J.: Nuovo Cimento **19**, 154 (1961)
     Goldstone J., Salam, A., Weinberg, S.: Phys. Rev. **117**, 965 (1962)
51.7 Kibble, T.W.B.: In Proc. Oxford Intl. Conf. Elementary Particles (1965)
     Gottfried, K., Weisskopf, V.F.: *Concepts of Particle Physics*, Vol. II (Oxford University Press, Oxford 1986) Sect. VI C

Chapter 52

52.1 Yamamoto, T.: J. Chem. Phys. **33**, 281 (1960)
52.2 Zubarev, D.N.: In *Nonequilibrium Statistical Thermodynamics*, ed. by P. Gray, Studies in Soviet Science (Consultants Bureau, New York 1974) Chap. IV

Chapter 53

53.1 Landau, L.D.: Sov. Phys. – JETP **35**, 70 (1959)
53.2 Rice, M.J.: Phys. Rev. **159**, 153 (1967); **162**, 189 (1967)
53.3 Schindler, A.I., Rice, M.J.: Phys. Rev. **164**, 759 (1967)

Additional Reading

Chapter 3

Kadanoff, L.P., Martin, P.C.: Ann. Phys. **24**, 419 (1963)
Martin, P.C., Schwinger, J.: Phys. Rev. **115**, 1342 (1959)

Chapter 4

Brown, R.: Philos. Mag. **4**, 161 (1828); **6**, 161 (1829)
Chandrasekhar, S.: Rev. Mod. Phys. **15**, 1 (1943)
Kubo, R., Toda, M., Hashitsume, N.: *Statistical Physics II*, Springer Ser. Solid-State Sci., Vol. 31 (Springer, Berlin, Heidelberg 1985) Chap. 1
Wang, Min Chen, Uhlenbeck, G.E.: Rev. Mod. Phys. **17**, 323 (1945)

Chapter 6

Kadanoff, L.P., Martin, P.C.: Ann. Phys. **24**, 419 (1963)
Kubo, R., Toda, M., Hashitsume, N.: *Statistical Physics II*, Springer Ser. Solid-State Sci., Vol. 31 (Springer, Berlin, Heidelberg 1985) Sects. 3.6, 4.3

Chapter 7

Forster, D.: *Hydrodynamic Fluctuations, Broken Symmetry, and Correlation Functions* (W.A. Benjamin, Reading, MA 1975) Sect. 3.2
Kubo, R., Toda, M., Hashitsume, N.: *Statistical Physics II*, Springer Ser. Solid-State Sci., Vol. 31 (Springer, Berlin, Heidelberg 1985) Sect. 4.3

Chapter 8

Callen, H.B., Welton, T.A.: Phys. Rev. B. **83**, 34 (1951)
Forster, D.: *Hydrodynamic Fluctuations, Broken Symmetry, and Correlation Functions* (W.A. Benjamin, Reading, MA 1975) Sect. 2.7
Kubo, R., Toda, M., Hashitsume, N.: *Statistical Physics II*, Springer Ser. Solid-State Sci., Vol. 31 (Springer, Berlin, Heidelberg 1985) Sect. 4.4

Chapter 9

Martin, P.C.: *Measurements and Correlation Functions* (Gordon and Breach, New York 1968) Chap. E
Pines, D.: *Elementary Excitations in Solids* (W.A. Benjamin, Reading, MA 1963) Sect. 2.5

Chapter 10

Forster, D.: *Hydrodynamic Fluctuations, Broken Symmetry, and Correlation Functions* (W.A. Benjamin, Reading, MA 1975) Sect. 3.3

Chapter 11

Forster, D.: *Hydrodynamic Fluctuations, Broken Symmetry, and Correlation Functions* (W.A. Benjamin, Reading, MA 1975) Sect. 2.9
Kubo, R., Toda, M., Hashitsume, N.: *Statistical Physics II*, Springer Ser. Solid-State Sci, Vol. 31 (Springer, Berlin, Heidelberg 1985) Sect. 3.7
Mori, H.: Prog. Theor. Phys. **33**, 423 (1965)

Chapter 12

Mori, H.: Prog. Theor. Phys. **33**, 423 (1965)
Mori, H.: Prog. Theor. Phys. **34**, 399 (1965)

Chapter 13

Forster, D.: *Hydrodynamic Fluctuations, Broken Symmetry, and Correlation Functions* (W.A. Benjamin, Reading, MA 1975) Chaps. 5,6
Nakajima, S.: Prog. Theor. Phys. **20**, 948 (1958)

Chapter 14

Forster, D.: *Hydrodynamic Fluctuations, Broken Symmetry, and Correlation Functions* (W.A. Benjamin, Reading, MA 1975) Chap. 6

Chapter 15

Zubarev, D.N.: In *Nonequilibrium Statistical Thermodynamics*, ed. by P. Gray, Studies in Soviet Science (Consultants Bureau, New York 1974)

Chapter 16

Kreuzer, H.J.: *Nonequilibrium Thermodynamics and Its Statistical Foundations* (Clarendon, Oxford 1981) Chap. 11

Chapter 17

Boltzmann, L.: Wien. Ber. **66**, 275 (1872)
Boltzmann, L.: Wien. Ber. **75**, 62 (1877)
Boltzmann, L.: Wiedemann Ann. **57**, 773 (1896)
Davies, P.C.W.: *The Physics of Time Asymmetry* (University of California Press, Berkeley, CA 1974)
Loschmidt, J.: Wien. Ber. **73**, 135 (1876)
Zermélo, E.: Wiedemann Ann. **57**, 485 (1896)
Zermélo, E.: Wiedemann Ann. **59**, 793 (1896)

Chapter 18

Groot, S.R. de, Mazur, P.: *Nonequilibrium Thermodynamics* (North-Holland, Amsterdam 1969)

Chapter 20
Groot, S.R. de, Mazur, P.: *Nonequilibrium Thermodynamics* (North-Holland, Amsterdam 1969) Chap. XI

Chapter 21
Thomson, W. (Lord Kelvin): Proc. R. Soc. Edinburgh **3**, 225 (1854)

Chapter 22
Groot, S.R. de, Mazur, P.: *Nonequilibrium Thermodynamics* (North-Holland, Amsterdam 1969) Chap. X
Laidler, K.J.: *Chemical Kinetics* (McGraw-Hill, New York 1965)

Chapter 23
Laidler, K.J.: *Chemical Kinetics* (McGraw-Hill, New York 1965)

Chapter 26
Langmuir, I.: J. Am. Chem. Soc. **40**, 1361 (1918)

Chapter 33
Born, M., Green, H.S.: Proc. R. Soc. London A **188**, 10 (1946); A **190**, 455 (1947)
Kirkwood, J.G.: J. Chem. Phys. **14**, 180 (1946)
Yvon, J.: Acta Sci. Ind., nos. 542, 543 (1937)

Chapter 34
Drude, P.: Ann. Phys. (Leipzig) **1**, 566 (1900)
Drude, P.: Ann. Phys. (Leipzig) **3**, 369 (1900)
Warren, J.L., Ferrell, R.A.: Phys. Rev. **117**, 1252 (1960)

Chapter 38
Onsager, L., Machlup, S.: Phys. Rev. **91**, 1505, 1512 (1953)

Chapter 40
Kohler, M.: Ann. Phys. (Leipzig) **40**, 601 (1941)
Kohler, M.: Z. Phys. **124**, 772 (1948)

Chapter 45
Götze, W.: Solid State Commun. **27**, 1393 (1978)

Chapter 46
Belitz, D., Gold, A., Götze, W.: Z. Phys. B **44**, 273 (1981)
Hikami, S.: In *Anderson Localization*, ed. by Y. Nagaoka, H. Fukuyama, Springer Ser. Solid-State Sci., Vol. 39 (Springer, Berlin, Heidelberg 1982) p. 15
Prelovšek, P.: In *Recent Developments in Condensed Matter Physics*, Vol. II, ed. by J.T. Devreese (Plenum, New York 1981)
Prelovšek, P.: Phys. Rev. B **23**, 1304 (1981)
Vollhardt, D., Wölfle, P.: Phys. Rev. Lett. **45**, 842 (1980)
Vollhardt, D., Wölfle, P.: Phys. Rev. B **22**, 4666 (1980)
Wegner, F.: In *Anderson Localization*, ed. by Y. Nagaoka, H. Fukuyama, Springer Ser. Solid-State Sci., Vol. 39 (Springer, Berlin, Heidelberg 1982) p. 8

Chapter 47
Bergmann, G.: Solid State Commun. **46**, 347 (1983)
Bergmann, G.: Phys. Rep. **107**, 1 (1984)
Khmelnitzskii, D.E.: Proc. of LT-17, Physica **126** B+C, 235 (1984)

Chapter 49
Schwabl, F.: Z. Phys. **246**, 13 (1971)

Chapter 50
Schwabl, F.: Z. Phys. **246**, 13 (1971)

Subject Index

Adiabatic compressibility 147
Adiabatic process 18
Adiabatic susceptibility 17, 56, 141
Adsorption 127
Affinity 110
Annihilation operators 217
Anomalous skin effect 173
Antiunitary operators 39
Attractor 121
Autocatalytic reaction 113, 115

Backscattering effects 231
BBGKY hierarchy 165, 280 (Ref. [33.9])
Belousov-Zhabotinski reaction 120
Bogoliubov inequality 265
Boltzmann equation 164, 169, 214, 202–210, 224
— hydrodynamic equations derived from 193–200
— linearization of 193
Bosons 158, 160, 268
Broken symmetry
— ferromagnetic system 266
— general definition 264
— and low frequency modes 264–268
— superfluids (superconductors) 267
Brownian motion
— and diffusion 184–187
— oscillator 13, 29, 81
— particle 13, 24, 25, 61, 69, 89, 187
— relaxator 22
Brusselator 119, 278 (Ref.[24.4])

Causality 10, 33, 60

Chaos 120
Chemical fluctuations 123
Chemical potential 108, 131, 142, 158
Chemical reactions 108, 112, 116, 271
Coherence length 247
Collision term (operator, rate) 159, 165, 193, 269, 174
Collision time 25, 71, 92, 178, 185, 187
Commutation relations 57, 217
Compressibility
— adiabatic 147
— isothermal 147
Conservation of
— energy 102, 143, 159, 164, 182, 193, 200
— momentum 143, 159, 164, 193, 200
— particle number 100, 143, 159, 164, 193
— probability 11
— spin 254
Conserved quantities
— energy density 102, 164, 182
— entropy density 102, 164, 182
— mass density 144
— momentum density 164, 182
— particle density 102, 164, 182
Continued fraction expansion 58
Continuity equation 101, 143, 159, 164, 193
Continuous phase transition 247–263
Correlation functions (matrices) 20, 29, 37, 217, 232
— advanced 34
— retarded 34

— in scattering 46–49
— velocity autocorrelation function 149
Creation operators 217
Critical exponents 241, 248–263
Critical point 247, 257, 258
Critical slowing down 247
Current (flux) density
— diffusion 100
— energy current 102, 143, 182
— entropy current 102, 143
— heat current 102, 143, 145
— mass current 144
— momentum flux (current) 182
— particle current 100, 143, 159, 164, 193

Debye frequency 228
Density fluctuations (response) 177
Desorption 127
Detailed balance 19, 35, 42, 49, 50, 52, 161, 163
Diamagnetism (Landau) 175
Diamagnetism in superconductors 157
Diffusion 95, 99, 195, 230
— and Brownian motion 184
— coefficient, constant 95, 195, 231, 239, 244–246
— equation 100, 195
Dirac picture 8
Dispersion relation 33
Drude theory 171, 175, 185, 225
Dufour effect 93
Dynamic scaling laws, *see* Scaling laws
Dynamical conductivity 169
Dynamical susceptibility 9, 56, 141, 218, 241

Eigenvalues (of collision term) 162, 163
Eigenvectors (of collision term) 162, 163, 193

Einstein relation (diffusion) 42, 100, 103, 172, 185, 231
Electric conductivity 155, 156, 219, 221
Electromagnetic fields 152, 165
Electron localization 237–243, 244
Energy dissipation 52
Enthalpy 74
Entropy 55, 78, 102, 143
Entropy increase 55, 78, 82
Entropy production 101, 103, 148, 201
Ergodicity 17, 56, 187

Fermi liquid 166, 210, 275
Fermi momentum (energy, surface) 169, 199, 210, 218
Fermions 158, 160, 210
Fick's law 99, 101, 184
Fluctuation-dissipation theorem 42, 123
Fokker-Planck equation 180–183, 188
Fountain effect 97, 253
Friction 22, 138
F-sum rule 57, 144

Generating functional 190
Gibbs ink parable 83
Goldstone bosons 268

Hamilton's equations 77
Heat transfer 23
Heat transport 95, 197
Heisenberg picture 7
Helium (superfluid) 97, 250
Hydrodynamic approximation (regime) 172, 177, 193–200, 231, 249, 255
Hydrodynamic equations 142
— derived from Boltzmann equation 193–200
— linearized 147, 193
— nonlinear (Navier-Stokes equation) 143–146
Hydrodynamic long time tails 149–151
Hydrodynamic modes 148, 264

Interaction picture 8

Internal energy 142
Irreversibility 3, 78
Isolated susceptibility 16, 18, 56, 242
Isothermal susceptibility 14, 18, 30, 41, 56, 63, 141

Kinetic coefficients 31, 59, 61, 76, 213, 257
Kinetic equations 1, 4, 31, 93, 193
Kinetic oscillations 117
Kramers-Kronig relations 34, 36
Kubo formula for conductivity 155
Kubo relaxation function,
    see Relaxation function

Lagrange parameters (multipliers) 8, 14, 60, 73
Landau damping 173
Landau-Placzek peak 140
Langevin equation 63, 69, 92, 188, 221
Lennard-Jones 6–12 potential 207
Light scattering 45–49
Limit cycles 119
Linear response theory 14–21, 79, 154
Liouville equation 11, 214, 215
Local equilibrium 101, 167, 177, 193, 194
Logistic equation 120
Long time tails 149
Loschmidt reversibility objection 86
Lotka-Volterra model 118
Low frequency response 59, 75, 81, 21

Markov processes 31, 62, 75
Master equation 159
Maxwell(-Boltzmann) distribution 186, 203
Maxwell demon 83
Mean free path 61, 162, 196
Mechano-caloric effect 95
Memory function (matrix) 27, 59, 62, 65, 73, 80, 92, 221
Mobility 31, 147, 163, 197
Mode coupling theory 230, 247, 257, 260

Moment expansion 56
Momentum flux tensor 143, 182
Mori formalism 66

Navier-Stokes equations 143–146, 214, 215
von Neumann equation 7, 55, 83, 215
Neutron scattering 45–49
Nonequilibrium phase transition 116
Nonergodic behavior 17, 18, 239
Nonlinear chemical reactions 116
Nonlinear kinetic equations 73, 77, 116
Nonlinear response theory 73
Nucleation 131
Nyquist formula 90

Ohm's law 105, 155, 219
Onsager's symmetry relations 31, 41, 93, 95, 98, 111
Ornstein-Zernicke equation 257, 258, 262
Oscillator
— Brownian 29
— overdamped 22, 257
— with thermal attenuation 137

Passivity 52, 53, 79
Peltier-effect 104, 106
Period doubling 120
Perturbation theory 162, 163, 221, 239, 269
Phase space 164
Phase transitions
— continuous 247–263
— nonequilibrium 116
Phonons 206, 227, 257, 265
Predator-prey model 118
Pressure 93, 97, 102, 133, 142–147
Progress variables 110, 123
Projection operators 66

Quasiparticles 66

Rate equations 112, 158–163
Rayleigh peak 148, 258

R-C element 23, 89, 90
Regression of fluctuations 20, 29
Relaxation function 15, 18, 59, 63, 123, 219, 220
Relaxation rate (time, frequency) 76, 91, 169, 226, 249, 257
Renormalization group 237, 247

Scalar product 16, 66, 68
Scaling laws
— anisotropic ferromagnets 258
— electron localization 241
— isotropic ferromagnets 254
— liquid-gas transition 259
— liquid helium 250
— structural phase transitions 257
— uniaxial antiferromagnets 257
— uniaxial ferromagnets 261
Scattering
— cross section 46, 49, 162
— electrons 20, 46, 49
— light 20, 47, 49
— neutrons 20, 47, 49
Schlögl model 117
Schrödinger picture 7
Schwarz inequality 201, 265
Screening 173
Second law of thermodynamics 94
Second sound 250–253
Seebeck effect 107
Soft mode transition 247
Soret effect 93
Sound 148, 177, 267
Spectral function 33
Spectral representation 33, 56, 80
Spin diffusion 254–258
Sticking 127, 135
Stochastic forces 63, 64, 74
Stokes formula 28, 150, 215
Stosszahlansatz 158
Stress tensor 143
Sum rules 57, 144
Superconductor 157
Superfluid 250, 267

Susceptibility
— adiabatic 17, 56
— dynamical 9, 56, 141, 218, 248
— electromagnetic 154
— isolated 16, 18, 56, 242
— isothermal 14, 18, 30, 56, 63, 141

Thermal conductivity 102, 197
Thermal expansion coefficient 21, 140, 147
Thermocouple 104
Thermodiffusion 102
Thermodynamic forces 31
Thermodynamic limit 215
Thermoelectric effect 104
Thermomechanical effect 93, 95, 97
Thomas-Fermi screening 173
Thomson effect 106
Time average 24, 187
Time-reversal invariance 38, 40, 50, 232, 234, 245
Time translation invariance 24, 34, 38, 40
Transition probability 159, 163
Transition rate 159, 163
Transport coefficients 193–200, 202–210
Transport theory 164–176, 193–210
Turbulence 121, 193
Two-fluid model 250, 267

Ultrasonic attenuation 173, 179, 259
Unitary transformation 7

Velocity autocorrelation function 149, 262
Viscosity 145, 178, 198, 209
Vlasov equation 166

Wiedemann-Franz law 209
Wiener-Khinchin theorem 44

Zermelo recurrence paradox 84
Zero sound 177–179
Zhabotinski reaction 120
Zwanzig-Mori formalism 66